Lecture Notes in Computer Science 12923

More information about this subseries at http://www.springer.com/series/7409

Christine Strauss · Gabriele Kotsis ·
A Min Tjoa · Ismail Khalil (Eds.)

Database and Expert Systems Applications

32nd International Conference, DEXA 2021
Virtual Event, September 27–30, 2021
Proceedings, Part I

 Springer

Editors
Christine Strauss
University of Vienna
Vienna, Austria

Gabriele Kotsis
Johannes Kepler University of Linz
Linz, Oberösterreich, Austria

A Min Tjoa
Vienna University of Technology
Vienna, Austria

Ismail Khalil
Johannes Kepler University of Linz
Linz, Austria

ISSN 0302-9743   ISSN 1611-3349   (electronic)
Lecture Notes in Computer Science
ISBN 978-3-030-86471-2   ISBN 978-3-030-86472-9   (eBook)
https://doi.org/10.1007/978-3-030-86472-9

LNCS Sublibrary: SL3 – Information Systems and Applications, incl. Internet/Web, and HCI

This Springer imprint is published by the registered company Springer Nature Switzerland AG
The registered company address is: Gewerbestrasse 11, 6330 Cham, Switzerland

Preface

The volume at hand represents the result of joint efforts of contributing researchers, reviewers, and organizers, and contains the papers presented at the 32nd International Conference on Database and Expert Systems Applications (DEXA 2021). This year, DEXA was held for the second time as a virtual conference during September 27–30, 2021, instead of in Linz, Austria, as originally planned. The decision to organize another virtual version of DEXA was driven by the intention to provide stable conditions for all DEXA participants and set a good example in temporarily suspending on-site meetings. We put our trust in the loyalty of DEXA community and look forward to personal DEXA meetings in 2022.

We are proud to report that authors from 43 different countries submitted papers to DEXA this year. The number of submissions was similar to those of the past few years. Our Program Committee conducted more than 500 reviews. We would like to sincerely thank our Program Committee members for their rigorous and critical, and at the same time motivating, reviews of these submissions. Based on the total number of accepted papers, we can report that the acceptance rate this year was 27%, a rate comparable to DEXA conferences of the last few years.

The conference program this year covered a wide range of important topics such as data management and analytics; consistency; integrity; quality of data; data analysis and data modeling; data mining; databases and data management; information retrieval; prediction and decision support; authenticity, privacy, security, and trust; cloud databases and workflows; data and information processing; knowledge discovery; machine learning; semantic web and ontologies; stream data processing; and temporal, spatial, and high dimensional databases.

We tried to follow our on-site face-to-face format. Thus, the authors of the accepted papers presented their research online using video conference software over four days. Presentations were performed live in 12 different thematic clusters structured as 15 sessions, each one with an assigned session chair. The scientific presentations, discussions, and question-and-answer time were all live and part of each session. As we were aware of time difference issues, for example, for participants from Australia or South American countries having to present or participate during unusual times of the day, we tried to minimize this inconvenience.

We would like to express our gratitude to the distinguished keynote speakers for illuminating us on their leading-edge topics: Elisa Bertino (Purdue University, USA) for her talk on "Privacy in the Era of Big Data, Machine Learning, IoT, and 5G", Amit Sheth (University of South Carolina, USA) for his talk on the third wave of AI, and Torben Bach Pedersen (Aalborg University, Denmark) for his talk on "Extreme-Scale Model-Based Time Series Management with ModelarDB".

In addition, we had a panel discussion on "Big Minds Sharing their Vision on the Future of AI" led by Bernhard Moser (SCCH, Austria), with Battista Biggio (University of Cagliari, Italy), Claudia Diaz (Katholieke Universiteit Leuven,

Belgium), Heiko Paulheim (University of Mannheim, Germany), and Olga Saukh (Complexity Science Hub, Austria).

As is the tradition of DEXA, all accepted papers were published in "Lecture Notes in Computer Science" (LNCS) and made available by Springer. Authors of selected papers presented at the conference will be invited to submit substantially extended versions of their conference papers for publication in special issues of international journals. The submitted extended versions will undergo a further review process.

The 32nd edition of DEXA featured six international workshops – three established ones and three brand-new ones – covering a variety of specific topics:

- The 12th International Workshop on Biological Knowledge Discovery from Data (BIOKDD 2021)
- The 5th International Workshop on Cyber-Security and Functional Safety in Cyber-Physical Systems (IWCFS 2021)
- The 3rd International Workshop on Machine Learning and Knowledge Graphs (MLKgraphs 2021)
- The 1st International Workshop on Artificial Intelligence for Clean, Affordable, and Reliable Energy Supply (AI-CARES 2021)
- The 1st International Workshop on Time Ordered Data (ProTime2021)
- The 1st International Workshop on AI System Engineering: Math, Modelling, and Software (AISys2021)

The success of the conference is due to the continuous and generous support of its participants and their relentless efforts. Our sincere thanks go to the dedicated authors, renowned Program Committee members, session chairs, organizing and steering committee members, and student volunteers who worked tirelessly to ensure the continuity and high quality of DEXA 2021.

We would also like to express our thanks to all institutions actively supporting this event, namely:

- Institute of Telekooperation, Johannes Kepler University Linz (JKU), Austria
- Software Competence Center Hagenberg (SCCH), Austria
- Web Applications Society (@WAS)

We hope you have enjoyed the conference! We are looking forward to seeing you again next year.

September 2021   Christine Strauss

Organization

Program Committee Chair

Christine Strauss   University of Vienna, Austria

Steering Committee

Gabriele Kotsis	Johannes Kepler University Linz, Austria
A Min Tjoa	Vienna University of Technology, Austria
Robert Wille	Software Competence Center Hagenberg, Austria
Bernhard Moser	Software Competence Center Hagenberg, Austria
Ismail Khalil	Johannes Kepler University Linz, Austria

Program Committee

Susan Ariel Aaronson	George Washington University, USA
Javier Nieves Acedo	Azterlan, Spain
Sonali Agarwal	IIIT, India
Hamid Aghajan	Ghent University, Belgium
Hans Akkermans	Vrije Universiteit Amsterdam, The Netherlands
Riccardo Albertoni	CNR-IMATI, Italy
Idir Amine Amarouche	USTHB, Algeria
Rachid Anane	Coventry University, UK
Mustafa Atay	Winston-Salem State University, USA
Sören Auer	Leibniz Universität Hannover, Germany
Juan Carlos Augusto	Middlessex University London, UK
Monica Barratt	RMIT University, Australia
Ladjel Bellatreche	LIAS, ENSMA, France
Nadia Bennani	LIRIS, INSA de Lyon, France
Karim Benouaret	Université Claude Bernard Lyon 1, France
Djamal Benslimane	Université de Lyon, France
Morad Benyoucef	University of Ottawa, Canada
Mikael Berndtsson	University of Skövde, Sweden
Catherine Berrut	LIG, Université Joseph Fourier, France
Vasudha Bhatnagar	University of Delhi, India
Didier Bigo	King's College London, UK
Steven Bird	Charles Darwin University, Australia
Ankur Singh Bist	KIET Ghaziabad, India
Joseph Bonneau	New York University, USA
Johan Bos	University of Groningen, The Netherlands
Athman Bouguettaya	University of Sydney, Australia
Olivier Bousquet	Google Brain, Zurich, Switzerland

Omar Boussai	ERIC Laboratory, France
Kevin Bowyer	University of Notre Dame, USA
Stephane Bressan	National University of Singapore, Singapore
Marcel Broersma	University of Groningen, The Netherlands
Axel Bruns	Queensland University of Technology in Brisbane, Australia
Jean Burgess	Queensland University of Technology in Brisbane, Australia
Maria Chiara Carrozza	Scuola Superiore Sant'Anna, Pisa, Italy
Lemuria Carter	University of New South Wales, Australia
Antonio Casilli	Télécom Paris, France
Pablo Castells	Universidad Autónonoma de Madrid, Spain
Carlos Castillo	Universitat Pompeu Fabra, Spain
Barbara Catania	Università degli Studi di Genova, Italy
Sharma Chakravarthy	University of Texas at Arlington, USA
Max Chevalier	IRIT, France
Chen-Fu Chien	National Tsing Hua University, Taiwan
Ruzanna Chitchyan	University of Bristol, UK
Soon Ae Chun	City University of New York, USA
David Clark	MIT Computer Science and Artificial Intelligence Lab, USA
Mark Coeckelbergh	University of Vienna, Austria
Diane Cook	Washington State University, USA
Alfredo Cuzzocrea	University of Calabria, Italy
Debora Dahl	Conversational Technologies, USA
Boyd Danah	Harvard University, USA
Jérôme Darmont	Université de Lyon, France
Trevor Darrell	University of California, Berkeley, USA
Soumyava Das	Teradata Labs, USA
Robert Davison	City University of Hong Kong, Hong Kong
Emiliano De Cristofaro	University College London, UK
Ronald Deibert	University of Toronto, Canada
Vincenzo Deufemia	University of Salerno, Italy
Roberto Di Pietro	Hamad Bin Khalifa University, Qatar
Juliette Dibie-Barthélemy	INRAE, France
Dejing Dou	University of Oregon, USA
Karen Douglas	University of Kent, UK
Brian Earp	Yale School of Medicine, USA
Johann Eder	University of Klagenfurt, Austria
Nicole Ellison	University of Michigan, USA
Suzanne Embury	University of Manchester, UK
Markus Endres	University of Passau, Germany
Sergio Escalera	University of Barcelona, Spain
Charles Ess	University of Oslo, Norway
Kevin Esterling	University of California, Riverside, USA
James Evans	University of Chicago, USA

Magdalena Hurtado	Arizona State University, USA
Ionut Iacob	Georgia Southern University, USA
Sergio Ilarri	University of Zaragoza, Spain
Abdessamad Imine	Loria, France
Yasunori Ishihara	Nanzan University, Japan
Ivan Izonin	Lviv Polytechnic National University, Ukraine
Peiquan Jin	University of Science and Technology of China, China
Deborah Johnson	University of Virginia, USA
Anne Kao	Boeing, USA
Dimitris Karagiannis	University of Vienna, Austria
Stefan Katzenbeisser	TU Darmstand, Germany
Anne Kayem	Hasso Plattner Institute, University of Potsdam, Germany
Deanna Kemp	University of Queensland, Australia
Faisal Khan	University of Calgary, Canada
Ewan Klein	University of Edinburgh, UK
Carsten Kleiner	University of Applied Science and Arts Hannover, Germany
Peter Knees	Vienna University of Technology, Austria
Henning Koehler	Massey University, New Zealand
Michal Kratky	VSB-Technical University of Ostrava, Czech Republic
Petr Kremen	Czech Technical University in Prague, Czech Republic
David Kreps	Stanford University, USA
Agnes Kukulska-Hulme	Open University, UK
Tahu Kukutai	University of Waikato, New Zealand
Josef Küng	Johannes Kepler University Linz, Austria
Nhien-An Le Khac	University College Dublin, Ireland
Lenka Lhotska	Czech Technical University in Prague, Czech Republic
Wenxin Liang	Chongqing University of Posts and Telecommunications, China
Chuan-Ming Liu	National Taipei University of Technology, Taiwan
Oscar Pastor Lopez	Universitat Politècnica de València, Spain
Hui Ma	Victoria University of Wellington, New Zealand
Qiang Ma	Kyoto University, Japan
Zakaria Maamar	Zayed University, UAE
Sanjay Madria	Missouri University of Science and Technology, USA
Elio Masciari	Federico II University, Italy
Brahim Medjahed	University of Michigan, USA
Jun Miyazaki	Tokyo Institute of Technology, Japan
Lars Moench	University of Hagen, Germany
Riad Mokadem	Paul Sabatier University, France
Anirban Mondal	University of Tokyo, Japan
Yang-Sae Moon	Kangwon National University, South Korea
Franck Morvan	IRIT, Paul Sabatier University, France
Cedric du Mouza	CNAM, France
Francesc Munoz-Escoi	Universitat Politècnica de València, Spain

Ismael Navas-Delgado	University of Malaga, Spain
Wilfred Ng	Hong Kong University of Science and Technology, Hong Kong
Marcin Paprzycki	Systems Research Institute, Polish Academy of Sciences, Poland
Dhaval Patel	IBM, USA
Clara Pizzuti	CNR-ICAR, Italy
Elaheh Pourabbas	CNR-ICAR, Italy
Uday Kiran Rage	University of Tokyo, Japan
Rodolfo Resende	Federal University of Minas Gerais, Brazil
Claudia Roncancio	Grenoble Alps University, France
Viera Rozinajova	Slovak University of Technology in Bratislava, Slovakia
Massimo Ruffolo	ICAR-CNR, Italy
Shelly Sachdeva	National Institute of Technology Delhi, India
Marinette Savonnet	University of Burgundy, France
Florence Sedes	IRIT, Paul Sabatier University, France
Nazha Selmaoui	University of New Caledonia, New Caledonia
Michael Sheng	Macquarie University, Australia
Patrick Siarry	Université de Paris 12, France
Tarique Siddiqui	Microsoft Research Lab, Redmond, USA
Gheorghe Cosmin Silaghi	Babes-Bolyai University, Romania
Hala Skaf-Molli	University of Nantes, LS2N, France
Srinivasa Srinath	IIITB, India
Bala Srinivasan	Monash University, Australia
Olivier Teste	IRIT, France
Stephanie Teufel	University of Fribourg, Switzerland
Jukka Teuhola	University of Turku, Finland
Jean-Marc Thevenin	IRIT, Université Toulouse I, France
A Min Tjoa	Vienna University of Technology, Austria
Vicenc Torra	University Skövde, Sweden
Traian Marius Truta	Northern Kentucky University, USA
Lucia Vaira	University of Salento, Italy
Ismini Vasileiou	De Montfort University, UK
Krishnamurthy Vidyasankar	Memorial University, Canada
Marco Vieira	University of Coimbra, Portugal
Piotr Wisniewski	Nicolaus Copernicus University, Poland
Ming Hour Yang	Chung Yuan Chritian University, Taiwan
Haruo Yokota	Tokyo Institute of Technology, Japan
Qiang Zhu	University of Michigan, USA
Yan Zhu	Southwest Jiaotong University, China
Ester Zumpano	University of Calabria, Italy

External Reviewers

Tooba Aamir
Amani Abusafia
Abdulwahab Aljubairy
Mohammed Bahutair
Andrea Baraldi
Nabila Berkani
Francesco Del Buono
Loredana Caruccio
Olivier De Casanove
Dipankar Chaki
Rachid Chelouah
Stefano Cirillo
Labbe Cyril
Matthew Damigos
Jonathan Debure
Abir Farouzi
Sheik Mohammad Mostakim Fattah
Angelo Ferrando
Lukas Fischer
Arnaud Flori
Jorge Galicia
María del Carmen Rodríguez Hernández
Akm Tauhidul Islam
Eleftherios Kalogeros
Julius Köpke
Bogdan Kostov
Cyril Labbe
Chuan-Chi Lai
Hieu Hanh Le
Xuhong Li

Ji Liu
Jin Lu
Qiuhao Lu
Jia-Ning Luo
Jorge Martinez-Gil
Ahcene Menasria
Niccolo Meneghetti
Quoc Hung Ngo
Daria Novoseltseva
Matteo Paganelli
Louise Parkin
Gang Qian
Subhash Sagar
Nadouri Sana
Chayma Sellami
Mohamed Sellami
Vladimir A. Shekhovtsov
Tao Shi
Hannes Sochor
Manel Souibgui
Sofia Stamou
Carlos Telleria-Orriols
Daniele Traversaro
Oscar Urra
Francesco Visalli
Shuang Wang
Yi-Hung Wu
Fa Yao Yin
Feng Yu
Eric Zhang

Organizers

Abstracts of Keynote Talks

Privacy in the Era of Big Data, Machine Learning, IoT, and 5G

Elisa Bertino

Samuel Conte Professor of Computer Science, Cyber2SLab, Director,
CS Department, Purdue University, West Lafayette, Indiana, USA

Abstract. Technological advances, such as IoT devices, cyber-physical systems, smart mobile devices, data analytics, social networks, and increased communication capabilities are making possible to capture and to quickly process and analyze huge amounts of data from which to extract information critical for many critical tasks, such as healthcare and cyber security. In the area of cyber security, such tasks include user authentication, access control, anomaly detection, user monitoring, and protection from insider threat. By analyzing and integrating data collected on the Internet and the Web one can identify connections and relationships among individuals that may in turn help with homeland protection. By collecting and mining data concerning user travels, contacts and disease outbreaks one can predict disease spreading across geographical areas. And those are just a few examples. The use of data for those tasks raises however major privacy concerns. Collected data, even if anonymized by removing identifiers such as names or social security numbers, when linked with other data may lead to re-identify the individuals to which specific data items are related to. Also, as organizations, such as governmental agencies, often need to collaborate on security tasks, data sets are exchanged across different organizations, resulting in these data sets being available to many different parties. Privacy breaches may occur at different layers and components in our interconnected systems. In this talk, I first present an interesting privacy attack that exploits paging occasion in 5G cellular networks and possible defenses. Such attack shows that achieving privacy is challenging and there is no unique technique that one can use; rather one must combine different techniques depending also on the intended use of data. Examples of these techniques and their applications are presented. Finally, I discuss the notion of data transparency – critical when dealing with user sensitive data, and elaborate on the different dimensions of data transparency.

Don't Handicap AI without Explicit Knowledge

Amit Sheth

University of South Carolina, USA

Abstract. Knowledge representation as expert system rules or using frames and variety of logics, played a key role in capturing explicit knowledge during the hay days of AI in the past century. Such knowledge, aligned with planning and reasoning are part of what we refer to as Symbolic AI. The resurgent AI of this century in the form of Statistical AI has benefitted from massive data and computing. On some tasks, deep learning methods have even exceeded human performance levels. This gave the false sense that data alone is enough, and explicit knowledge is not needed. But as we start chasing machine intelligence that is comparable with human intelligence, there is an increasing realization that we cannot do without explicit knowledge. Neuroscience (role of long-term memory, strong interactions between different specialized regions of data on tasks such as multimodal sensing), cognitive science (bottom brain versus top brain, perception versus cognition), brain-inspired computing, behavioral economics (system 1 versus system 2), and other disciplines point to need for furthering AI to neuro-symbolic AI (i.e., hybrid of Statistical AI and Symbolic AI, also referred to as the third wave of AI). As we make this progress, the role of explicit knowledge becomes more evident. I will specifically look at our endeavor to support human-like intelligence, our desire for AI systems to interact with humans naturally, and our need to explain the path and reasons for AI systems' workings. Nevertheless, the variety of knowledge needed to support understanding and intelligence is varied and complex. Using the example of progressing from NLP to NLU, I will demonstrate the dimensions of explicit knowledge, which may include, linguistic, language syntax, common sense, general (world model), specialized (e.g., geographic), and domain-specific (e.g., mental health) knowledge. I will also argue that despite this complexity, such knowledge can be scalability created and maintained (even dynamically or continually). Finally, I will describe our work on knowledge-infused learning as an example strategy for fusing statistical and symbolic AI in a variety of ways.

Extreme-Scale Model-Based Time Series Management with ModelarDB

Torben Bach Pedersen

Aalborg University, Denmark

Abstract. To monitor critical industrial devices such as wind turbines, high quality sensors sampled at a high frequency are increasingly used. Current technology does not handle these extreme-scale time series well, so only simple aggregates are traditionally stored, removing outliers and fluctuations that could indicate problems. As a remedy, we present a model-based approach for managing extreme-scale time series that approximates the time series values using mathematical functions (models) and stores only model coefficients rather than data values. Compression is done both for individual time series and for correlated groups of time series. The keynote will present concepts, techniques, and algorithms from model-based time series management and our implementation of these in the open source Time Series Management System (TSMS) ModelarDB. Furthermore, it will present our experimental evaluation of ModelarDB on extreme-scale real-world time series, which shows that that compared to widely used Big Data formats, ModelarDB provides up to 14x faster ingestion due to high compression, 113x better compression due to its adaptability, 573x faster aggregation by using models, and close to linear scale-out scalability.

Big Minds Sharing their Vision on the Future of AI (Panel)

Panelists

Battista Biggio, University of Cagliari, Italy
Claudia Diaz, Katholieke Universiteit Leuven, Belgium
Heiko Paulheim, University Mannheim, Germany
Olga Saukh, Complexity Science Hub, Austria

Moderator

Bernhard Moser, Software Competence Center Hagenberg and Austrian Society
for Artificial Intelligence, Austria

Abstract. While we are currently mainly talking about narrow AI systems, in the future, neural networks will increasingly be combined with graph-based and symbolic-logical approaches (3rd wave of AI).

How will this technological trend affect the key issues of security such as integrity protection or privacy protection, and environmental impact? In this context, in this interactive panel discussion, technology experts will discuss current and envisioned challenges to AI from the research perspective of their respective fields.

Contents – Part I

Data Mining

Databases and Data Management

Information Retrieval

Prediction and Decision Support

Contents – Part II

Knowledge Discovery

Machine Learning

Semantic Web and Ontologies

Temporal, Spatial, and High Dimensional Databases

Big Data

Reference Architecture for Running Large Scale Data Integration Experiments

Michał Bodziony[1] and Robert Wrembel[2]([⊠])

[1] IBM Poland, Software Lab Kraków, Kraków, Poland
michal.bodziony@pl.ibm.com
[2] Poznan University of Technology, Poznań, Poland
robert.wrembel@cs.put.poznan.pl

Abstract. This paper contributes a reference architecture of a reusable infrastructure for scientific experiments on data processing and data integration. The architecture is based on containerization and is integrated with an external machine learning cloud service to build performance models.

Keywords: Data integration · Data processing · Hybrid cloud · Containerization · Performance testing · Software architecture · Experimentation

1 Introduction

In recent years, the volume of collected data increases faster than ever in the past. There is a high demand for well optimized, high performing, and reliable data processing and data integration solutions, e.g., in data warehouse, data lake, and data science architectures. Although there exist numerous solutions in these areas, the need for further research seems to be bigger than ever. The 3Vs of big data need excessive scientific experiments to workout better performance, scalability, security, and more trustworthy solutions for data processing.

In a hybrid cloud architecture, an infrastructure necessary to perform experiments can be highly distributed. A subject of an experiment can be a component of the cloud or can be provided as a service, but more often it is necessary to have it on premise to access performance metrics at a low level of a technology stack. To this end, monitoring resources consumption, I/O, and operating system performance indicators is necessary. Integrating all monitoring components and tooling required for an experiment is often a challenging integration project itself.

In general, the majority of experiments share a life-cycle similar to the one shown in Fig. 1. Each experiment has to be carefully designed and then a necessary infrastructure has to be set up, cf. the *prepare* step. Experiments are executed based on specifications, aka. *recipes*, to produce experimental data. These data are then analyzed and conclusions are drawn, which may result in

© Springer Nature Switzerland AG 2021
C. Strauss et al. (Eds.): DEXA 2021, LNCS 12923, pp. 3–9, 2021.
https://doi.org/10.1007/978-3-030-86472-9_1

some tuning of a design and recycle of all the steps. Typically, data integration experiments require expensive resources. Therefore, it is important to figure out in which phases such resources are required and allocate them only when needed.

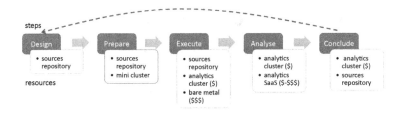

Fig. 1. The life-cycle of a scientific experiment

A few contributions to automating experiments have been proposed in scientific publications so far. [8] and [5] address the problem of running experiments on biological data. In [8] the authors proposed workflow templates, which represent types of a task (services) to run. A workflow template is instantiated with concrete tasks. The paper provides a general description of the solution, without providing details on the implementation. In [5], a method for defining, deploying, and running scientific workflows on biological data was proposed. Such workflows can have an arbitrary structure, like in our approach. [7] contributes a Python library for experimenting with recursive inter-network architectures (which are based on inter-process communication). The library allows to programmatically define a network topology and other parameters of the network. [2] describes a distributed framework based on containers for building a fog of IoT devices and scheduling works of heterogeneous IoT applications. Finally, the state of the art analysis of approaches to designing environments for continuous experimentation is available in [1].

In recent years, a common trend is observed to support performance tuning of systems and software pieces by machine learning (ML) algorithms, e.g., [3,4,6]. Such algorithms require large volumes of test data to learn reliable performance models. Thus, excessive experimental evaluations are needed to provide performance data, to feed ML algorithms.

In this short paper we propose a **software architecture for automatically deploying, running, collecting, and analyzing performance data**. Our solution differs from the aforementioned ones in the following: (1) it includes a ML component to build performance models and (2) it is based on virtualization by means of containerization (Docker). The architecture has been designed to fulfill the major **requirement** - to support full automatization of all tasks in an experiment's pipeline, i.e., all components of the architecture are automatically deployable either in a cloud or on-premise environment. The automatization encompasses: (1) setting up multiple test environments that differ with various parameters, e.g., the number of nodes, memory, CPUs, (2) deploying software components to be evaluated, (3) running experiments with various parameters,

(4) monitoring the execution of the experiments, (5) gathering and persistently storing performance data, (6) preparing the performance data for ML algorithms, and (7) running the ML algorithms. The architecture has been designed at a logical level (cf. Sect. 2) and at an implementation level (cf. Sect. 3). To the best of our knowledge, it is the first architecture that is able to fully automatize the full pipeline of scientific experiments.

2 Reference Architecture: Logical Level

In our reference architecture, all services are distributed across the cloud or on-premise infrastructure. At the logical level, there are the following groups of services: *Core, Analytics, External SaaS*, and *Experiment subject*. Each such group is a separate deployment unit, which can be provisioned on a single or multiple nodes.

Experiment subject represents all operational services and data sources that are subjects of experiments. This is the most frequently changing part of experiments' architecture, as different experiments are focused on testing different components (e.g., databases, ETL engines, or other data integration components). They use different data sets, with different scale factors and configurations of data sources. Nevertheless, a still significant portion of these services can be reusable. Services to initialize data sets, generate data, and monitor resources consumption can be continuously developed to provide a good toolkit for subsequent experiments.

Core provides: (1) an entry point for experiments and (2) the orchestration of all experiments, accessed via GUIs and APIs. These services are reusable in a broad scope of experiments without any changes required.

Analytics includes a data repository and services needed to store, (pre-)process, and analyze data collected during experiments. This component stores all results of all executed experiments.

External SaaS integrates all external services that may be needed for experiments, e.g., a repository of recipies, a repository of experimental results, data preparation, machine learning, and variety of analytical functions. *Recipes repository* is a persistent and versioned repository for documents that describe all experiments to be run. Recipes are maintained like any other code. These services do not require any maintenance, but the services have to be picked up carefully to ensure long lasting support, backward compatibility, and general reliability.

3 Reference Architecture: Implementation Level

The aforementioned logical architecture has been implemented by means of encapsulated functionalities, deployed as Docker containers. We outline the implementation for a use-case on testing performance of user defined functions (UDFs) in data processing workflows (DPWs). In such applications, UDFs are

typically treated as black boxes, i.e., their semantics and performance characteristics are unknown, which prevents from optimizing execution plans of DPWs.
The goal of UDFs testing is to collect their performance characteristics to further build ML models that can determine a type of an executed UDF (thus to
open a black box), based on these characteristics.

The experiments for collecting performance characteristics of UDFs were
conducted with Spark UDFs on top of the PostgreSQL database. All services
in the architecture were packed in **Docker containers** and deployed on an
OpenShift cluster. The implementations of the logical components (cf. Sect. 2)
were deployed on dedicated clusters, as presented in Fig. 2.

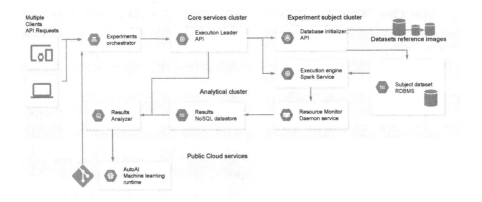

Fig. 2. The implementation components of the logical reference architecture

The **Experiment subject** service from the logical architecture (cf., Sect. 2)
is implemented by *Database initializer, Execution engine, Resource monitor*, and
Subject dataset. Database initializer prepares and maintains databases used for
experiments. In particular, it (1) generates PostgreSQL databases of requested
sizes, using the TPC-DS[1] generator; (2) prepares a fresh new database by cloning
the volumes snapshot generated in the previous step; (3) warms-up a database
if needed - some experiments require databases to be in a state other than
initial (e.g., triggered lazy initialization, pre-filled caches, collected statistics).
Execution engine makes available an environment to run experiments. In the
described use-case, UDFs were executed on Apache Spark. The Spark cluster was
deployed with Apache Spark operator for Kubernetes (spark-on-k8s-operator).
UDFs were specified by recipes stored in the Git repository and were executed
on data from the TPC-DS benchmark, stored in PostgreSQL. *Resource monitor*
is implemented as a service with a RESTfull interface to be used by *Execution
engine*. Resource monitoring by *Resource monitor* can be done either at an operating system (OS) level or at the Spark level. At the OS level raw characteristics

---

[1] http://www.tpc.org/tpcds/.

of CPU usage, memory allocations, network traffic, and I/O are monitored. At the Apache level some performance metrics of this engine are collected. All the metrics are periodically uploaded into *Results datastore*. *Subject dataset* stores in a relational database (in the current implementation PostgreSQL) datasets used by experiments.

The **Core** service is implemented by *Experiments orchestrator* and *Execution leader*. The first service manages experiments, each of which is described by a recipe stored in a repository. From a set of workers, it assigns a service called *Execution leader* that is responsible for communicating with other components of the architecture.

The **Analytics** service is implemented by *Results analyzer* and *Results datastore*. *Results analyzer* prepares data collected from experiments in order to pass them to the *AutoAI* service for analysis. Data being analyzed are stored in *Results datastore*. Since experiments are of different natures and their results are of heterogeneous formats, a NoSQL database (MongoDB) is used. It also stores the results of analyzes.

The **External SaaS** service is composed of *Git repository* and *AutoAI*. *Git repository* stores recipies, described in YAML files. Recipes contain all the information necessary to execute and analyse experiments. To start an experiment via *Experiment orchestrator*, the only required input is a link to a recipe in *Git repository*. This way, an exact context of an experiment execution is versioned. A recipe can be composed of two sections, namely: *Execution specification* and *Analyzes specification*. *Execution specification* contains all the information needed by *Execution engine* to execute UDFs in Spark (a UDF implementation, sequences of executions, the number of re-tries, the set of indicators to be monitored, datasets specification). *Analyzes specification* defines a type of processing to be run on data collected by *Resource monitor*. The specification includes a sequence of actions, like data pre-processing for ML, model training, model evaluation, hyper-parameter tuning, and model re-training. The *AutoAI* service[2] in IBM Cloud, is used to automatically train multiple models, evaluate them, tune hyper-parameters, and return both a model and its assessment, as the result of the whole experiment's cycle.

Technology Stack. The following technology stack was proposed to implement the architecture: (1) *Docker containerization* - an operating system level virtualization, which is resource efficient and ensures exactly the same conditions for subsequent software re-executions; (2) *Kubernetes containers orchestration* - a system for automating application deployment, scaling, and management on multiple nodes; (3) *OpenShift container platform* - platform as a service (PaaS) for containerized software managed by Kubernetes; (4) *Bare metal as a service* (MaaS) - hardware that can be configured in a flexible way with web GUI or API; typically available within minuets for experiments, often with customized operating system, middle-ware, and cluster management software.

---

[2] https://www.ibm.com/cloud/watson-studio/autoai.

Experimental Evaluation. The architecture for evaluating the performance characteristics of UDFs was deployed on two OpenShift clusters. Deploying the full architecture in a cloud performed as follows: (1) the *Core* service on a 3-node cluster was provisioned within less than 5 min, (2) the *Analytics* service on another 3-node cluster was provisioned also within less than 5 min, (3) the *Experiment subject* on another 3-node cluster was provisioned within less than 4 h; most of the time was spent on data sets images generation. The time can be easily reduced by increasing the number of nodes in a cluster, since the data generation process scales linearly w.r.t. the number of nodes. Having completed this experimental evaluation, we can conclude that the architecture proved its flexibility in defining a parameterized environment for experiments.

4   Summary

In this paper we contributed an architecture for conducting scientific experiments. Unlike, other solutions in this field, our architecture is built with chunks encapsulated in containers. Moreover, it integrates with a ML service to build performance models. The architecture enables a high level of re-usability of its components for various (even very different) experiments. As soon as the experiments are finished, the most expensive part of the infrastructure can be scaled down to zero. The architecture was deployed in the cloud to test performance of user defined functions. This application proved that the whole architecture could be deployed quickly and fully automatically.

Acknowledgement. The work of Robert Wrembel is partially supported by IBM Shared University Reward 2019.

References

1. Auer, F., Ros, R., Kaltenbrunner, L., Runeson, P., Felderer, M.: Controlled experimentation in continuous experimentation: knowledge and challenges. Inf. Softw. Technol. **134**, 106551 (2021)
2. Deng, Q., Goudarzi, M., Buyya, R.: Fogbus2: a lightweight and distributed container-based framework for integration of IoT-enabled systems with edge and cloud computing. In: International Workshop on Big Data in Emergent Distributed Environments (BiDEDE) @SIGMOD. ACM (2021)
3. Hernández, Á.B., Pérez, M.S., Gupta, S., Muntés-Mulero, V.: Using machine learning to optimize parallelism in big data applications. Future Gener. Comput. Syst. **86**, 1076–1092 (2018)
4. Herodotou, H., et al.: Starfish: a self-tuning system for big data analytics. In: CIDR, pp. 261–272 (2011)
5. Majithia, S., Walker, D.W., Gray, W.A.: Automating scientific experiments on the semantic grid. In: McIlraith, S.A., Plexousakis, D., van Harmelen, F. (eds.) ISWC 2004. LNCS, vol. 3298, pp. 365–379. Springer, Heidelberg (2004). https://doi.org/10.1007/978-3-540-30475-3_26

6. Popescu, A.D., Ercegovac, V., Balmin, A., Branco, M., Ailamaki, A.: Same queries, different data: can we predict runtime performance? In: ICDE Workshops, pp. 275–280 (2012)
7. Vrijders, S., Staessens, D., Capitani, M., Maffione, V.: Rumba: a Python framework for automating large-scale recursive internet experiments on GENI and FIRE+. In: IEEE Conference on Computer Communications Workshops, pp. 324–329 (2018)
8. Wroe, C., et al.: Automating experiments using semantic data on a bioinformatics grid. IEEE Intell. Syst. **19**(1), 48–55 (2004)

Subgroup Discovery with Consecutive Erosion on Discontinuous Intervals

Reynald Eugenie[(⊠)] and Erick Stattner[(⊠)]

LAMIA Laboratory, Université des Antilles, Pointe-à-Pitre Cedex, France
{reynald.eugenie,erick.stattner}@univ-antilles.fr

Abstract. The subgroup discovery problem aims to identify a subset of objects which exhibit interesting characteristics according to a quality measure defined on a target attribute. In this paper, we propose a new optimized approach, called SD-CEDI, which originality consists of extracting subgroups defined on discontinued attribute intervals. The intuition behind this approach is that disjoint intervals allow refining the definition of subgroups and therefore the quality of the subgroups identified. For this purpose, the approach we propose models the search space through a hypercube and aims to slice this hypercube to highlight subgroups. Unlike recent methods that also exploit the concept of hypercubes, the originality of our approach relies on the way it performs the slicing in order to bring out discontinuous attribute intervals. The good performances of the approach are demonstrated by comparing, qualitatively and quantitatively, the results with the main algorithms that are the references in the domain.

Keywords: Data science · Subgroup discovery · Knowledge extraction · Algorithm

1 Introduction

Numerous subgroup discovery techniques have been proposed. The vast majority of these approaches make the assumption that optimal subgroups emerge from continuous intervals on attributes. Thus, the main contribution of the literature for extracting subgroups aims to search for those defined on continuous attribute intervals. That result in the identification of groups defined over wider intervals thus containing some irrelevant objects that may degrade the quality function.

In this paper, we focus on the subgroup discovery problem and we propose a new approach, called SD-CEDI, which originality consists of searching for subgroups defined on discontinuous attribute intervals. The intuition behind this approach is that disjoint intervals allow refining the definition of subgroups and therefore the quality of the subgroups identified. For this purpose, the approach we propose models the search space through a hypercube and aims to slice this hypercube to highlight subgroups. Unlike recent methods that also exploit the concept of hypercubes, the originality of our approach relies on the way it performs the slicing in order to bring out discontinuous attribute intervals.

© Springer Nature Switzerland AG 2021
C. Strauss et al. (Eds.): DEXA 2021, LNCS 12923, pp. 10–21, 2021.
https://doi.org/10.1007/978-3-030-86472-9_2

In this work, we detail the SD-CEDI algorithm we propose and the optimizations it integrates during the searching process. The efficiency of the approach is demonstrated by comparing the results with the main algorithms that are the references in the domain. Thus, by applying our approach to the benchmark of the four datasets traditionally used in the field, we show qualitatively and quantitatively the good performances of the approach.

The paper is organized as follows. Section 2 reviews the main related works. Section 3 formally defines the concept of subgroups and describes the approach we propose. Section 4 is devoted to performances based on experimental results. Section 5 concludes and presents the future directions.

2   Related Works

Subgroup Discovery is a descriptive data analysis technique method of data mining which aims to extract a group of transactions with the highest "quality" defined on an attribute or a group of attributes according to a quality function. The attributes used in the quality function are called target variables.

In [1], Atzmueller et al. presented an overview of the main quality function in the domain for each type of target variable. Thus, the quality function which they presented for the numerical target in their paper, being one of the most relevant, will be used in the rest of the paper:

$$q_\alpha(P) = n^\alpha(m_P - m_0) \tag{1}$$

In this equation, q_α is the quality function, P is the subgroup, n is the number of elements in the subgroup, $\alpha \in [0; 1]$ is a parameter to adjust the weight of n in the final result, and finally m_P and m_0 are respectively the means of the target value calculated on the subgroup and the whole dataset.

Regarding the subgroup discovery methods, Herrera et al. [5] introduce a classification according to three main families: (1) **Extensions of classification algorithms**, based on algorithms which search rules to separate classes, and include notably the pioneer of the subgroup discovery (EXPLORA [6] and MIDOS [8]), (2) **Evolutionary algorithms for extracting subgroups**, which allows extracting fuzzy rules by using bio-inspired methods such as genetic algorithms (as MESDIF [3] for instance) and (3) **Extensions of association algorithms**, which particularity stands out in the possibility to use many types of target variable (binary, nominal, numerical, etc.)

In [2], Atzmueller et al. proposed SD-MAP, an algorithm based on FP-Tree which became a reference to the domain for years.

Recently, Millot et al. proposed OSMIND [7], that is able to extract optimal subgroups with better quality than SD-MAP, even without discretization, through a fast but exhaustive search. However, the previous methods have their limitation about the format of their subgroup, which can only be defined on continuous intervals.

In our recent works, we have proposed DISDi [4], that is the first attempt at using discontinuous intervals in its subgroup description, and was even able to surpass OSMIND in some situations.

In this paper, we propose to deepen the principle of discontinuous intervals in subgroup discovery through the SD-CEDI algorithm, by using consecutive erosion on a hypercube in order to extract the best subgroup with discontinuous intervals.

3   SD-CEDI Algorithm

In order to understand the tasks which have to be done by SD-CEDI, it is necessary to understand the principles of the Subgroup Discovery.

Thus, in this section, the tasks of Subgroup Discovery will be formally defined, then the concept of SD-CEDI will be detailed.

3.1   Preliminaries

Let D be a dataset constituted with a set of attributes A and a set of transactions T such as: $\forall t \in T, \forall a \in A, \exists v$ such that $t[a] = v$. The dataset D can also be seen as a hypercube where each attribute a is considered as a dimension on the cube. The search for subgroup discovery aims to find interesting patterns which describe a particular set of objects according to a quality function q_α (see Eq. 1).

In [2], Atzmueller et al. pinpoints 4 properties needed in order to conduct a Subgroup Discovery: (1) A quality function q_α that evaluate the quality of the subgroup, (2) a target variable a_{target} (or a group of target variables) used in the quality function, (3) a subgroup description language in which the format of the subgroup description is defined and (4) a search strategy, essentially the concept of the algorithm used for the extraction.

Obviously, the target variable depends on the dataset; nevertheless in this study we always work with non-discretized numerical variable as targets. As for the description language, we define a subgroup g as a set of objects called "selectors", s_i composed of a left part and a right part. The left part is an attribute a_i in A, while the right part is an interval or a set of intervals leading to a restriction on the value upon the a_i attribute.

For the quality function, we consider the function which is commonly used for the numeric target quality function [1] described previously in the Eq. 1.

At last, the search strategy used the principle of a FP-Tree which specificity will be explained further.

More formally, for each attribute a_i in A, the values Min_i and Max_i are defined respectively as the minimal and maximal value of the dataset on the attribute a_i.

Definition 1: Selector. A selector s_j is defined as a couple of objects (l_j, r_j) with $l_j = a_i \in A$ and r_j such as

$$r_j = \bigcup_{1 \to nb_I}^{k} [min_k; max_k] \tag{2}$$

where $min_1 \geq Min_i$, $max_{nb_I} \leq Max_i$ and $max_k < min_{k+1}$.

We can then define two functions: lh, the function which return the left part of a given selector and rh which return its right part, namely $lh : s_j \to l_j$ and $rh : s_j \to r_j$. We note S the set of all possible selectors.

Definition 2: Extent of a Selector. For every selector s_j, we can define its extent $ext(s_j)$ as the transactions of T for which the value of the attribute $lh(s_j)$ is in the range of $rh(s_i)$.

$$ext(s_j) = \{t \in T \text{ such as } t[lh(s_j)] \in rh(s_j)\} \tag{3}$$

Definition 3: Extent of a Subgroup. The extent $ext(g)$ of a subgroup g composed by the selectors $s_1, s_2, ..., s_l$ is the intersection of the extent of its selectors.

$$ext(g) = \bigcap_{1 \to l}^{k} ext(s_k) \tag{4}$$

Property 1: Closer Upper-Bound. For a subgroup g, a closer upper-bound $ub(g)$ can be identified as the highest value theoretically reachable with the combination of the transactions in $ext(g)$. This bound can be found by the formula:

$$ub(g) = max \left(q_\alpha(Top_1), q_\alpha(Top_2), ..., q_\alpha(Top_{\|ext(g)\|}) \right) \tag{5}$$

with Top_n the set of the n transactions of $ext(g)$ with the highest target value.

Proof. Let consider a set of transactions g_{small} with nb_{small} elements and sum_{small} the sum of the targets value of its transactions. For two transactions t_1 and t_2 with val_1 and val_2 as their value on the target variable, with $val_1 \geq val_2$ we can calculate the score of the new sets created by adding t_1 and t_2 to g_{small}, respectively g_1 and g_2:

$$val_1 \geq val_2 \Rightarrow sum_{small} + val_1 \geq sum_{small} + val_2$$

$$\Rightarrow mean_1 = \frac{sum_{small} + val_1}{nb_{small} + 1} \geq \frac{sum_{small} + val_2}{nb_{small} + 1} = mean_2$$

$$\Rightarrow q_\alpha(g_1) = n^\alpha (mean_1 - m_0) \geq n^\alpha (mean_2 - m_0) = q_\alpha(g_2)$$

With this property, using any subgroup g allows creating the best subsets of elements maximizing the quality function in the Eq. 1 for each possible size. Thus, the highest value determines the best score that any subset of g can achieve.

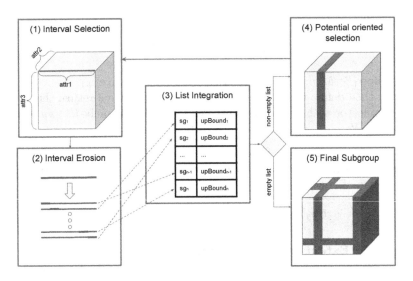

Fig. 1. 5 steps of the SD-CEDI algorithm

3.2   SD-CEDI

The algorithm SD-CEDI that presented performs the search for subgroups in 5 main steps, as depicted in Fig. 1: (i) **Interval Selection:** Selection of the first intervals available for an erosion in the current subgroup. (ii) **Interval Erosion:** Erosion of the selected interval and creation of the corresponding subgroups. (iii) **List Integration:** Addition of the new subgroups candidate in the list of candidates ordered by their upper bound. (iv) **Potential Oriented Selection:** Selection of a subgroup with the best potential in the list. If there is one, repetition of step 1 with the new subgroup. Otherwise, go to step 5. (v) **Final Subgroup:** Return the best subgroup found during the search if there are not more candidates.

More precisely, the SD-CEDI algorithm is detailed in Algorithm 1. SD-CEDI uses successive erosion in order to merge to the optimal solution. Thus, the first step is to identify the first subgroup which includes every transaction (Algorithm 1 - lines 2 and 3). From there, SD-CEDI can recursively extract all of the other subgroups of interest through the erosion (Algorithm 1 - line 7). Usually, this kind of erosion should be represented by a graph, as erosion of different intervals may lead to the same result. For instance, let's consider an ordered set $\{a, b, c, d, e\}$, the intervals $it_1 = [a; d]$ and $it_2 = [b; e]$. A third interval, $it_3 = [b; d]$ can be obtained with either a left erosion of it_1 or a right erosion of it_2.

In order to extract the list of sub-selectors without redundancies, additional information was added to the interval with which they are composed. Thus, the intervals which can be eroded exist in 4 states, according to the Fig. 2.

Three parameters have to be considered during the erosion: the lock on the left bracket (preventing the left erosion), the lock of the right bracket (preventing

---

Algorithm 1. SD-CEDI

Require: A : list of attributes, A_{target} : target variable, T : list of transactions
Ensure: $best_{sg}$: Subgroup with the highest score
1: $orderedList \leftarrow \{\}$
2: $first_{sg} \leftarrow completeBreakableSg(A, T)$
3: $curSg \leftarrow first_{sg}$
4: $best_{sg} \leftarrow first_{sg}$
5: $bestScore \leftarrow 0$
6: **while** $curSg \neq NULL$ **do**
7:      $sub_{sg} \leftarrow EROSION(curSg)$
8:      $orderedList.remove(curSg)$
9:      **for** $newSg \in sub_{sg}$ **do**
10:         **if** $newSg.getScore() > bestScore$ **then**
11:            $best_{sg} \leftarrow newSg$
12:            $bestScore \leftarrow newSg.getScore()$
13:         **end if**
14:         **if** $newSg.getUpperBound() > bestScore$ **then**
15:            $orderedList.addByUpperBound(newSg)$
16:         **end if**
17:      **end for**
18:      $curSg \leftarrow orderedList.takeFirst()$
19: **end while**
20: **return** $best_{sg}$

---

the right erosion) and the possibility of breaking the interval into the union of two intervals. At last, by considering that the erosion of a current subgroup can only be done on the first unlocked interval of the fist unlocked selector, the search space can be reduced from a graph, in which a subgroup can be the result of many different erosions, to a tree in which each node had only one parent. The erosion process is detailed in the Algorithm 2.

In order to find the best subgroup in this tree, SD-CEDI will go through the 5 steps presented earlier:

During the **(i) Interval selection** (Algorithm 2 - lines 2 to 11), SD-CEDI will identify, in the current subgroup, the interval which has to be eroded, which is the first interval which is not completely locked in all the selectors.

The interval will then be identified as one of the four possible types shown in Fig. 2, extract the corresponding subintervals and create the new subgroup candidate in the **(ii) Interval Erosion** step (Algorithm 2 - lines 12 to 25).

It can also be noted that each candidate is tested in order to become the best subgroup if they show a better result than the current best (Algorithm 1 - lines 10 to 13).

At the **(iii) List Integration**, the candidates which were generated will be tested: to optimize the exploration of the tree, SD-CEDI uses the Property 1. With this property, the method is able to prune the branches of the tree when the closer upper bound of their subgroup cannot reach the score of the best current subgroup (Algorithm 1 - lines 14 to 15).

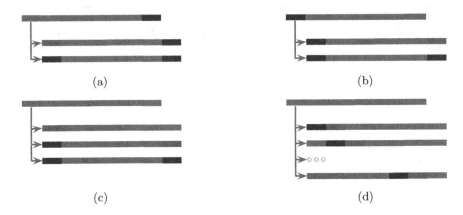

Fig. 2. Four kinds of interval erosion in the SD-CEDI algorithm: (a) left lock, (b) right lock, (c) no lock and (d) breakable, with blue for locks, red for removed elements and green for removable elements (Color figure online)

Furthermore, the closer upper-bound property is better exploited by finding a subgroup with a high score sooner, which expose two problems with the classic exploration of the tree. The subgroup with good score tends to be the results of many erosion and thus be positioned deep in the tree. The Breadth First Search is then inadequate with the closer upper-bound property. However, the Depth First Search also shown his flaws as the algorithms may invest too much time on subgroups with a low and unreachable closer upper-bound if the current best subgroup has a low score.

To overcome this limitation, SD-CEDI perform a **(iv) Potential Oriented selection**: each node of the tree which has to be tested is ordered by its closer upper-bound, as seen in the line 15 of the Algorithm 1. This method allows SD-CEDI to find the high scored subgroup faster, and then increase the effect of the pruning throw the upper bound.

At last, the **(v) Final Subgroup** extracted by SD-CEDI is returned at the end of the Algorithm 1, line 20.

4   Experimental Results

The performances of SD-CEDI have been evaluated on 4 datasets traditionally used as a benchmark for evaluating performances of subgroup discovery algorithms. These datasets, that come from the Bilkent repository[1], are the following: (i) Airport (AP), which contains air hubs in the United States as defined by the Federal Aviation Administration. (ii) Bolt (BL), which gathers data from an experiment on the effects of machine adjustments at the time to count bolt. (iii) Body data (Body), which represents data on body Temperature and Heart Rate. (iv) Pollution (Pol), which are data on pollution of cities.

---

[1] http://pcaltay.cs.bilkent.edu.tr/DataSets/.

Algorithm 2. EROSION

Require: *Subgroup* : sg
Ensure: sub_{sg} : List of the eroded subgroups created from sg
 1: $sub_{sg} \leftarrow \{\}$
 2: $indexSel \leftarrow 1$
 3: **while** $sg.isSelectorLock(indexSel)$ **do**
 4:     $indexSel \leftarrow indexSel + 1$
 5: **end while**
 6: $sel \leftarrow sg.getSelector(indexSel)$
 7: $indexInter \leftarrow 1$
 8: **while** $sel.isIntervalLock(indexInter)$ **do**
 9:     $indexInter \leftarrow indexInter + 1$
10: **end while**
11: $inter \leftarrow sel.getInterval(indexInter)$
12: **if** inter.isLockLeft() **then**
13:     $newSubInters \leftarrow inter.erodLeft()$
14: **else if** inter.isLockRight() **then**
15:     $newSubInters \leftarrow inter.erodRight()$
16: **else**
17:     $newSubInters \leftarrow inter.erodNoLock()$
18:     **if** inter.isBreakable() **then**
19:         $newSubInters.addAll(inter.erodBreak())$
20:     **end if**
21: **end if**
22: **for all** curInter $\in$ newSubsInters **do**
23:     $newSg \leftarrow sg.replaceInterval(curInter, indexSel, indexInter)$
24:     $sub_{sg}.add(newSg)$
25: **end for**
26: **return** sub_{sg}

The number of attributes as well as the number of transactions varies from a dataset to another, as shown in Table 1.

Table 1. Description of the datasets

	Airport	Bolt	Body	Pollution
Nb. attributes	5	7	2	15
Nb. transactions	135	40	130	60

We have compared SD-CEDI to the three main algorithms that are references in the domain: SD-MAP [2], OSMIND [7] and DISDi [4]. For each dataset, all attributes are numeric attributes and have not been discretized beforehand. Such an approach, without discretization, is interesting as the discretization process may not be an intuitive operation due to the multiple existing methods to do so. Thus observing these results under those circumstances may reveal the actual capacity of the subgroup discovery algorithm in front of raw data. Finally, note that in the case of the DISDi algorithm, all tests have been done with the β parameter - that defines the minimal size of the selector - fixed at the value which optimizes the score.

4.1   Best Quality on Raw Dataset

In a first step, we have studied the quality score of the extracted subgroups for each algorithm. Figure 3 describes the score of each method on the datasets of the benchmark, normalize by the maximum score obtained. The comparison of the best subgroup extracted by SD-CEDI is always used as a reference.

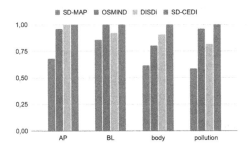

Fig. 3. Comparison of the quality value of the best subgroup identified

First of all, we can observe that for all datasets SD-CEDI always provides the best score.

In the case of SD-MAP, one of the reference algorithms, the best result are shown on the BL dataset, where it reaches 86% of the score of SD-CEDI. Nevertheless, if the dataset became more complex either by the number of attributes or the number of transactions, a larger gap is observed ranging from 59% to the pollution dataset to 68% to the AP dataset.

There is a major interest in the comparison with the other two algorithms, as SD-CEDI shares a similar search strategy with OSMIND but extends the description of selectors with discontinuous intervals, and inversely DISDi used a FP-Tree for the search strategy but also search discontinuous intervals.

The 4 datasets can be categorized in two types with their structure: (1) the few attribute number with high number of transactions (AP and Body) and the (2) high attribute number with low transaction number (BL and Pollution).

For the first type of dataset, the use of discontinuous intervals seems to be more effective, as both DISDi and SD-CEDI are able to extract subgroups with the best score possible. On the other hand, the algorithm that uses consecutive erosion shows better result with the second type of dataset, with OSMIND and SD-CEDI reaching the best score on this situation.

These results allow SD-CEDI to highlight the good performances of the approach we propose regarding the score of the extracted

subgroup. Indeed, we could observe that in all configurations SD-CEDI always identify subgroups having a score greater or equal to the best known approaches.

4.2 Best Quality on Resized Datasets

The second step of the analysis focuses on the evolution of the gain provided by SD-CEDI when the size of the datasets varies from the values 10, 20, 50 and 100 transactions when the dataset was large enough. The gain is evaluated through the equation below:

$$GAIN_{SD-CEDI} = \frac{Score_{SD-CEDI} - Score_{other}}{Score_{other}}$$

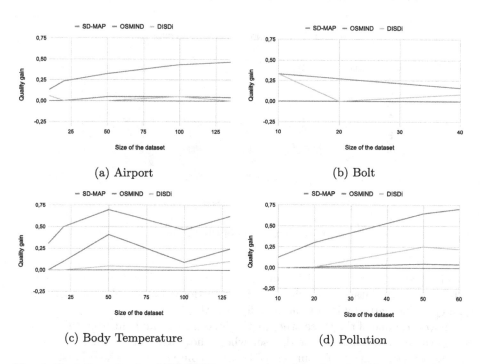

(a) Airport

(b) Bolt

(c) Body Temperature

(d) Pollution

Fig. 4. Evolution of gain of SD-CEDI compared to the other algorithms according to the dataset size

The Fig. 4 illustrates the evolution of the gain on the different dataset sizes. As previously observed, the results are significant since SD-CEDI is able to extract the subgroups with the best scores in each configuration. In accordance with the precedent section, it can be observed that SD-Map constantly shows the worst results. Except for the Bolt dataset, SD-CEDI provides a gain which either increase (AP and Pollution) or oscillating stably around a value (Body).

In the case of OSMIND, SD-CEDI provides a subgroup with a score 6% higher on average, but the contribution of our method stands out in the body dataset, where the average gain rises up to 17%. Compared to DISDi, the quality of the subgroup extracted by SD-CEDI are 8% better, reaching 12% on the pollution dataset and even 14% on the BL dataset.

These results confirm our previous observations. Indeed, we observe that SD-CEDI extract subgroup with better score quality than its pairs regardless of the size of the dataset, which confirms the good performances of the approach.

4.3   Time Comparison

In the last part, we focus on the runtime to extract subgroups. The Fig. 5 displays the computation time taken by each algorithm.

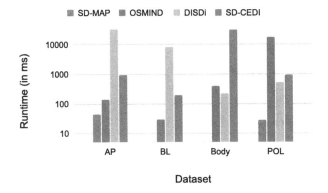

Fig. 5. Comparison of the runtime

The first observation which can be made is on SD-MAP. Indeed, it is the fastest on every dataset, especially on BL and Body, but we observed that the quality of the patterns is not the best for SD-MAP.

However, none of the three other methods seems to stand out from the others. In the three cases, there is one dataset where they take more than 10 s, and no special pattern seems to favour one of them in comparison to the others.

Thus regarding the calculation time, these results suggest that the approaches are relatively equivalent.

5   Conclusion

In this paper, we have addressed the subgroup discovery problem and we have presented SD-CEDI, a new approach which the originality consists of extracting subgroups defined discontinuous intervals. The main idea behind the approach is to model the search space through a hypercube and to slice this hypercube to highlight subgroups. The results of the experiments, conducted on a benchmark datasets, show the good performances of SD-CEDI to always extract an equivalent or higher subgroups than the main algorithms of the domain that only focus on continuous intervals.

As perspectives, it may be interesting to parallelize the method in order to scaling up for more complex datasets. Another interesting track is to focus on a way to express the quality of a subgroup in regard to the description of its attributes, thus defining a new quality measure which takes into account the discontinuity of the intervals.

References

1. Atzmueller, M.: Subgroup discovery. Wiley Interdiscipl. Rev. Data Min. Knowl. Discov. **5**(1), 35–49 (2015)
2. Atzmueller, M., Puppe, F.: SD-Map – a fast algorithm for exhaustive subgroup discovery. In: Fürnkranz, J., Scheffer, T., Spiliopoulou, M. (eds.) PKDD 2006. LNCS, vol. 4213, pp. 6–17. Springer, Heidelberg (2006). https://doi.org/10.1007/11871637_6
3. Berlanga, F., Del Jesus, M.J., González, P., Herrera, F., Mesonero, M.: Multiobjective evolutionary induction of subgroup discovery fuzzy rules: a case study in marketing. In: Perner, P. (ed.) ICDM 2006. LNCS, vol. 4065, pp. 337–349. Springer, Heidelberg (2006). https://doi.org/10.1007/11790853_27
4. Eugenie, R., Stattner, E.: DISDI: discontinuous intervals in subgroup discovery (2021)
5. Herrera, F., Carmona, C.J., González, P., Del Jesus, M.J.: An overview on subgroup discovery: foundations and applications. Knowl. Inf. Syst. **29**(3), 495–525 (2011)
6. Klösgen, W.: Explora: a multipattern and multistrategy discovery assistant. In: Advances in Knowledge Discovery and Data Mining, pp. 249–271 (1996)
7. Millot, A., Cazabet, R., Boulicaut, J.F.: Optimal subgroup discovery in purely numerical data. In: Lauw, H., Wong, R.W., Ntoulas, A., Lim, E.P., Ng, S.K., Pan, S. (eds.) PAKDD 2020. LNCS, vol. 12085, pp. 112–124. Springer, Cham (2020). https://doi.org/10.1007/978-3-030-47436-2_9
8. Wrobel, S.: An algorithm for multi-relational discovery of subgroups. In: Komorowski, J., Zytkow, J. (eds.) PKDD 1997. LNCS (LNAI), vol. 1263, pp. 78–87. Springer, Heidelberg (1997). https://doi.org/10.1007/3-540-63223-9_108

Fast SQL/Row Pattern Recognition Query Processing Using Parallel Primitives on GPUs

Tsubasa Ohara, Qiong Chang$^{(\boxtimes)}$, and Jun Miyazaki$^{(\boxtimes)}$

School of Computing, Tokyo Institute of Technology, Tokyo, Japan
ohara@lsc.cs.titech.ac.jp, {q.chang,miyazaki}@c.titech.ac.jp

Abstract. SQL/Row Pattern Recognition (SQL/RPR), a row matching query processing for sequence data stored in a database, has been standardized in SQL:2016. So far, many studies have focused on developing technology to perform SQL/RPR for large-scale sequence data, such as stock chart records and system logs. However, due to the large amount of data and complex calculation problems, the processing speeds of current methods are not sufficient. In this paper, we propose a fast SQL/RPR pattern-recognition method that uses parallel primitives on GPUs. This method improves on the processing speed of PostgreSQL, a commonly used RPR method, by 7.1 to 22.6 times in different use cases.

Keywords: GPGPU · Parallel primitives · Big data processing · SQL/RPR

1 Introduction

With the recent development of information technology, a large amount of data is generated every day. Sequence data, which is data continuously arranged along the time axis such as stock chart records and system logs, is one of the most common forms of data storage. It is usually arranged by row in a database, with multiple rows of data showing relationships and changing trends. Users can perform a matching task from the sequence data in accordance with a pre-specified pattern, such as "the stock price goes down" followed by "the stock price goes up". Thus, important feature points (e.g. inflection points) can be extracted, thereby affecting users' decisions (e.g. bottom fishing).

This kind of row-pattern matching task on sequence data is called *SQL/Row Pattern Recognition* (SQL/RPR) [9]. Because a large number of applications require pattern recognition for sequence data, SQL/RPR has been standardized in SQL:2016 and divided into four steps: 1) grouping the target sequence data set, 2) sorting the data in each group in accordance with user requirements, 3) transforming target row patterns on the basis of regular expression, and 4) matching target row patterns with each group. This four-step process is suitable for various types of sequence data and provides support for data analysis in many

© Springer Nature Switzerland AG 2021
C. Strauss et al. (Eds.): DEXA 2021, LNCS 12923, pp. 22–34, 2021.
https://doi.org/10.1007/978-3-030-86472-9_3

applications. However, the operations in each step are computationally expensive, especially for large-scale data sets. Therefore, accelerating the processing speed of SQL/RPR is a challenging task.

Over the past decade, the GPU, a programmable processor originally designed for accelerating computer graphics and image processing, has become compatible with many particular frameworks that provide access to general purpose computing (GPGPU). In contrast to CPUs, current GPUs consist of thousands of smaller cores, which give applications their high computing capability. Mars [7] was proposed to implement the MapReduce framework [5], and is an efficient and simple parallel programming model, on GPUs. However, implementing SQL/RPR effectively with MapReduce is difficult. Fortunately, some GPU programming libraries of parallel algorithm primitives are available, such as ModernGPU [4] and CUDPP [2,6]. These primitives are important building blocks for a wide variety of algorithms such as gather, scatter, scan, sort, split, and reduction, and building data structures such as trees and summed-area tables. As a result, these parallel primitives are suitable for and can accelerate the SQL/RPR four-step process.

In this paper, we propose a GPU-based framework for accelerating SQL/RPR. The main contribution of our work is to map the four steps of SQL/RPR onto GPUs and use the corresponding parallel primitives to perform each step at a high processing speed.

The remainder of the paper is organized as follows. Section 2 describes related work. In Sects. 3 and 4, the mechanism of SQL/RPR and the proposed method are described. The experimental results are discussed in Sect. 5, and are followed by concluding remarks in Sect. 6.

2 Related Work

2.1 Implementation of SQL/RPR Based on Spark SQL

Nakabasami et al. [10] proposed a method consisting of sequence and row filtering to reduce the processing cost of row pattern matching for sequence data. Sequence filtering is a method that groups the data with the PARTITION BY clause. Then, it removes the unmapped groups from the row pattern matching in accordance with conditions defined by the DEFINE clause. The conditions may include row pattern variables such as PREV() and NEXT(). Row filtering is a method that keeps only the rows that match the conditions of the row pattern variable defined by the DEFINE clause, plus a few rows before and after, and removes all other rows. Using the above methods, the number of target rows for row pattern matching and the overall matching cost will be greatly reduced.

2.2 SQL Query Processing on GPUs

PG-Strom [3] is an extension module designed for PostgreSQL, and it accelerates SQL aggregation queries and batch processing for large data sets by using GPUs.

PG-Strom consists of a code generator, that automatically generates a GPU code from SQL instructions, and an execution engine, that executes SQL workloads in parallel on GPUs. Moreover, for tasks suitable for GPU processing, it provides a function that automatically replaces PostgreSQL in the background to improve the processing capability of the entire system.

He et al. [8] proposed algorithms for relational database operators on GPUs. They defined eight essential functions to efficiently implement algorithms for major relational operators, such as selection, projection, join, and aggregation. Due to these algorithms, query processing can be accelerated by 2 to 27 times.

3 SQL/Row Pattern Recognition

This section describes the mechanism of SQL/RPR standardized in SQL:2016. Figure 1 shows an example of a trading database. This database has three attributes: item, trade date, and price. Figure 2 shows an example SQL/RPR query to the database.

Table

item	tradedate	price
item1	2021-02-05	152.00
item1	2021-02-15	142.00
item1	2021-02-17	142.00
item1	2021-02-18	138.00
item1	2021-02-19	144.00
item1	2021-02-21	150.00
item1	2021-02-22	152.00
item2	2021-02-13	335.00
item2	2021-02-14	329.00
item2	2021-02-19	329.00
item2	2021-02-20	326.00
item2	2021-02-22	320.00
item2	2021-02-25	326.00
item2	2021-02-27	328.00
item3	2021-02-05	148.00
item3	2021-02-10	153.00
item3	2021-02-13	151.00
item3	2021-02-20	326.00

```
SELECT *
FROM Table
    MATCH_RECOGNIZE
    (
        PARTITION BY item
        ORDER BY tradedate
        MEASURES
            CLASSIFIER() AS classy,
            A.tradedate AS startdate,
            A.price AS startprice,
            LAST(B.tradedate) AS bottomdate,
            LAST(B.price) AS bottomprice,
            LAST(C.tradedate) AS enddate,
            LAST(C.price) AS endprice,
            MAX(U.price) AS maxprice
        ALL ROWS PER MATCH
        AFTER MATCH SKIP PAST LAST ROW
        PATTERN (A B+ C+)
        SUBSET U = (A, B, C)
        DEFINE
            B AS B.price < PREV(B.price),
            C AS C.price > PREV(C.price)
    )
```

Fig. 1. Example of database　　　　**Fig. 2.** Example of SQL/RPR query

In this query, a V-shaped pattern, in which the transaction price decreases one or more times followed by one or more increases, is extracted. To use the SQL/RPR function, the MATCH_RECOGNIZE clause is used, and a series of clauses is then defined under it to extract the corresponding pattern, including PARTITION BY, ORDER BY, MEASURES, ONE ROW PER MATCH, ALL ROWS PER MATCH, AFTER MATCH SKIP, PATTERN, SUBSET, and DEFINE.

- PARTITION BY: Groups rows with the same specified value by an attribute.
- ORDER BY: Sorts rows in each group by the attribute specified in the PARTITION BY clause.
- MEASURES: Defines sorted rows on the basis of row pattern measurement conditions.
- ALL ROWS PER MATCH: Outputs one row for each extracted pattern.
- AFTER MATCH SKIP: Specifies start of next pattern matching location, after extracting a non-empty pattern.
- PATTERN: Defines the pattern to be extracted using a regular expression with a row pattern variable and quantity specifier, such as $\{+,*\}$.
- SUBSET: Defines a set list of row pattern variables.
- DEFINE: Defines the matching conditions of row pattern variables, including the functions PREV() and NEXT(), and aggregation functions such as AVG(), MAX(), and COUNT().

4   Implementation of SQL/RPR on GPUs

In this section, we propose a method for accelerating SQL/RPR with GPUs. As shown in Fig. 3, our method builds on the four-step process flow mentioned in Sect. 1 with an additional step at the beginning and end of the process.

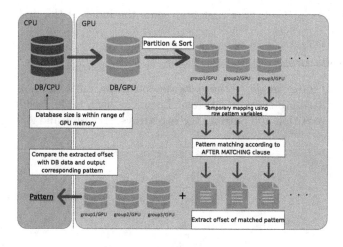

Fig. 3. Workflow of proposed method

In our method, first the SQL/RPR processing is transferred from the CPU-database to GPU memory. After that, the data is partitioned into several groups and sorted by respective specified item. Next, a row pattern variable is temporarily mapped to each row in each group on the basis of the DEFINE clause, called *temporary mapping*. Then, pattern matching is performed for each group in parallel in accordance with the options defined by the AFTER MATCH SKIP clause. This allows the GPU to produce an offset of the matched pattern. In our acceleration method, the offset is then sent back to the CPU database where the results are generated and displayed on the host (CPU).

In the first step, when the CPU database is transferred to the GPU, each attribute is arranged as an array. This includes the ID attribute, which is called the *ID array*. Moreover, in this research, it is assumed that the size of the CPU database is within the range that can be stored in GPU memory (GPU database).

4.1   Partition

During the partition phase, the data is grouped by attributes with the same value (called *partition attributes*). In our research, a hash table is required to perform the partition for the GPU database. The value of a partition attribute is set as a *key*, and the *ID* transferred from the CPU database is set as a *value*. Thus, both are inserted into the hash table in pairs. In our implementation, the hash table parallel primitive in CUDPP is used to perform the grouping for each value of the partition attributes. It outputs two kinds of arrays: one is the *ID* array that represents a series of groups organized as:

$$Array_{ID} = (group1\_1, group1\_2, ..., group2\_1, ..., group3\_1, \cdots); \qquad (1)$$

the other is the *SegHead* array that represents the index of the first element of each group as:

$$Array_{SegHead} = (index\ of\ group2\_1,\ index\ of\ group3\_1, \cdots). \qquad (2)$$

Figure 4 shows an example of the database partition on a GPU. In this case, the partition attribute is set as an *item*. Then, the hash table parallel primitive completes the partition and outputs the two arrays, $Array_{ID}$ and $Array_{SegHead}$, for each group.

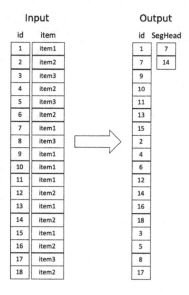

Fig. 4. Example of partition

4.2 Sorting

Sorting each partitioned group requires not only the two outputs from Sect. 4.1 but also an attribute specified by the user, called the *sorting attribute*. Since sorting must be done independently within each group, the segmented sort parallel primitive in ModernGPU is used. It uses two kinds of arrays as inputs: *Array_A*, which is sorted for each segment, and *Array_B*, which defines the interval between any two segments. In our implementation, *Array_A* is set as the array that holds the sorting attribute, and *Array_B* is set as the $Array_{SegHead}$. Then, *Array_A* is sorted in accordance with *Array_B* and updated with the sorting results. At the same time, a new array *Array_C*, which is stored with the index of the updated *Array_A* is also generated.

4.3 Pattern Definition Using Regular Expression

Inside of SQL/RPR, the matching begins by performing temporary mapping using row pattern variables in accordance with the DEFINE clause. As each row is mapped, the system also checks for possible matching patterns. If a pattern is matched, the temporary mapping of row pattern variables becomes permanent; if not, the matching is cancelled. This process is performed for all rows. However, in our implementation, the ordering of the temporary mapping and pattern checking is performed differently. First, temporary mapping is performed on all rows; then, the rows are checked for possible matching patterns. Each process is performed in parallel by GPU threads.

```
AFTER MATCH SKIP PAST LAST ROW
PATTERN (A+ B+)
DEFINE    A AS PREV (A.Price + A.Tax) < 100
          B AS PREV (B.Price) >= 100
```

Use Case 1

```
AFTER MATCH SKIP PAST LAST ROW
PATTERN (A+ B+)
DEFINE    A AS NEXT (A.Price + A.Tax) < 100
          B AS NEXT (B.Price) >= 100
```

Use Case 2

```
AFTER MATCH SKIP PAST LAST ROW
PATTERN (A+ B+)
DEFINE    A AS A.Price < 200
          B AS B.Price + B.Tax > FIRST (A.Price) + 200
```

Use Case 3

```
AFTER MATCH SKIP PAST LAST ROW
PATTERN (A* B+ C)
DEFINE    A AS A.Price + A.Tax > 200
          B AS B.Price < 100
          C AS C.Tax < MAX (A.Tax)
```

Use Case 4

```
AFTER MATCH SKIP PAST LAST ROW
PATTERN ( (A | B) + C )
DEFINE    A AS A.Price >= 150
          B AS B.Price < 100
          C AS CASE
                  WHEN PREV (CLASSIFIER ()) = 'A'
                  AND C.Price >= 100 AND C.Price <130
                  THEN 1
                  WHEN PREV (CLASSIFIER ()) = 'B'
                  AND C.Price >= 130 AND C.Price <150
                  THEN 1
                  ELSE 0
```

Use Case 5

Fig. 5. Use cases

Because the definition of the matching condition of row pattern variables is performed by combining multiple functions specified in SQL:2016, the number of patterns in the DEFINE clause is limitless. In our experiment, the five use cases shown in Fig. 5 are used to evaluate our method. These use cases generally use the functions specified in SQL:2016, which are similar to the DEFINE clauses of the original use cases. The temporary mapping of each row in each group must be done in parallel. Furthermore, the *ID* array and the *SegHead* array output in Sect. 4.1 are also required to avoid a cross-row during temporary mapping. Use cases 3 and 4 contain row pattern variables in the DEFINE clause that cannot be matched in temporary mapping, such as First() and aggregation functions. In this case, temporary mapping can only be performed using suitable functions, such as *A* in use case 3 and *A, B* in use case 4. The row pattern variables that are not suitable for temporary mapping are moved to the next step, such as *B* in use case 3 and *C* in use case 4.

4.4 Pattern Matching

During the pattern matching step, CUDA-grep [1] is used to extract patterns of row pattern variables from the temporary mapping data. It matches multiple regular expression patterns with the target data in parallel. However, in our implementation, one regular expression pattern, defined in the PATTERN clause, is matched with multiple partitioned groups. Therefore, CUDA-grep is adjusted to use multiple threads of GPUs in order to perform pattern matching for multiple pieces of target data in parallel.

Next, use case 4 is used to introduce how to perform matching when row pattern variables are not suitable for temporary mapping. Here, the target data is defined in the PATTERN clause: $(A * B + C)$. First, temporary mapping is performed for row pattern variables A and B. Then, matching is performed on all rows in each partitioned group until pattern $(A * B+)$ is matched. After that, for the row where $(A * B+)$ is matched to its previous row and no row pattern variable itself is mapped, the MAX() function is performed. The purpose of the MAX() function is to return the highest *Tax* value among all rows that are temporarily mapped as row pattern variable A. In our implementation, MaxReduce of ModernGPU is used, which is a parallel primitive that outputs the largest value in an input array and makes it possible to calculate the conditions for mapping row pattern variable C to each row.

After pattern matching is completed, an offset array is output to show which rows are matched. In this array, each pattern is assigned a sequential pattern number that is added to all rows included in the corresponding pattern. In addition, the pattern numbers are independent for each group, which means that the last number shows the total number of matched patterns in that group. The offset array is compared with the CPU-database and the matched patterns are output as the final results.

5   Evaluation

In this section, we compare and evaluate our method with a method that implements SQL/RPR on PostgreSQL.

5.1   Method Used for Comparison

There are five tasks in the PostgreSQL-based method used for comparison.

Partition&Sort: The input data is partitioned into several groups and sorted within each group using the PARTITION BY and ORDER BY clause.

Mapping RPV: Temporary mapping of row pattern variables is performed for each row on the basis of the DEFINE clause.

Concat RPV: The row pattern variables temporarily mapped are concatenated for each group and converted into a long character string as shown in Fig. 6.

Pattern Matching: Pattern matching is performed on the concatenated row pattern variable attributes for each group. These attributes are converted into a table with two attributes: 1) *id*, which is a serial number starting from 1, and 2) *is_match*, which is a pattern number assigned in accordance with the pattern appearance position.

Collecting Results: The final result is generated based on the table output from the *Pattern Matching* and *Mapping RPV* tasks.

item	tradedate	price	RPV
item1	2021-02-05	152.00	A
item1	2021-02-15	142.00	B
item1	2021-02-17	142.00	A
item1	2021-02-18	138.00	B
item1	2021-02-19	144.00	C
item1	2021-02-21	150.00	C
item1	2021-02-22	152.00	C
item2	2021-02-13	335.00	A
item2	2021-02-14	329.00	B
item2	2021-02-19	329.00	A
item2	2021-02-20	326.00	B
item2	2021-02-22	320.00	B
item2	2021-02-25	326.00	C
item2	2021-02-27	328.00	C
item3	2021-02-05	148.00	A
item3	2021-02-10	153.00	C
item3	2021-02-13	151.00	B
item3	2021-02-20	326.00	C

item	RPVS
item1	ABABCCC
item2	ABABBCC
item3	ACBC

Fig. 6. Example of Concat RPV Task

All tasks, except for the *Pattern Matching* task, were performed using the SQL queries. Since the *Pattern Matching* task cannot be implemented by SQL query alone, it was implemented in C++.

5.2   Experiment Setup

The architecture of the system we used is shown in Table 1. We tuned the buffer parameters of PostgreSQL used as the method for comparison. In this experiment, the execution time was measured from the time when data was read into the CPU database, to the time when the calculation result was written back to the memory on the CPU. Thus, the data transfer time between CPU and GPU was included.

Table 1. Architecture of machine

CPU	Intel Core i7-6800K (3.4 GHz, 6 cores)
RAM	32 GB
GPU Series	NVIDIA TITAN X (Pascal)
CUDA Cores	3584
GPU Basic Clocks	1.53 GHz
GPU Memory Size	12 GB
CUDA	CUDA8.0
OS	CentOS 7.7
DBMS	PostgreSQL 9.2

A test table was designed for the CPU database first. It had four attributes: item, trade date, price, and tax. Artificial data was used to be read into the test table. The data for each attribute was random and followed a uniform distribution. We used three databases with overall sizes of 870 MB, 1.7 GB, and 10 GB. The numbers of records in the three databases are around 1.74×10^7, 3.47×10^7, and 2.08×10^8, respectively. In addition, there were ten types of data for the item attribute. Figure 7 shows the queries used in our experiment, and the definitions of the AFTER MATCH SKIP, PATTERN and DEFINE clauses are the same as the use cases shown in Fig. 5.

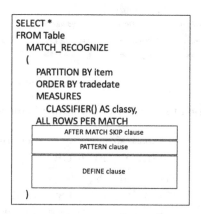

Fig. 7. Queries used in experiment

5.3 Experimental Results

The execution time of our proposed method was measured for each task. The tasks included *copyHostToDevice, Partition, Sort, Mapping RPV, Pattern Matching,* and *copyDeviceToHost. copyHostToDevice* is the transferring of data from CPU to GPU, and *copyDeviceToHost* is the reverse. In addition, we measured the execution time of the method used for comparison for each task described in Sect. 5.1.

The execution times for use case 1 are shown in Fig. 8. The time of each task and the total time of all tasks were measured for the different database sizes. The processing speed of our method was about 7.1–16.9 times faster than the method for comparison. Moreover, in all other use cases, the speed of our method was 7.5–22.6 times faster than the method for comparison.

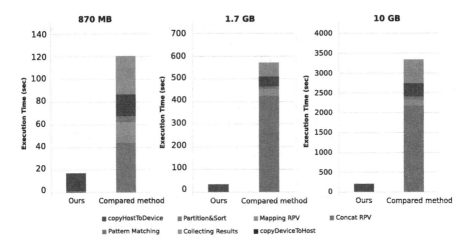

Fig. 8. Comparison of execution time for different data set sizes (use case 1)

Table 2. Execution time of Pattern Matching in different use cases

Execution time (s)	Database size			Performance (times)
	870 MB	1.7 GB	10 GB	
Use case 1	15.617	31.228	186.957	7.1–16.9
Use case 2	15.213	30.325	181.821	7.5–17.8
Use case 3	15.635	31.093	186.157	12.0–22.6
Use case 4	14.411	28.637	172.212	10.5–22.5
Use case 5	15.221	30.291	180.606	10.3–20.2

As shown in Fig. 8, with our method, the *Pattern Matching* task took the most time, and there was a big gap in the execution times between the *Pattern Matching* task and others. Except for *Pattern Matching*, all other tasks had almost no effect on the overall execution time of the system and could be ignored. Table 2 shows the execution time of the *Pattern Matching* task for each use case. Among them, the execution time of use case 4 showed a slight difference from the other use cases due to its different extraction pattern. Nevertheless, the execution time of all tasks with the method for comparison increased as the database size increased as shown in Fig. 8. Compared with our method, the execution time of each step was higher and could not be ignored. This also fully proves the effectiveness of our acceleration method.

Tables 3 and 4 show the execution time of each task and the total time of all tasks when the number of groups was changed from 10 to 10,000. The database size was 870 MB, and use case 1 was used for the AFTER MATCH SKIP, PATTERN, and DEFINE clauses.

Table 3. Execution time of proposed method for different group numbers

Execution time (s)		Group number		
		10	1000	10000
Task	copyHostToDev	0.30	0.30	0.30
	Partition	0.03	0.03	0.03
	Sort	1.00	0.99	0.97
	Mapping RPV	0.001	0.001	0.001
	Pattern Match	15.62	12.41	8.15
	copyDevToHost	0.080	0.081	0.076
Total time		17.03	13.81	9.53

Table 4. Execution time of method for comparison for different group numbers

Execution time (s)		Group number		
		10	1000	10000
Task	Partition & Sort	44.13	84.77	103.97
	Mapping RPV	17.89	17.66	17.91
	Concat RPV	5.37	5.08	4.63
	Pattern match	19.38	19.71	18.58
	Collect results	34.04	38.10	35.23
Total time		120.81	165.33	180.31

As shown in these two tables, the execution time of the *Pattern Matching* task in our proposed method notably decreased as the number of groups increased. The reason is that processing with a high degree of parallelism was performed as the number of groups increased, leading to a decrease in execution time. In comparison, for the other method, the execution time of the *Partition&Sort* task increased as the number of groups increased due to the larger amount of partition calculations. In accordance with this result, the larger the number of partitioned groups, the more effective the proposed method was with high parallelism.

To measure the performance by a GPU-supported RDB, we have also tested a few queries with the latest version of PostgreSQL (PostgreSQL 12) and PG-Strom, because PG-Strom can work only with PostgreSQL 12. The execution time of PG-Strom and that of PostgreSQL 12 were almost the same in most cases, which means that the GPU of PG-Strom could not contribute to SQL/RPR queries. It is noteworthy that the performance of PostgreSQL 12 was roughly two times higher than that of PostgreSQL 9. Nevertheless, their efficiency was far lower than our proposed.

6   Conclusion

In this paper, we proposed a fast SQL/RPR query execution method that uses parallel primitives on GPUs. In an experiment, SQL/RPR was implemented on PostgreSQL as a method for comparison. The performance was evaluated by comparing the two methods by using SQL/RPR use case queries. The results show that it is possible to efficiently perform SQL/RPR query processing on GPUs by combining technologies such as parallel primitives. In addition, the proposed method achieved a processing speed of about 7.1–22.6 times faster than the method for comparison.

In the future, a compiler that can automatically generate a GPU code for the DEFINE clause is expected to be built. In addition, supporting multiple options that exist in the AFTER MATCH SKIP clause and a fast SQL/RPR method for a database that exceeds the GPU memory size are also worthy of challenge.

References

1. Cuda-grep. https://github.com/bkase/CUDA-grep. Accessed 20 Jan 2021
2. Cudpp. https://github.com/cudpp/cudpp. Accessed 20 Jan 2021
3. Pg-strom. https://heterodb.github.io/pg-strom/. Accessed 20 Jan 2021
4. Sean baxter: Moderngpu 2.0. https://github.com/moderngpu/moderngpu. Accessed 20 Jan 2021
5. Dean, J., Ghemawat, S.: MapReduce: simplified data processing on large clusters. Commun. ACM **51**(1), 107–113 (2008)
6. Harris, M., Sengupta, S., Owens, J.D.: GPU Gems 3 – Parallel Prefix Sum (Scan) with CUDA (Chap. 39), pp. 851–876. Addison Wesley, Boston (2007)
7. He, B., Fang, W., Luo, Q., Govindaraju, N.K., Wang, T.: Mars: a mapreduce framework on graphics processors. In: 17th International Conference on Parallel Architectures and Compilation Techniques, PACT 2008, Toronto, Ontario, Canada, 25–29 October 2008, pp. 260–269. ACM (2008)
8. He, B., et al.: Relational query coprocessing on graphics processors. ACM Trans. Database Syst. **34**(4), 21:1–21:39 (2009)
9. ISO/IEC JTC 1/SC 32: Information technology - database languages - SQL technical reports - part 5: row pattern recognition in SQL. Technical report, ISO/IEC (2016)
10. Nakabasami, K., Kitagawa, H., Nasu, Y.: Optimization of row pattern matching over sequence data in spark SQL. In: Hartmann, S., Küng, J., Chakravarthy, S., Anderst-Kotsis, G., Tjoa, A., Khalil, I. (eds.) DEXA 2019. LNCS, vol. 11706, pp. 3–17. Springer, Cham (2019). https://doi.org/10.1007/978-3-030-27615-7_1

Scalable Tabular Metadata Location and Classification in Large-Scale Structured Datasets

Kazi Islam and Michael Gubanov[(✉)]

Department of Computer Science, Florida State University,
Tallahassee, FL 32306, USA
{km20fr,gubanov}@cs.fsu.edu

Abstract. Tabular metadata (i.e. attribute names) location and classification is a *fundamental* problem for large-scale structured corpora. Web tables [24], CORD-19 [35], have thousands to millions of tables, but often have missing or incorrect labels for rows (or columns) with attribute names (e.g. *Last Name*). Missing or incorrect metadata labels [19] prevent or at least significantly complicate the fundamental data management tasks such as *query processing, data integration, indexing*, and many other. Different sources position metadata rows/columns differently inside a table, which makes its reliable identification challenging.

In this work we describe a scalable, hybrid two-layer Deep- and Machine-learning based ensemble, combining Long Short Term Memory (LSTM) and Naive Bayes Classifier to accurately identify Metadata-containing rows or columns in a table. We have performed an extensive evaluation on several structures datasets, including an ultra large-scale dataset containing more than 15 million tables coming from more than 26 thousands of sources to justify scalability and resistance to heterogeneity, stemming from a large number of sources. We observed superiority of this two-layer ensemble, compared to the recent previous approaches and report an impressive 81.53% accuracy at scale.

Keywords: Hierarchical metadata · Metadata classification · Web table

1 Introduction

Large-scale structured datasets are becoming ubiquitous. WDC [24], CORD-19 [35], Census Buerau [1] are just a few examples. Such corpora exhibit a wealth of useful structured data usually originating from hundreds to millions of different sources. Each source represents even the tables of the same category (e.g. Songs, COVID vaccines side-effects) differently, hence efficiently accessing or deriving insights from such heterogeneous and large-scale data is extremely complicated [2,13,16,17,27,28,30,31]. Not only that, but also some sources use only *relational* tables, other sources may use tables of different *non-relational* formats.

© Springer Nature Switzerland AG 2021
C. Strauss et al. (Eds.): DEXA 2021, LNCS 12923, pp. 35–50, 2021.
https://doi.org/10.1007/978-3-030-86472-9_4

For example, it is very common that information about a particular product is stored differently by different Websites with different attributes and formatting. The problem becomes worse when the table attributes are *hierarchical* (i.e. several attributes nested inside another one).

Because of the table format and metadata variety, efficient metadata annotation and identification is still a subject of ongoing research [4,19,23]. Since tabular data can be stored not only in a relational database, but also in CSV format, as a spreadsheet, etc., there have been related research focused on CSV structure detection [7,23] and Web table metadata annotation [10]. Although these systems showed promising performance, the evaluation sets used in their experiments have relatively small number of sources, hence are very *homogeneous* (i.e. do not exhibit high variety of tabular and metadata representations by contrast to heterogeneous datasets composed from many sources that we used to evaluate our approach). Each source has a liberty to choose the table and metadata format, hence an algorithm, which works good for one source, usually performs much worse for tables from another source unless the formats are significantly similar. To prove that the approach is robust for diverse sources, the evaluation sets should be composed from as many sources as possible.

In this work, we describe and evaluate a hybrid Deep- and Machine-Learning ensemble to classify metadata rows/columns in a table. To gauge generality, we have evaluated it on two large-scale collections of tables - CORD-19 [35] and Web Data Commons (WDC) [24], which have more than 88 thousand and 15 million tables in English respectively. In both cases, there are hundreds to hundred thousands of different sources, storing data and forming tables in different ways. We have designed an ensemble combining Long Short Term Memory (LSTM) [20] based Recurrent Neural Network (RNN) [21] and a Naive Bayes Classifier. Most of the previous approaches are limited to only on table cell-level analysis, whereas we have take into account the cell context (i.e. the whole tuple or a column). Our approach takes the context of the cell into account along with the position of the cell and the surroundings of that cell in the table. The first-layer model - RNN is order-sensitive as order matters for the terms inside a cell (i.e. *First Name* is different from *Name First*). However, the second-layer model is order-insensitive as the order of attribute values does not matter in a tuple. For example, a tuple having *artist* and then *album* value is the same as vice versa.

Finally, we designed an algorithm that using our hybrid metadata classification ensemble can distinguish different kinds of metadata in a table - row/column-based or hierarchical. To summarize, the main contributions of this paper are the following:

1. A novel two-layer ensemble comprised of an RNN and Naive Bayes models for metadata detection. On the first layer, the RNN model performs analysis of an input cell and encodes the context. In the second step, the feature vectors composed of encodings of all cells in a tuple or a column are classified with the Naive Bayes classifier combined with the decision-tree. The output is a binary label indicating whether a tuple/column contains metadata or just regular data as well as whether it is of hierarchical or plain type.

2. A Web-scale training and evaluation infrastructure that we architected and which is essential for experiments with large-scale datasets.
3. Extensive evaluation on several Web-scale datasets with tables coming from thousands to millions of different sources from a wide variety of domains.

The rest of the paper is structured as follows. First we define the terminology that we used throughout the paper. Then we describe the methodology, followed by the metadata classification ensemble description. After it we explain the large-scale evaluation architecture and experimental study on large-scale datasets with comparative evaluation against the previous approaches. We finish with the related work discussion and conclusion.

2 Definitions

Relational tables, defined in [8], have the following properties: values are atomic, each column has values of same type, each column has unique name. An example of relational table is illustrated in Fig. 1(a).

Def_1: <u>Cell</u> is a data value (i.e. can be a number, string, etc.) found at the intersection of a row and a column in a table. A relational table has C * R cells total, where C number of columns and R number of rows.

Def_2: <u>Metadata</u> is a sequence of the *attribute* names of a table, found in a row or a column. Metadata can be represented as a row - Fig. 1(a), or a column - Fig. 1(c)

Def_3: We call <u>non-relational</u> here the tables that have *hierarchical/nested* metadata. For example, a metadata row or column may have nested metadata, such as in Fig. 1(b) - "Name" column is divided into two columns having "First name" and "Last name" separately.

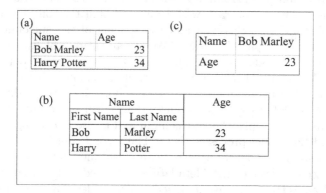

Fig. 1. Vertical and horizontal alignment of Metadata (a, c); hierarchical metadata (b).

2.1 Non-relational Table Representation

In Fig. 1, three different types of tables that our ensemble is working with are shown - two relational tables (a, c) and a non-relational table having hierarchical

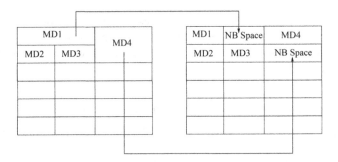

Fig. 2. Tables with hierarchical metadata and their structural representation. MD denotes Metadata.

Metadata (b). In Fig. 2 we illustrate how we convert non-relational tables with hierarchical metadata into the relational format by adding a non-breaking space in the corresponding cells to bring it to a valid relational table format. We converted both the WDC and CORD-19 non-relational tables like that to the relational format used by the ensemble.

3   Methodology

In this work, we describe a scalable ensemble of Machine- and Deep-learning models for Metadata identification in tables. Our ensemble can classify apart Metadata from data rows and columns. In previous works, researchers worked mostly on a single cell level, that is they have analyzed table cells, their position, data value types, and the surroundings such as blank cells among the neighbouring cells [4,11,19,23]. By contrast, we take into account the contexts for all cells in a tuple/column, being classified and the spacing of the surroundings. The model architecture is depicted in Fig. 3. We use an entire table tuple or column as the input of our model and the model predicts whether it contains Metadata or not. We do not check every table tuple or column, because we never saw metadata stored in the middle, at the very bottom or as a rightmost table column, but if the dataset specifics requires that, nothing prevents that. Hence, we take the first table tuple, the first column and the second tuple as input for the model and the model processes them. The second tuple is taken as input, because for non-relational tables with *hierarchical* metadata, the second row contains the second layer of the Metadata as depicted in Fig. 1(b). Finally, to post-filter false-positives, we have designed a custom decision-tree model, based on the number of spaces in the column or row. Algorithm 1 encodes this process step by step in pseudo-code for clarity.

In Algorithm 1 code, we see that it takes two table rows and one column as input and stores the predicted value for each input. Then it generates the final decision by using the decision-tree, based on the spaces. As any model, it has false-positives, which we mitigate with this space-based logic. For example, if both the first and second rows of a table are classified as Metadata rows,

---

Algorithm 1: Classifying Metadata row/columns of a table.

Input:

row_1 = the first table row

row_2 = the second row

$column_1$ = the first column

Output: Metadata position

1 y_1 = the ensemble's classification of row_1

2 y_2 = the ensemble's classification of row_2

3 y_3 = the ensemble's classification of $column_1$

4 s_1 = number of spaces in row_1

5 s_2 = number of spaces in row_2

6 s_1 = number of spaces in $column_1$

7 **if** $y_1 = 1$ *and* $y_2, y_3 = 0$ **then**

8 | return *Metadata in the first row only*;

9 **else if** $y_3 = 1$ *and* $y_1, y_2 = 0$ **then**

10 | return *Metadata on first column only*

11 **else if** $y_1 = 1$ *and* $y_2 = 1$ *and* $s_2 > 0$ **then**

12 | return *Hierarchical metadata*

13 **else**

14 | return *No metadata*;

15 **end**

---

but there are no blank cells in either row, then the second row cannot be the second row of a hierarchical Metadata. If the first and second rows has the same number of cells without any spaces and it can not be hierarchical metadata of a non-relational table.

3.1   Ensemble Architecture

In this section, we describe the two-layer ensemble we have designed for tabular Metadata classification and the ideas behind its design. The first layer consists of a Recurrent Neural Network (RNN) [21] model containing LSTM [20] units for the analysis of a table cell. The number of rows and columns are different for each table and also the number of terms inside each cell is different. Keeping this in mind, we have designed our RNN model to process one cell of a tuple or column at a time. Each cell may contain different number of terms as its value and the order matters in this case. For example, "First Name" and "Name First" are different values both syntactically and semantically. Hence, we decided to use the LSTM-based model, which is *order-sensitive*, to predict the probability of a cell having Metadata. Although, it is understood that many common terms, some of them referring to *data types* are used in the Metadata cells, such as "Time", "Date", "Name", etc., a purely rule-based classification would not be accurate enough at scale given enormous variety of sources in large-scale datasets, all having different naming conventions, even for the data types. Moreover, the whole Metadata cells as components forming a tuple or a column are *order-insensitive*. For example, suppose there are two columns in a table - "Name" and "Age". It does not matter if we swap "Name" with "Age" in a tuple, the tuple still remains the same, retaining its semantics. Hence, the second layer that we added (Naive Bayes Classifier) is order-insensitive. After feeding the cell to the RNN model, we have taken the output of an intermediate dense layer as an encoding of the cell and for a whole row or column, we concatenated all such encodings to form the feature vector, input to the Naive Bayes Classifier. Figure 3 illustrates the ensemble architecture.

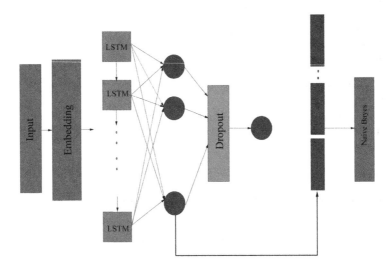

Fig. 3. Hybrid ensemble architecture used for Metadata classification. The RNN is on the first layer, followed by the Naive Bayes on the second layer. The circles represent Dense units [33] and blank boxes before Naive Bayes classifiers are the output feature vectors of the Dense layer that represent cell encodings.

Feature Space: We have used 100'000 dimensional feature space, i.e. 100K English terms in our vocabulary that we have selected by taking all terms from our datasets, sorting by frequency and cutting off the noise words and spam [34]. Increasing the dimensionality further led to significantly slower training time, which would prevent or make the experiments much more difficult (see Sect. 4 for configuration of our cluster).

Feature Vectors: First the term sequence in a cell is converted into a feature vector, encoded by one-hot encoding. The maximum number of terms in a cell is 734 for our datasets. Each cell is converted to a collection of 734 vectors, each vector corresponding to a term using zero-padding. We did not use any pre-trained word embedding, rather we have trained our local embedding using our vocabulary and datasets. The dimensionality of our dense embedding layer is 50, which helps reduce the dimensionality of input vector space (100K) significantly and alleviates sparsity. We have used keras embedding layer in our implementation. The vectors output of the embedding layer are passed to the input of the LSTM units. We have used 60 LSTM units in this case followed by a fully connected layer, containing 256 dense nodes. Then we have added a *dropout* layer, usually used to avoid overfitting and finally, the output layer has one dense unit with the sigmoid function as the activation function, which outputs the probability of the input cell is a Metadata cell, however it does not generate the final decision for an entire tuple or column.

To generate the final label, the first layer, first, processes all cells of a tuple or column sequentially, before the second layer starts processing and producing the final decision. Although the individual cells contain sequences of terms that are order-sensitive as discussed, the order of each cell within the row or column does not matter. So we need an order-insensitive model to process an order-insensitive sequence of cells.

After training the RNN model, for each cell, we extract the output of the intermediate dense layer that contains 256 dense units. This layer produces an encoding for each input cell. We have taken each cell's encoding, concatenated them and used it to form the input feature vector for the Naive Bayes Classifier. For example, if there are 4 cells in a row, each cell will have a feature vector of 256 dimensions that of the intermediate dense layer. For a tuple, concatenating the individual cell's feature vector, we'll have a vector of $4 * 256 = 1024$ dimensions as the input feature vector for the Naive Bayes Classifier.

The final output is binary, whether the input sequence of cells, i.e. tuple or column is a Metadata row/column or not. The output of the Naive Bayes Classifier is then used with the decision-tree rules in Algorithm 1 to produce the final output having the Metadata type.

3.2 Large-Scale Evaluation Architecture

In this work, we have used two different datasets. One is named CORD-19 [35] and the other is Web Data Commons [24], a large-scale corpus having Web Tables from different sources on different topics. The CORD-19 dataset is a novel collection of papers related to COVID-19. Tables used in the papers are stored in a JSON file having HTML format and they have used the structured representation shown in Fig. 2. We have extracted the HTML formatted tables to JSON. Given mostly medical tables in CORD-19, there are tables of all three formats, metadata on left, metadata on top and hierarchical metadata. Most tables having hierarchical metadata have metadata on top. From the CORD-19 papers, we were able to extract 88777 tables.

Another large-scale dataset that we have used in this work is WDC [24]. This dataset consists of more than 100M Web tables, including information about their source URL, table type, text before and after the table, where more than 15M tables are in English language. The tables are from a variety of domains, including scientific, news articles, product information and many other. This dataset is a very robust, large-scale dataset with significant representation of tables from many domains, so it is very attractive choice for evaluating our models, scalability, and generalizability across domains. Concerning metadata position, we have found six types of tables in total: relational, metadata on left, metadata on top, hierarchical metadata on left, hierarchical metadata on top. In total there are more than 100 million tables from 265804 different sources.

4   Experimental Study

In this section, we first describe the training and test sets construction followed by the experimental evaluation on several large-scale datasets.

Hardware: We run all experiments on a cluster of 4 machines, each having 4 Intel Xeon 2.4 GHz 40-core CPUs, from 192 GB to 1 TB of RAM, 10 TB disk space each, interconnected with a 1 GB Ethernet.

Software: For implementing the RNN model, we have used Keras, a popular python library for deep learning, having Tensorflow framework as the backend. The second step i.e. the Naive Bayes classifier is implemented using Scikit-learn, a popular machine learning library for python. Throughout the experimentation and implementation, we have used python as the programming language.

4.1   Training the Models

Our ensemble has 2 layers that we trained separately. On our large-scale datasets, we are able to achieve at best 82.76% accuracy for classifying Metadata on top and 78% for hierarchical Metadata on top. These are the best results in class, to the best of our knowledge, given the scale, a huge number of sources, and a variety of domains represented in the datasets.

Main Parameter Settings: In the first layer, for LSTM model, we have used 60 LSTM units and an embedding layer where the dense embedding dimensionality of 50. LSTM units are unidirectional and order sensitive. The input feature vector is composed of a collection of terms in a cell and it is first encoded using *one-hot* encoding and then is passed through the embedding layer that reduces the dimensionality to 50. After the LSTM units, we have added a fully connected layer with 256 nodes. We have used rectified linear activation function for all nodes. We are aware that such models tend to overfit even on a large training set. To alleviate that we have used a dropout layer after the dense layer with a value of 0.2, known to be a balance between dropping too many features and degree of overfitting. Lastly, there is a node having *sigmoid* as the activation function.

First we have trained the LSTM model as a binary classifier i.e. to classify a cell being a Metadata cell or not. We have taken the output feature vector of the dense layer having 256 nodes as an encoding of the input cell, i.e. the dimensionality of an encoded vector is 256. For a tuple or column, there are several cells. We have concatenated their encodings to form an encoding for a tuple/column. This feature vector is the input of the Multinomial Naive Bayes Classifier, which is also a binary classifier that predicts the input tuple or column being Metadata or not.

Dimensionality: Each table has a different number of columns and rows, hence a different number of cells. Each cell has a different number of terms, some of them can be blank, e.g. with hierarchical Metadata or just missing values. The largest cell had 764 terms among the datasets we have used. So after tokenization, we had to *pad* each cell with zeros to align all cells to 734 dimensionality. As discussed above, the input we have used in the Naive Bayes model is the concatenated encodings of all cells in the tuple or column. The widest table has 36 cells in one row. Other vectors, having less cells were amended with zero matrices.

4.2   Training Data

First, we have selected only English tables from the WDC dataset as the dataset is multilingual and we wanted to experiment first on the English subset, which is more than 15 million tables. From our experiments, we have observed that one source maintains a common format for most of its tables. So if we compose the training set from one source it will be biased to that source. Hence, we have uniformly sampled at random an equal number of tables from all sources and constructed the training set containing 570K tables. We used space-based logic to find the Metadata rows/columns to form the training set. Per the representation illustrated in Fig. 2, tables having hierarchical metadata have spaces inside the Metadata row. However, this is not a 100% guarantee, sometimes there is a blank space purposefully in the leftmost column. So we first pre-selected all tables having more than one space in the top row. Second, we pre-selected the well-formed relational tables that should not have blank spaces neither in the first row *nor* the column. After that, we took samples from these subsets and checked the table type and Metadata location manually to ensure there are 500 correct positive labels for each kind of metadata.

We amended the positively labeled training instances with the same number of negative (regular data rows and columns, without Metadata) instances by sampling tables from the entire dataset and taking the second last row from each table in the sample. To ensure the training set is balanced, we made sure there is equal number of positively and negatively labeled instances.

4.3   Test Data

We have constructed three separate test sets - for Metadata in the first row, in the first column, and hierarchical Metadata. To ensure heterogeneity of our test sets, we have first uniformly sampled the tables from the entire dataset, excluding the tables used for the training set. In total, we had 6,785 different sources for the three test sets. Then we annotated 500 tables for each test set and the same number without metadata to ensure a balanced test set.

5   Evaluation

Here we evaluate the accuracy of our trained ensemble to recognize Metadata on two large-scale tabular corpora - WDC [24] and CORD-19 [35] as well as four other popular corpora used in the recent related work [6,7,10,23]. We have used *accuracy* as defined in Eq. 1 below as the metric. *TP, TN, FP, FN* denote the true positives and negatives.

$$Accuracy = \frac{TP + TN}{TP + TN + FP + FN} \tag{1}$$

Table 1 illustrates Metadata classification accuracy for three Metadata types evaluated on WDC [24] and CORD-19 [35] datasets. The accuracy varies, depending on the dataset size and number of sources. The CORD-19 dataset contains only medical tables extracted from scientific publications on COVID-19. Although the scientific papers are from different sources, there might be an inherent similarity of formats common among tables in this domain, hence the accuracy is slightly better than that for WDC [24], which is also much larger and has more sources.

Table 1. Classification accuracy evaluation.

Dataset	Metadata type	Accuracy
Cord-19	Metadata in the first row	83.76%
	Metadata in the leftmost first column	82.9%
	Hierarchical metadata	84.35%
WDC	Metadata in the first row	79.87%
	Metadata in the leftmost first column	76.47%
	Hierarchical metadata	81.53%

In recent related work [6,7,10,23], the authors evaluated their Metadata classification models on much smaller datasets, composed from much fewer sources. By contrast, CORD-19 and especially WDC are *ultra large-scale heterogeneous* datasets and we would like to highlight high accuracy and generalizability of our approach in such challenging context.

Purely cell-based approaches that consider and classify just a single table cell in isolation [6,7,10,23] without treating them sequentially as a row or a column, unlike our approach, do not scale and generalize very well across domains, datasets, and sources. We implemented a naive standard cell-level classification including Random Forest classifier as in [23] and evaluated it on CORD-19 dataset. Its accuracy was 65% (for Metadata in the first row) compared to our 83.7%, which yields an impressive ≈19% delta in accuracy.

Our approach performs approximately the same on all datasets used recent works [6,7,10,23] on location and classification of the components of verbose

CSV files, such as Metadata, Header, Group, Data, Notes with a slight delta in F-measure. Our work is most similar to the header detection in these files (i.e. Metadata in our terminology). We picked 3 largest and most heterogeneous datasets, used by the recent studies to comparatively evaluate our approach. Other two datasets were not available for public access as well as are manually crawled and post-processed, which makes them more customized and less attractive for comparison. Table 2 compares these and our datasets used for comparative evaluation by size and the number of sources. Our F-1 score is a slightly higher on DeEx (83% vs. 80.7%) and SUAS datasets (97% vs. 96%) and a bit lower on CIUS dataset (95% vs. 97.2%). The heterogeneity of our training and test sets is much higher, because they are composed from a large number of sources, which significantly affect generalizability of our trained ensemble as justified by the comparative evaluation against a regular cell-based approach on our large-scale datasets ($\approx$19% delta in accuracy, see this Section above for the detailed description). I.e. instead of using many tables from one or several large sources for training and evaluation, we have uniformly at random taken tables from a wide variety of sources both for training and evaluation purposes to ensure the trained model is more robust and naturally resistant to heterogeneity as well as making the evaluation set much more challenging.

Table 2. Comparison of the datasets used for evaluation by size (number of tables) and the number of sources. For our datasets, we count just the English subset of WDC and the CORD-19 dataset.

Related work	Dataset size	Number of sources
Christodoulakis *et al.* [7]	4500	2
Fang *et al.* [10]	255	2
Zhe *et al.* [6]	1.1 Million	Not mentioned
Lan *et al.* [23]	1345	5
Our datasets [24, 35]	**More than 15 million**	**More than 26512**

Except many *fundamental* data management activities, naturally dependent on metadata such as *query processing, data integration, warehousing, replication* and many, another important application of Metadata location and classification is distinguishing *relational* and *non-relational* tables. In our datasets, we had both kinds of tables and we have evaluated performance of our ensemble on this task as well. Although the number of sources is ultra large, it performs remarkably well on WDC. We were more interested in *Recall*, because of the large number of sources and our goal to evaluate *generalizability* of our trained ensemble in context of many sources. *Precision* was $\approx$75% for both datasets, but we thought *Recall* is more interesting as in context of thousands of sources it charctarizes the ability to recognize new representations and resistance to heterogeneity at scale, which is a big challenge in practice [14, 15, 18, 22]. For classifying relational tables we observed Recall - 79.9% on WDC and 94.6% on CORD-19 and for non-relational tables on WDC - 76.5% and 96.7% on CORD-19. We would like to highlight stellar Recall on CORD-19 and lower, but still remarkable Recall on WDC due to the vast scale and unprecedented heterogeneity of this Web-scale dataset.

Table 3. Accuracy of our hybrid ensemble, evaluated on the datasets used by the related work. Here MD abbreviates to Metadata; "w/o NB" means the results are calculated by using only the first-layer RNN model; "with NB" means the ensemble is two-layer as usual with NB as a second layer. The best results are in bold.

Dataset	Algorithm	MD on top	MD on left	Hierarchical MD
Cord-19 [35]	w/o NB	77.97%	80.75%	83.43%
	with NB	**83.7%**	**82.9%**	**84.3%**
Lehmberg *et al.* [24]	w/o NB	67.3%	69.8%	74.3%
	with NB	**78.8%**	**76.4%**	**81.5%**
DeEx [12]	w/o NB	**83.7%**	91.8%	81.9%
	with NB	78.8%	**95.73%**	**82.36%**
SAUS [12]	w/o NB	78.5%	89.3%	**91.4%**
	with NB	**87.4%**	**95.7%**	82.6%

We finish by discussing Table 3, which illustrates the comparative evaluation results of our ensemble for the same classification task of three different kinds of Metadata - Metadata on Top, Metadata on Left and Hierarchical Metadata. This experiment is performed on 4 different datasets and the accuracy of locating these metadata types is presented.

From these last evaluation results, we can infer that using the order-insensitive Naive Bayes classifier as a second layer generally improves the Metadata classification performance compared to just using the Deep Learning RNN model alone. In most cases the ensemble performs better, because analyzing only the cell content, which is done in the first layer is insufficient. Analyzing the cell composition in a tuple or a column in an order-insensitive way improves performance. In only two cases for the SAUS Hierarchical metadata and DeEx regular Metadata on top, just the first layer performs better, because in these case the Metadata rows have many numeric values and Naive Bayes is classically term-based.

6   Related Work

Accurate, scalable, generalizable Metadata location and classification in Web Tables, CSV files, tables, extracted from scientific publications, other large-scale structured corpora is starting gaining momentum in the Data Management and Science communities [6, 7, 10, 23]. It is due to both fundamental nature of Metadata and it being of critical importance for many data management tasks on structured data and the fact that large-scale structured datasets are becoming ubiquitous, lack accurate Metadata labeling, and are, at the same time invaluable assets full of useful information.

Here, we presented a hybrid, two-layer Machine- and Deep-Learning ensemble for location and classification of Metadata rows, columns as well as more complex hierarchical Metadata in tables. Several recent works study discriminating between relational and non-relational tables, which is related to metadata classification [5,37]. Except being capable to do the same and at scale, our approach is also useful for precise Metadata rows or columns location and Metadata type identification. In [7], authors proposed Pytheas, a line classification system for a CSV files. Using fuzzy logic, it determines whether a field is data or not and based on the rules, it detects table border and hence it can differentiate table from metadata to some extent. Their algorithm uses two phases for the task, offline (training) phase and online table discovery (inference) phase. In the offline phase, the algorithm learns the weights for the rule set and in the inference phase, using fuzzy logic it computes confidence value for each line whether it belongs to data or not data. They have evaluated their algorithm on two manually annotated datasets containing a total of 4500 CSV files with a recall value of 95.7%. Although the recall is high, the evaluation set size is much smaller and less heterogeneous (composed only from two different sources) than the Web-scale corpora, we used to evaluate our approach. Size and the number of sources used for training and validation sets drastically affect classification performance at scale and in presence of heterogeneity [11,22].

The authors in [6] proposed rule-assisted active learning for header spreadsheet property detection including differentiating header from data to reduce human annotation. They have used a hybrid iterative learning framework, with and without user-provided rules for learning property detection of a CSV document. At first, a human labeler labels the spreadsheet properties manually and the models are trained based on the human labeling activity. They have worked with two different sources of data, a web-crawled dataset containing 1.1M sheets and Web400 data containing 400 sheets. Authors have shown 89% accuracy for header detection in a CSV file. Again the validation set size is much less compared to what we have used and the number of sources is not specified by the authors. Table 2 compares the datasets including the one used by the authors.

The authors in [10] used Random Forest classifier to detect and classify table headers. They have proposed two heuristic strategies to separate data and the header. As a baseline, they have used the first row and first table columns as default headers. Their evaluation set contains only 255 tables, which is remarkably small compared to all recent works and our datasets. They have achieved 92% accuracy on their test set, which does not unfortunately mean it will remain the same when the validation set becomes larger and more heterogeneous, even slightly.

To finalize the discussion, the authors in a very recent work [23] performed line and cell classification in verbose CSV. In this work, authors have focused on CSV structure detection and for this purpose, they have detected various cell types like metadata, header, group, data, derived, notes etc. Our work is similar to the part of their work on cell classification, more precisely, the header cell classification. For cell classification, they have used content, contextual and computational features of the cell. That is they have analyzed the number of empty cells, position of row or column, is there any empty column besides the column being analyzed, block size, data type etc. This analysis is, however, is purely cell-based and does not take into account compositional features of cells when they become tuples or columns, which we do. Our ensemble is also two-layered, where the first layer takes into account term order inside cells and on the second layer the cells are order-insensitive. This idea has been proven valuable and sound for structured data in context of set-based relational model and is absent in [23] as well as all other recent works to the best of our knowledge.

Lastly, the authors trained a multi-class random forest classifier and evaluated performance of their system on 5 different datasets. Their evaluation results are good compared to the related works, but the evaluation set is again much smaller than ours. Moreover, the number of sources of their datasets is very limited whereas we have evaluated our system on very large-scale, heterogeneous datasets.

7   Conclusion

Here, we presented a hybrid, two-layer Machine- and Deep-Learning ensemble for location and classification of Metadata rows, columns as well as more complex hierarchical Metadata in large-scale structured datasets. For cell-level analysis in the first layer, we used a Recurrent Neural Network (RNN) model with LSTM units. The cell-level analysis is order-sensitive as the cell content - a sequence of terms is order-sensitive. On the second layer, which analysis the composition of several cells into a tuple or a column, we have used the Naive Bayes classifier, which is order-insensitive in accord with order-insensitivity of a tuple or a column regarding the cell order. Except this architectural novelty, in the recent previous work, the authors used a small number of sources and relatively small-scale datasets to evaluate their approaches. We have evaluated scalability and generalizability our approach on a very large-scale heterogeneous dataset. We have specifically constructed our training and evaluation sets to contain tables from thousands of sources. Tables are represented very differently depending on a source, so an algorithms trained on one source, having high accuracy one or several sources, does not usually generalize and works very poorly on thousands of other sources [3,9,11,22,25,26,29,32,36]. That is why rule-based or regular Machine-learning based approaches would be incapable of inferring additional source-dependent features, unlike Deep-Learning. Our work achieves high accuracy on two large-scale datasets, WDC and CORD-19 that confirms scalability and generalizability in context of a large number of sources.

References

1. Census bureau. https://www.census.gov/data/datasets.html
2. Alexe, B., et al.: Simplifying information integration: object-based flow-of-mappings framework for integration. In: Castellanos, M., Dayal, U., Sellis, T. (eds.) BIRTE 2008. LNBIP, vol. 27, pp. 108–121. Springer, Heidelberg (2009). https://doi.org/10.1007/978-3-642-03422-0_9
3. Braunschweig, K., Thiele, M., Lehner, W.: From web tables to concepts: a semantic normalization approach. In: Johannesson, P., Lee, M.L., Liddle, S.W., Opdahl, A.L., López, Ó.P. (eds.) ER 2015. LNCS, vol. 9381, pp. 247–260. Springer, Cham (2015). https://doi.org/10.1007/978-3-319-25264-3_18
4. Cafarella, M.J., Halevy, A., Wang, D.Z., Wu, E., Zhang, Y.: WebTables: exploring the power of tables on the web. In: VLDB (2008)
5. Cafarella, M.J., Halevy, A., Zhang, Y., Wang, D., Wu, E.: Uncovering the relational web. In: WebDB (2008)
6. Chen, Z., Dadiomov, S., Wesley, R., Xiao, G., Cory, D., Cafarella, M., Mackinlay, J.: Spreadsheet property detection with rule-assisted active learning. In: CIKM. ACM (2017)
7. Christodoulakis, C., Munson, E.B., Gabel, M., Brown, A.D., Miller, R.J.: Pytheas: pattern-based table discovery in CSV files. In: PVLDB, July 2020
8. Codd, E.F.: A relational model of data for large shared data banks. In: CACM. vol. 13, no. 6, June 1970
9. Dong, X.L.: Challenges and innovations in building a product knowledge graph. In: KDD (2018)
10. Fang, J., Mitra, P., Tang, Z., Giles, C.L.: Table header detection and classification. In: AAAI, vol. 26, no. 1, July 2012
11. Gentile, A.L., Ristoski, P., Eckel, S., Ritze, D., Paulheim, H.: Entity matching on web tables: a table embeddings approach for blocking. In: EDBT (2017)
12. Gol, M.G., Pujara, J., Szekely, P.: Tabular cell classification using pre-trained cell embeddings. In: ICDM (2019)
13. Gubanov, M.: Hybrid: a large-scale in-memory image analytics system. In: CIDR (2017)
14. Gubanov, M.: Polyfuse: a large-scale hybrid data fusion system. In: ICDE (2017)
15. Gubanov, M., Priya, M., Podkorytov, M.: CognitiveDB: an intelligent navigator for large-scale dark structured data. In: WWW (2017)
16. Gubanov, M., Pyayt, A.: READFAST: high-relevance search-engine for big text. In: ACM CIKM (2013)
17. Gubanov, M., Pyayt, A.: Type-aware web search. In: EDBT (2014)
18. Gubanov, M.N., Popa, L., Ho, H., Pirahesh, H., Chang, J.-Y., Chen, S.-C.: IBM UFO repository: object-oriented data integration. In: VLDB (2009)
19. Hancock, B., Lee, H., Yu, C.: Generating titles for web tables. In: WWW. ACM, New York (2019)
20. Hochreiter, S., Schmidhuber, J.: Long short-term memory. Neural Comput. 9(8), 1735–1780 (1997)
21. Jain, L.C., Medsker, L.R.: Recurrent Neural Networks: Design and Applications, 1st edn. CRC Press Inc., Boca Raton (1999)
22. Khan, R., Gubanov, M.: WebLens: towards interactive large-scale structured data profiling. In: CIKM. ACM (2020)
23. Jiang, L., Vitagliano, G.: Structure detection in verbose CSV files. In: EDBT, March 2021

24. Lehmberg, O., Ritze, D., Meusel, R., Bizer, C.: A large public corpus of web tables containing time and context metadata. In: Bourdeau, J., Hendler, J., Nkambou, R., Horrocks, I., Zhao, B.Y. (eds.) WWW (2016)

25. Limaye, G., Sarawagi, S., Chakrabarti, S.: Annotating and searching web tables using entities, types and relationships (2010)

26. Mulwad, V., Finin, T., Joshi, A.: Generating linked data by inferring the semantics of tables. In: VLDS, CEUR Workshop. CEUR-WS.org (2011)

27. Ortiz, S., Enbatan, C., Podkorytov, M., Soderman, D., Gubanov, M.: Hybrid.json: high-velocity parallel in-memory polystore JSON ingest. In: IEEE Bigdata (2017)

28. Podkorytov, M., Soderman, D., Gubanov, M.N.: Hybrid.poly: an interactive large-scale in-memory analytical polystore. In: ICDM Workshops, pp. 43–50. IEEE Computer Society (2017)

29. Ritze, D., Bizer, C.: Matching web tables to DBpedia - a feature utility study. In: EDBT (2017)

30. Simmons, M., Armstrong, D., Soderman, D., Gubanov, M.: Hybrid.media: high velocity video ingestion in an in-memory scalable analytical polystore. In: IEEE Bigdata (2017)

31. Soderman, S., Kola, A., Podkorytov, M., Geyer, M., Gubanov, M.: Hybrid.AI: a learning search engine for large-scale structured data. In: WWW (2018)

32. Subramanian, A., Srinivasa, S.: Semantic interpretation and integration of open data tables. In: Sarda, N.L., Acharya, P.S., Sen, S. (eds.) Geospatial Infrastructure, Applications and Technologies: India Case Studies, pp. 217–233. Springer, Singapore (2018). https://doi.org/10.1007/978-981-13-2330-0_17

33. Uhrig, R.: Introduction to artificial neural networks. In: IECON, vol. 1, pp. 33–37 (1995)

34. Villasenor, S., Nguyen, T., Kola, A., Soderman, S., Gubanov, M.: Scalable spam classifier for web tables. In: IEEE Big Data (2017)

35. Wang, L.L., Lo, K., et al.: The covid-19 open research dataset. ArXiv (2020)

36. Wang, N., Ren, X.: Identifying multiple entity columns in web tables. Int. J. Softw. Eng. Knowl. Eng. **28**(3), 287–310 (2018)

37. Wang, Y., Hu, J.: A machine learning based approach for table detection on the web. In: WWW 2002, pp. 242–250. ACM, New York (2002)

Unified and View-Specific Multiple Kernel K-Means Clustering

Yujing Zhang[ID], Siwei Wang[ID], and En Zhu[✉][ID]

College of Computer, National University of Defense Technology,
Changsha 410073, China
enzhu@nudt.edu.cn

Abstract. Multiple kernel clustering (MKC), as an important tool for handling multi-view non-linear data, has attracted notable attention among data mining and machine learning communities. The key issue of MKC is to obtain a more accurate and appropriate kernel similarity for clustering from the given multiple kernel library. However, existing MKC methods only capture the unified or consistent information while the view-specific individual structures have been ignored to degrade clustering performance. In this paper, we propose a novel multiple kernel k-means clustering method (UVSMKC), where the *unified* and *view-specific* information is seamlessly explored under multiple kernel setting. Different form traditional framework, we formulate the provided multi-view kernels with a *unified* low-rank similarity matrix, multiple *view-specific* matrices and the respective noise matrices. Furthermore, we design a four-step alternate algorithm to solve the proposed optimization problem. Comparing to the state-of-the-art multiple kernel clustering methods, the experimental results on six multiple kernel benchmark datasets validate the effectiveness of our proposed UVSMKC, showing promising ability to capture consensus and view-specific information.

Keywords: Multiple kernel clustering · Multi-view clustering · Data clustering

1 Introduction

Multi-view clustering is of great importance in the unsupervised data learning tasks which has received great attention among data mining society. To effectively deal with non-linearly separable data, kernel k-means is proposed to transform original data to appropriate space. However existing data are collected from multiple sources which gives rise to multiple kernel k-means clustering in reality. Given a group of pre-defined kernel matrices, multiple kernel clustering (MKC) optimizes the available information to classify data items with similar structures or patterns into the same group [1, 4–6, 8, 10, 12, 14–16, 19, 21]. By following traditional multiple kernel learning framework, MKC optimizes the optimal coefficients for the pre-defined kernel matrices [1, 3, 5, 10, 13–15, 20]. In [1],

C. Strauss et al. (Eds.): DEXA 2021, LNCS 12923, pp. 51–62, 2021.
https://doi.org/10.1007/978-3-030-86472-9_5

a three-step alternate algorithm is put forward to seamlessly obtain clustering result, kernel coefficients and dimension reduction. The work in [14] proposes a multiple kernel k-means clustering algorithm with a matrix-induced regularization term to reduce the redundancy of the selected kernels. Furthermore, the local kernel alignment criterion has been applied to multiple kernel learning to enhance the clustering performance in [10]. On the contrary, late fusion based multiple kernel clustering strategy seeks to exploit the complementary information in kernel partition space to reach consensus on partition level [18]. In [18], late fusion alignment is firstly proposed to maximally align the multiple base partitions with the consensus partition and enjoys considerable algorithm acceleration and satisfactory clustering performance.

Although the aforementioned methods have improved MKC from massive aspects, most of them only capture the consistent information to obtain a unified kernel matrix while the view-specific individual structures have been ignored. In fact, individual view may contain independent local information since they are collected from different sides. Moreover, each kernel matrix may consist of various noise. Therefore, how to jointly optimize consistent and view-specific information still keeps to be unsolved.

In this paper, we propose a novel multiple kernel k-means clustering method (UVSMKC), where the *unified* and *view-specific* information are seamlessly explored under multiple kernel setting. Different from traditional framework, we formulate the provided multi-view kernels with a *unified* low-rank similarity matrix, multiple *view-specific* matrices and the respective noise matrices. Furthermore, we design a four-step alternate algorithm to solve the proposed optimization problem. Extensive experiments on six multiple kernel benchmark datasets are conducted to evaluate the effectiveness of the proposed method. As demonstrated, the proposed algorithm enjoys superior clustering performance with better clustering structure, in comparison with the state-of-the-art multiple kernel clustering methods.

The contributions of this paper are summarized as follows,

- Different from traditional multiple kernel k-means strategy, we propose a novel MKKM method termed as Unified and View-specific Multiple Kernel K-means Clustering (UVSMKC). Unlike existing methods seeking for optimal kernel coefficients, we propose to divide multiple kernel matrices into the shared similarity, view-specific structures and individual noises.
- The proposed UVSMKC integrates multi-view information with unified and view-specific structures for clustering task. It simultaneously optimizes the low-rank consistent kernel similarity, view-specific matrix and the respective noise information. By introducing ℓ_2 and $\ell_{2,1}$ regularization term to the respective matrices, the noises can be further eliminated to better recover the clustering structures.
- An optimization algorithm is designed to efficiently tackle the resultant problem. By the virtue of it, UVSMKC shows clearly superior clustering performance and structure advantages in comparison with state-of-the-art methods.

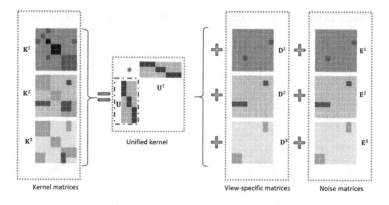

Fig. 1. The flow chart of the proposed Unified and View-specific Multiple Kernel Clustering. Given 3 kernel matrices, our method pursues a consistent kernel matrix, presented as $\mathbf{U} * \mathbf{U}^\top$, a set of view-specific matrices $\{\mathbf{D}^v\}_{v=1}^3$ and noise matrices $\{\mathbf{E}^v\}_{v=1}^3$. $\mathbf{U}$ will be used as input to the k-means clustering method to generate the final clustering result.

The rest of this paper is organized as follows. Section 2 outlines the related work of multiple kernel clustering. Section 3 introduces the proposed optimization objective and the four-step alternate algorithm with its convergence and the computational complexity. Section 4 shows the experiment results with evaluation. Section 5 concludes the paper.

2 Background

In this section, we briefly review the most related work, including kernel k-means (KKM) and multiple kernel k-means (MKKM).

2.1 Kernel k-means (KKM)

As an important task in unsupervised learning, clustering can categorize data into different groups according to the similarity of the samples. k-means is one of the most classic clustering methods, and able to classify a given data set into a certain number of clusters in a simple and easy way. The main idea is to determine the cluster according to the cluster centers, which makes finding optimal cluster centers vital. Firstly, we just initialize the cluster centers randomly. It is noteworthy that different locations cause different results. That is the reason we need to repeat the experiment multiple times. Next, each point from the data set is regrouped into the cluster with the nearest cluster centers, then we need to re-calculate k new centers of the clusters resulting from the previous step. Repeat the steps until centers do not move again.

Unfortunately, k-means algorithm lacks the ability to deal with nonlinear data sets. Next, kernel k-means (KKM) algorithm is proposed to solve the problem. KKM applies the same trick as k-means but with one improvement that for the measurement of sample distance, kernel method is used instead of the Euclidean distance.

Suppose there is a data set $\mathbf{X} = [\boldsymbol{x}_1, \boldsymbol{x}_2, \cdots, \boldsymbol{x}_n] \subseteq \mathcal{R}^{d \times n}$, and $\phi_p(\cdot) : \boldsymbol{x} \in \mathcal{X} \mapsto \mathcal{H}$ which non-linearly maps $\boldsymbol{x}$ into a higher dimensional Hilbert space $\mathcal{H}$. It can be mathematically expressed as follows,

$$\min_{\mathbf{Z}^{n \times k}} \sum_{i=1}^{n} \sum_{c=1}^{k} \mathbf{Z}_{ic} \| \phi(\boldsymbol{x}_i) - \boldsymbol{\mu}_c \|_2^2, \quad \text{s.t.} \sum_{c=1}^{k} \mathbf{Z}_{ic} = 1, \tag{1}$$

where $\mathbf{Z} = [\mathbf{Z}_{ic}]_{n \times k}$ and $\mathbf{Z} \in \{0, 1\}^{n \times k}$ is the indicator matrix, $n_c = \sum_{i=1}^{n} \mathbf{Z}_{ic}$ is the number of the cluster and $\boldsymbol{\mu}_c = 1/n_c \sum_{i=1}^{n} \mathbf{Z}_{ic} \phi(\boldsymbol{x}_i)$ is the center of cluster c. Specifically, Eq. 1 is equivalent to the following one,

$$\min_{\mathbf{Z}^{n \times k}} \text{Tr}(\mathbf{K} - \mathbf{L}^{1/2} \mathbf{Z}^{\top} \mathbf{K} \mathbf{Z} \mathbf{L}^{1/2}), \quad \text{s.t.} \, \mathbf{Z} \mathbf{1}_k = \mathbf{1}_n, \tag{2}$$

where $\mathbf{L} = \text{diag}\left[\frac{1}{n_1}, \cdots, \frac{1}{n_k}\right]$, and $\mathbf{K}$ is the kernel matrix, defined as $\mathbf{K}_{ij} = \phi(\boldsymbol{x}_i)^{\top} \phi(\boldsymbol{x}_j)$. Unfortunately, this problem is NP-hard. To simplify the problem, note that $\mathbf{H} = \mathbf{Z} \mathbf{L}^{1/2}$, which represents normalized clustering assignment. Equation 2 can then be rewritten as follows,

$$\min_{\mathbf{H} \in \mathcal{R}^{n \times k}} \text{Tr}(\mathbf{K} - \mathbf{H}^{\top} \mathbf{K} \mathbf{H}), \quad \text{s.t.} \, \mathbf{H}^{\top} \mathbf{H} = \mathbf{I}_k. \tag{3}$$

The optimal solution is calculated by $\mathbf{H} = \mathbf{U}_K \mathbf{Q}$, where $\mathbf{U}_K$ is concatenated by the eigenvectors of $\mathbf{K}$ involved with k largest eigenvalues $\lambda_1 \geq \cdots \geq \lambda_k$ and $\mathbf{Q}$ is an arbitrary orthogonal matrix [7].

2.2   Multiple Kernel k-means (MKKM)

Let $\mathcal{X} = [\boldsymbol{x}_1, \boldsymbol{x}_2, \cdots, \boldsymbol{x}_n]$ be a collection of n samples, $\{\mathbf{K}^p\}_{p=1}^{m}$ be a set of pre-calculated kernel matrices. The optimal kernel matrix $\mathbf{K}_\alpha$ denotes $\sum_{p=1}^{m} \alpha_p^2 \mathbf{K}^p$, where $\sum_{p=1}^{m} \alpha_p = 1, \alpha_p \geq 0$. MKKM jointly optimizes the linear combination and the clustering partition matrix $\mathbf{H}$ by solving problem in Eq. 4.

$$\min_{\mathbf{H}, \alpha} \text{Tr}(\mathbf{K}_\alpha (\mathbf{I}_n - \mathbf{H} \mathbf{H}^{\top})), \text{s.t.} \, \mathbf{H} \in \mathcal{R}^{n \times k}, \mathbf{H}^{\top} \mathbf{H} = \mathbf{I}_k, \tag{4}$$

where $\mathbf{I}_k$ is an identity matrix with size $k \times k$. The optimization problem in Eq. 4 can be solved by alternately updating $\mathbf{H}$ and α:

i) **Optimizing H given α.** With coefficients α fixed, the optimization problem in Eq. 4 can be reduced to a simple kernel k-means clustering optimization problem shown in Eq. 3, and $\mathbf{H}$ is calculated by taking the k eigenvectors having the largest eigenvalues of $\mathbf{K}_\alpha$.

ii) **Optimizing α given H.** With $\mathbf{H}$ fixed, α can be optimized via solving the following quadratic programming with linear constraints,

$$\min_{\alpha} \sum_{p=1}^{m} \alpha_p^2 \text{Tr}(\mathbf{K}^p (\mathbf{I}_n - \mathbf{H} \mathbf{H}^{\top})), \quad \text{s.t.} \, \alpha^{\top} \mathbf{1}_m = 1, \, \alpha_p \geq 0, \tag{5}$$

which admits a closed-form solution.

MKKM is a classic method in multiple kernel clustering, there are many variants improving it from different aspects. The work in [2] studies nonlinear relationship, adopting the manifold adaptive kernel, instead of the original kernel, to integrate the local manifold structure of kernels. Multiple kernel clustering with corrupted kernels [11] proposes a scheme to address the problem of considerable corrupted kernels, where each given kernel is adaptively adjusted according to its corresponding error matrix. These variants are both good at handling outliers and noise, however they tend to ignore the view-specific information.

3   The Proposed Method

In this section, we firstly revisit kernel k-means, and then discuss a variant of the formula. Starting from the variant, we introduce our method at the second part. Finally, we propose a simple algorithm to solve the problem.

3.1   A Variant of Kernel k-means

We introduce a variant of the formula 3 as follows,

$$\min_{\mathbf{U}} \frac{1}{2} \|\mathbf{E}\|_F^2, \quad \text{s.t.} \ \mathbf{U}^\top \mathbf{U} = \mathbf{I}_k, \mathbf{K} = \mathbf{U}\mathbf{U}^\top + \mathbf{E}. \tag{6}$$

It seems that there is no direct connection between Eq. 3 and Eq. 6. We will prove the equivalence of the two formulas below.

Theorem 1. *Given a kernel matrix* $\mathbf{K}$*, the optimal solution of Eq. 3* $\mathbf{H}^*$ *and Eq. 6* $\mathbf{U}^*$ *satisfy the following equation* $\mathbf{H}^* = \mathbf{U}^*$*.*

Proof. Notice that $\text{Tr}(\mathbf{K} - \mathbf{H}^\top \mathbf{K}\mathbf{H}) = \text{Tr}(\mathbf{K}) - \text{Tr}(\mathbf{H}^\top \mathbf{K}\mathbf{H}) = \text{Tr}(\mathbf{K}) + \frac{1}{2}\{\|\mathbf{K} - \mathbf{H}\mathbf{H}^\top\|_F^2 - \text{Tr}(\mathbf{K}^\top \mathbf{K} + \mathbf{H}\mathbf{H}^\top \mathbf{H}\mathbf{H}^\top)\}$. It is obvious that $c = \text{Tr}(\mathbf{K}) - \text{Tr}(\mathbf{K}^\top \mathbf{K} + \mathbf{H}\mathbf{H}^\top \mathbf{H}\mathbf{H}^\top)$ is a constant. Let $\mathbf{E}$ represent $\mathbf{K} - \mathbf{H}\mathbf{H}^\top$, and take it into the equation. Then, we have $\text{Tr}(\mathbf{K} - \mathbf{H}^\top \mathbf{K}\mathbf{H}) = c + \frac{1}{2}\|\mathbf{E}\|_F^2$, which means Eq. 3 is equivalent to $\min_{\mathbf{H}} \frac{1}{2}\|\mathbf{E}\|_F^2, \text{s.t.} \ \mathbf{H}^\top \mathbf{H} = \mathbf{I}_k, \mathbf{K} = \mathbf{H}\mathbf{H}^\top + \mathbf{E}$.

These gives us a new angle to discuss kernel k-means: it can be thought of as a simple noise reduction process. In the next part, our method is proposed to improve this process.

3.2   Unified and View-Specific Multiple Kernel Clustering

Apply the single-view version variant into multi-view, and the new formula is written as follows,

$$\min_{\mathbf{U}} \sum_{v=1}^{m} \|\mathbf{E}^v\|_F^2, \quad \text{s.t.} \ \mathbf{U}^\top \mathbf{U} = \mathbf{I}_k, \mathbf{K}^v = \mathbf{U}\mathbf{U}^\top + \mathbf{E}^v. \tag{7}$$

In Eq. 7, the noise matrix is treated as a dependent variable, changing as $\mathbf{H}$ updating. And there goes the problem: when obtaining a consensus $\mathbf{H}$, $\{\mathbf{E}^v\}_{v=1}^{m}$

is carrying noise and specific information from each kernel. It is not appropriate to handle noise and view-specific information without distinction. Inspired by [17], we break the kernel matrices down into three parts: consensus part, specific part and noise part, as the Fig. 1 shows. It comes down to an equation as follows,

$$\mathbf{K}^v = \mathbf{U}\mathbf{U}^\top + \mathbf{E}^v + \mathbf{D}^v, \quad \forall v \in [1, m]. \tag{8}$$

Then, we argue that $\{\mathbf{E}^v\}_{v=1}^m$ is carrying redundant information, while $\{\mathbf{D}^v\}_{v=1}^m$ is carrying kernel-specific information. That is why we should treat them as independent variables. To obtain ideal low-rank consensus matrix, we use $\mathbf{U}\mathbf{U}^\top$ to denote it. In addition, we use F-norm and $\ell_{2,1}$-norm to ensure the sparseness of $\{\mathbf{D}^v\}_{v=1}^m$ and row sparseness of $\{\mathbf{E}^v\}_{v=1}^m$ respectively. Our idea can be mathematically expressed as follows,

$$\min_{\mathbf{U},\{\mathbf{D}^v\}_{v=1}^m,\{\mathbf{E}^v\}_{v=1}^m} \sum_{v=1}^m \{\|\mathbf{D}^v\|_F^2 + \lambda \|\mathbf{E}^v\|_{2,1}\}, \tag{9}$$

$$\text{s.t.} \, \forall v \in [1, m], \mathbf{K}^v = \mathbf{U}\mathbf{U}^\top + \mathbf{D}^v + \mathbf{E}^v, \mathbf{U}^\top \mathbf{U} = \mathbf{I}_k.$$

From Eq. 9, the proposed model treats $\mathbf{U}$, $\{\mathbf{D}^v\}_{v=1}^m$ and $\{\mathbf{E}^v\}_{v=1}^m$ as equal. In this way, we utilize kernel-specific information and redundant information to improve common information filtering.

3.3   Optimization

Given one of the constraints contains three variables, we find it hard to solve the problem in Eq. 9 directly. Fortunately, this problem can be solved by Augmented Lagrange Multiplier(ALM). By adding penalties to the original formula, we can obtain augmented Lagrange function as follows,

$$\min_{\mathbf{U},\{\mathbf{D}^v\}_{v=1}^m,\{\mathbf{E}^v\}_{v=1}^m,\{\mathbf{W}^v\}_{v=1}^m,\mu} \sum_{v=1}^m \left\{ \frac{\mu}{2} \|\mathbf{K}^v - \mathbf{U}\mathbf{U}^\top - \mathbf{D}^v - \mathbf{E}^v\|_F^2 \right.$$
$$\left. + \|\mathbf{D}^v\|_F^2 + \lambda \|\mathbf{E}^v\|_{2,1} + \langle \mathbf{W}^v, \mathbf{K}^v - \mathbf{U}\mathbf{U}^\top - \mathbf{D}^v - \mathbf{E}^v \rangle \right\}, \tag{10}$$

$$\text{s.t.} \, \mathbf{U}^\top \mathbf{U} = \mathbf{I}_k.$$

where $\{\mathbf{W}^v\}_{v=1}^m$ is a set of Lagrange multipliers, μ is a penalty parameter, and $\langle *, * \rangle$ denotes Euclidean space inner product of two matrices. We design a simple algorithm to solve this problem by alternatively updating $\mathbf{U}$, $\{\mathbf{D}^v\}_{v=1}^m$, $\{\mathbf{E}^v\}_{v=1}^m$.

update U With $\{\mathbf{D}^v\}_{v=1}^m, \{\mathbf{E}^v\}_{v=1}^m$ being fixed, the problem in Eq. 10 can be rewritten as follows,

$$\min_{\mathbf{U}} \sum_{v=1}^m \{\frac{\mu}{2} \|\mathbf{K}^v - \mathbf{U}\mathbf{U}^\top - \mathbf{D}^v - \mathbf{E}^v\|_F^2 + < \mathbf{W}^v, \mathbf{K}^v - \mathbf{U}\mathbf{U}^\top - \mathbf{D}^v - \mathbf{E}^v > \}, \tag{11}$$

$$\text{s.t. } \mathbf{U}^\top \mathbf{U} = \mathbf{I}_k.$$

The Eq. 11 can then be converted into Eq. 12.

$$\min_{\mathbf{U}} \text{Tr}(\mathbf{U}^\top \mathbf{M} \mathbf{U}), \tag{12}$$

$$\text{s.t. } \mathbf{U}^\top \mathbf{U} = \mathbf{I}_k,$$

where $\mathbf{M} = \mu(\mathbf{D}^v + \mathbf{E}^v - \mathbf{K}^v) - \mathbf{W}^v$. Obviously, the optimal of the problem in Eq. 12 can be obtained by calculating the k eigenvectors corresponding to the smallest k eigenvalues of $\mathbf{M}$.

update $\mathbf{D}^v$ With $\mathbf{U}$ and $\{\mathbf{E}^v\}_{v=1}^m$ being fixed, the Eq. 10 is equivalent to

$$\min_{\mathbf{D}^v} \sum_{v=1}^m \left\{ \|\mathbf{D}^v\|_{\mathrm{F}}^2 + \frac{\mu}{2} \left\| \mathbf{K}^v - \mathbf{U}\mathbf{U}^\top - \mathbf{D}^v - \mathbf{E}^v + \frac{\mathbf{W}^v}{\mu} \right\|_{\mathrm{F}}^2 \right\}. \tag{13}$$

By taking derivation of Eq. 13 with respect to E^v to zero, we can get the closed-form solution as Eq. 14.

$$\mathbf{D}^v = \frac{\mathbf{W}^{v\top} - \mu\mathbf{U}\mathbf{U}^\top - \mu\mathbf{E}^{v\top} + \mu\mathbf{K}^v}{\mu + 2}. \tag{14}$$

update$\{\mathbf{E}^v\}_{v=1}^m$ With $\mathbf{U}$ and $\{\mathbf{D}^v\}_{v=1}^m$ being fixed, the problem in Eq. 10 is equivalent to

$$\min_{\mathbf{E}^v} \sum_{v=1}^m \left\{ \frac{\mu}{2} \left\| \mathbf{K}^v - \mathbf{U}\mathbf{U}^\top - \mathbf{D}^v - \mathbf{E}^v + \frac{\mathbf{W}^v}{\mu} \right\|_{\mathrm{F}}^2 + \lambda \|\mathbf{E}^v\|_{2,1} \right\} \tag{15}$$

The problem in Eq. 15 can be divided into sub-problems by taking only one row of $\mathbf{E}^v$ into consideration each time. Then, the sub-problems can be written as,

$$\min_{e} \sum_{v=1}^m \left\{ \frac{\mu}{2} \|e - a\| + \lambda\|e\|_2 \right\}, \tag{16}$$

where e is a row of $\mathbf{E}^v$, a is the corresponding row of $\mathbf{A}$, and $\mathbf{A} = \mathbf{U}\mathbf{U}^\top + \mathbf{D}^v - \mathbf{K}^v - \frac{\mathbf{W}^v}{\mu}$. By taking derivation of Eq. 16 with respect to x to zero, we can obtain the closed-form solution as follows.

$$e = \begin{cases} \dfrac{(2\|a\|_2 - \frac{\mu}{2\lambda})a}{2\|a\|_2}, & \text{if } 2\|a\|_2 \geq \dfrac{\mu}{2\lambda}, \\[3mm] \mathbf{0}, & \text{if } 2\|a\|_2 < \dfrac{\mu}{2\lambda}. \end{cases} \tag{17}$$

where $\mathbf{0}$ denotes zero matrix.

4   Experiments and Analysis

In this section, we evaluate the clustering performance and convergence of the proposed algorithm on six benchmark data sets.

Algorithm 1. Unified and View-specific Multiple Kernel K-means Clustering

---

Input: Set of given kernel matrices $\{\mathbf{K}^v\}_{v=1}^m$, cluster number k, pre-defined parameterλ.

Initialize: $\{\mathbf{D}^v\}_{v=1}^m$, $\{\mathbf{E}^v\}_{v=1}^m$, $\{\mathbf{W}^v\}_{v=1}^m$, μ.

Output: Performing spectral clustering on $\mathbf{U}$.

1: **while** not convergence **do**
2:      update $\mathbf{U}$ by solving Eq. 12.
3:      update $\{\mathbf{D}^v\}_{v=1}^m$ by solving Eq. 13.
4:      update $\{\mathbf{E}^v\}_{v=1}^m$ by solving Eq. 15.
5:      update $\{\mathbf{W}^v\}_{v=1}^m$ by $\mathbf{W}^v = \mathbf{W}^v + \mu\left(\mathbf{K}^v - \mathbf{U}\mathbf{U}^\top - \mathbf{E}^v\right)$.
6:      update μ.
7: **end while**
8: **return** result.

4.1   DataSets

We conduct experiments on six widely-adopted benchmark datasets, they are Plant, Mfeat[1], AR10P[2], YALE[3], CCV[4], Caltech101-5[5]. In Table 2, detailed information about these data sets is presented. In particular, the number of samples, kernels and categories of these data sets is shown in the list.

4.2   Compared Algorithms

A wide range of the state-of-the-art clustering approaches are employed to compare with the proposed algorithm. (1) **Average Multiple Kernel k-means (A-MKKM).** AMKKM conducts clustering on the average of original kernel. (2) **Best Single View Kernel k-means (B-KKM).** BKM takes the best clustering performance based on only one view, and is often used as a benchmark to measure the effectiveness of multi-view clustering. (3) **Multiple Kernel k-means (MKKM).** MKKM alternatively performs kernel k-means and then calculates the optimal linear combination of the base kernel matrices. (4) **Co-regularized multi-view spectral clustering (CRSC)** [9]. Based on spectral clustering, CRSC proposes a framework to find the common cluster membership by co-regularizing the clustering hypotheses. (5) **Robust kernel k-means using $\ell_{2,1}$ norm (RMKKM)** [3]. Based on $\ell_{2,1}$ norm, RMKKM simultaneously finds the optimal clustering result, the cluster membership and the best linear combination of base kernels. (6) **Robust Multiple Kernel k-means (RMSC).** RMSC recovers a common low-rank transition probability matrix based on transition probability matrices for every view, and perform Markov chain clustering on it. (7) **Localized Data Fusion for Kernel -means clustering with Application to Cancer Biology (LMKKM)** [5]. LMKKM

---

[1] http://archive.ics.uci.edu/ml/datasets/Multiple+Features.
[2] http://featureselection.asu.edu/old/.
[3] http://cvc.yale.edu/projects/yalefaces/yalefaces.html.
[4] http://www.ee.clumbia.edu/ln/dvmm/CCV/.
[5] https://www.vision.caltech.edu/archive.html.

Table 1. The ACC, NMI and Purity comparison of different clustering algorithms on six benchmark data sets. The best results are red, and the results that are comparable to the best are bold. Besides, we calculated average rank by three criteria respectively.

Datasets	A-MKKM	SB-KKM	MKKM	Co-trian	RMKKM	RMSC	LMKKM	MKKM-MR	MKKM-LKA	Proposed
ACC(%)										
Plant	60.21	51.91	56.38	60.21	55.00	53.62	44.79	52.55	50.32	63.30
Mfeat	95.20	86.00	66.75	95.30	73.70	96.60	94.90	92.55	96.65	**95.90**
AR10P	38.46	43.08	40.00	38.46	30.77	30.77	40.77	39.23	27.69	56.15
YALE	52.12	56.97	52.12	56.97	56.36	58.03	53.33	**60.00**	46.67	60.61
CCV	19.98	20.23	18.29	19.98	16.76	16.29	20.17	20.86	18.35	23.65
Caltech101_5	36.67	36.86	28.63	36.08	32.75	33.73	**38.24**	**38.04**	32.16	38.24
Average Rank	5.17	4.83	7.33	4.50	7.50	6.00	5.33	4.33	7.67	1.33
NMI(%)										
Plant	25.54	17.19	20.02	25.54	19.43	23.18	13.79	21.65	21.46	28.69
Mfeat	89.83	75.79	60.84	90.02	73.05	**92.64**	89.68	85.90	92.70	**91.12**
AR10P	37.27	42.61	39.53	39.82	26.62	27.87	41.67	40.11	24.72	51.12
YALE	57.72	58.42	54.16	57.69	59.32	57.58	56.60	62.87	53.51	**61.38**
CCV	17.06	17.84	15.04	17.06	12.42	13.77	16.86	18.71	16.52	**18.57**
Caltech101_5	70.64	70.55	65.97	70.60	66.76	68.93	**71.28**	71.08	67.18	71.52
Average Rank	4.50	5.33	8.33	4.33	8.00	6.17	5.83	3.50	7.00	1.67
Purity(%)										
Plant	60.21	56.38	56.38	60.21	55.00	59.47	54.86	58.72	56.60	63.30
Mfeat	95.20	86.00	66.75	95.30	77.45	**96.60**	94.90	92.55	96.65	**95.90**
AR10P	39.23	43.08	40.00	39.23	32.31	33.08	40.77	39.23	28.46	56.15
YALE	53.94	57.58	52.73	57.58	58.18	57.24	53.94	60.00	49.09	62.42
CCV	23.56	23.62	22.41	23.56	20.79	20.08	24.36	24.88	22.90	25.58
Caltech101_5	38.24	38.04	29.80	37.65	33.92	34.90	**39.41**	39.02	33.92	40.00
Average Rank	4.67	5.00	8.00	4.33	7.83	6.17	5.17	4.00	7.00	1.33

obtains sample-specific information by calculating combined kernel in a localized way. (8) **Multiple kernel k-means with Matrix-induced Regularization (MKKM-MR)** [14]. MKMR proposes a matrix-induced regularization and plays a good role in redundancy elimination. (9) **Multiple Kernel Clustering with Local Kernel Alignment Maximization(MKKM-LKA)** [10]. MKKM-LKA requires the clustering algorithm to focus on sample pairs closed enough to be together and avoids sample pairs that far apart involving similarity evaluation.

Table 2. Detailed information of used data sets

Data sets	#Samples	#Kernels	#Clusters
Plant	940	69	4
Mfeat	2000	12	10
AR10P	130	6	10
YALE	165	5	15
CCV	6773	3	20
Caltech101-5	510	48	102

4.3   Experimental Setup

All base kernels are firstly centered and scaled so that, for every α and i, $\mathbf{K}_\alpha$ $(x_i, x_i) = 1$. The number of clusters is given for every data set, and we set it as the ground truth. In addition, the clustering performance are judged by three widely recognized criteria, including clustering accuracy (ACC), normalized mutual information (NMI) and purity (PUI). For the proposed algorithm, the trade-off parameter λ is chosen from $[10^{-5}, ..., 10^5]$. We repeat each experiment 50 times for all algorithm to avoid the effect of the random initialization and take the best result. Our experiments are conducted on desktop computer with Intel i9-9900K CPU @ 3.60 GHz × 16 and 64 GB RAM.

4.4   Clustering Performance

Table 1 lists the ACC, NMI, Purity comparison of the above algorithm on six data sets. The best results are written in red, and the results comparable to the best are bold. Based on the results, we find that our algorithm almost has the best performance on all given data sets: Plant, AR10P, YALE, ORL and Caltech101-5. Actually, it outperforms the second best algorithm 3.09%, 13.7%, 0.61%, 2.79%, 0.3% on ACC respectively. As for Mfeat, our method is not distinguishable from the best one. The NMI and purity in the table also show the proposed algorithm outperforms other multiple kernel clustering algorithms on most data sets. Furthermore, it is interesting that A-MKKM and SB-KKM have better performance than other methods on some data sets.

Particularly on the dataset Plant, the other methods are not as good as A-MKKM, while our method outperforms it. Given that A-MKKM does better than SB-KKM on Plant, we infer that it is necessary to combine information from every kernel to score higher than A-MKKM on the three criteria. This is the evidence that our method is able to combine effective information from different kernels. Meanwhile, on data set ORL, SB-KKM outperforms all methods other than ours. According to SB-KKM being better than A-MKKM on ORL, we find that our method has the ability to remove redundant information of each kernel. In summary, these results verify that the proposed algorithm can help improve clustering performance, compared to existing state-of-the-art approaches.

4.5   Kernel Structure and Parameter Sensitivity Study

To visualize the clustering result of our method, we show the affinity matrix in Fig. 2(a). There are ten blocks on the diagonal, which means our method manage to recognize every cluster. However, the color of these ten blocks is a bit mottled, indicating that our method can identify samples from the same cluster, but there are still deficiencies. Meanwhile, there are darker yet recognizable blocks on both sides of the diagonal. That shows affinity matrix captures the similarity between different clusters and might have a bad influence on clustering performance. As Fig. 2(b) shows, our method converges quickly within a few iterations. This verifies the convergence of our algorithm.

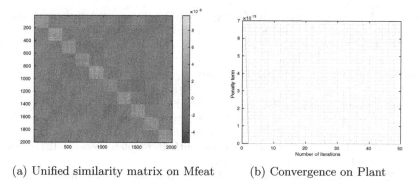

(a) Unified similarity matrix on Mfeat   (b) Convergence on Plant

Fig. 2. An illustration of the learned affinity matrix on Mfeat and the number of our method until convergence on dataset Plant.

5   Conclusion

In this article, we propose a novel multiple kernel clustering method named Unified and View-specific Multiple Kernel Clustering, which takes kernels down to unified, view-specific and noise matrices. We also introduce an algorithm to solve the Augmented Lagrange function of the original problem. Experiment results show that our method outperforms the compared state-of-art methods in clustering effects. In the future, we will consider utilize local information to improve our model.

References

1. Chen, J., Zhao, Z., Ye, J., Liu, H.: Nonlinear adaptive distance metric learning for clustering. In: Proceedings of the 13th ACM SIGKDD International Conference on Knowledge Discovery and Data Mining, pp. 123–132. ACM (2007)
2. Du, L., Zhang, H., Ren, X., Lv, X.: Manifold adaptive multiple kernel k-means for clustering. arXiv preprint arXiv:2009.14389 (2020)
3. Du, L., et al.: Robust multiple kernel k-means using l21-norm. In: Twenty-Fourth International Joint Conference on Artificial Intelligence (2015)
4. Gönen, M., Alpaydin, E.: Localized multiple kernel learning. In: Proceedings of the 25th International Conference on Machine Learning, pp. 352–359. ACM (2008)
5. Gönen, M., Margolin, A.A.: Localized data fusion for kernel k-means clustering with application to cancer biology. In: Advances in Neural Information Processing Systems, pp. 1305–1313 (2014)
6. Huang, H.C., Chuang, Y.Y., Chen, C.S.: Multiple kernel fuzzy clustering. IEEE Trans. Fuzzy Syst. **20**(1), 120–134 (2012)
7. Jegelka, S., Gretton, A., Schölkopf, B., Sriperumbudur, B.K., von Luxburg, U.: Generalized clustering via kernel embeddings. In: Mertsching, B., Hund, M., Aziz, Z. (eds.) KI 2009. LNCS (LNAI), vol. 5803, pp. 144–152. Springer, Heidelberg (2009). https://doi.org/10.1007/978-3-642-04617-9_19
8. Kang, Z., et al.: Partition level multiview subspace clustering. Neural Netw. **122**, 279–288 (2020)

9. Kumar, A., Rai, P., Daume, H.: Co-regularized multi-view spectral clustering. In: Advances in Neural Information Processing Systems, pp. 1413–1421 (2011)
10. Li, M., Liu, X., Wang, L., Dou, Y., Yin, J., Zhu, E.: Multiple kernel clustering with local kernel alignment maximization. In: International Joint Conference on Artificial Intelligence, pp. 1704–1710 (2016)
11. Li, T., Dou, Y., Liu, X., Zhao, Y., Lv, Q.: Multiple kernel clustering with corrupted kernels. Neurocomputing **267**, 447–454 (2017)
12. Liang, W., et al.: Multi-view spectral clustering with high-order optimal neighborhood laplacian matrix. IEEE Trans. Knowl. Data Eng. (2020)
13. Liu, X., et al.: Multiple kernel k-means with incomplete kernels. IEEE Trans. Pattern Anal. Mach. Intell. 1–14 (2019)
14. Liu, X., Dou, Y., Yin, J., Wang, L., Zhu, E.: Multiple kernel k -means clustering with matrix-induced regularization. In: Thirtieth AAAI Conference on Artificial Intelligence, pp. 1888–1894 (2016)
15. Liu, X., et al.: Optimal neighborhood kernel clustering with multiple kernels. In: Thirty-First AAAI Conference on Artificial Intelligence (2017)
16. Liu, X., et al.: Late fusion incomplete multi-view clustering. IEEE Trans. pattern Anal. Mach. Intell. (2018)
17. Luo, S., Zhang, C., Zhang, W., Cao, X.: Consistent and specific multi-view subspace clustering. In: Proceedings of the AAAI Conference on Artificial Intelligence, vol. 32 (2018)
18. Wang, S., et al.: Multi-view clustering via late fusion alignment maximization. In: IJCAI, pp. 3778–3784 (2019)
19. Wu, J., Liu, H., Xiong, H., Cao, J., Chen, J.: K-means-based consensus clustering: a unified view. IEEE Trans. Knowl. Data Eng. **27**(1), 155–169 (2014)
20. Xia, R., Pan, Y., Du, L., Yin, J.: Robust multi-view spectral clustering via low-rank and sparse decomposition. In: Twenty-Eighth AAAI Conference on Artificial Intelligence (2014)
21. Zhang, P., et al.: Consensus one-step multi-view subspace clustering. IEEE Trans. Knowle. Data Eng. (2020)

Data Analysis and Data Modeling

Augmented Lineage: Traceability of Data Analysis Including Complex UDFs

Masaya Yamada[1,2(✉)], Hiroyuki Kitagawa[1,2(✉)] [iD], Toshiyuki Amagasa[1] [iD], and Akiyoshi Matono[2] [iD]

[1] University of Tsukuba, Tsukuba, Ibaraki, Japan
`yamada@kde.cs.tsukuba.ac.jp`, {`kitagawa,amagasa`}`@cs.tsukuba.ac.jp`
[2] National Institute of Advanced Industrial Science and Technology, Koto-ku, Tokyo, Japan
`a.matono@aist.go.jp`

Abstract. Data lineage allows information to be traced to its origin in data analysis by showing how the results were derived. Although many methods have been proposed to identify the source data from which the analysis results are derived, analysis is becoming increasingly complex both with regard to the target (e.g., images, videos, and texts) and technology (e.g., AI and machine learning). In such complex data analysis, simply showing the source data may not ensure traceability. Analysts often need to know which parts of images are relevant to the output and why the classifier made a decision. Recent studies have intensively investigated interpretability and explainability in the machine learning (ML) domain. Integrating these techniques into the lineage framework will greatly enhance the traceability of complex data analysis, including the basis for decisions. In this paper, we propose the concept of *augmented lineage*, which is an extended lineage, and an efficient method to derive the augmented lineage for complex data analysis. We express complex data analysis flows using relational operators by combining user defined functions (UDFs). UDFs can represent invocations of AI/ML models within the data analysis. Then we present an algorithm to derive the augmented lineage for arbitrarily chosen tuples among the analysis results. We also experimentally demonstrate the efficiency of the proposed method.

Keywords: Data traceability · Data lineage · User Defined Function · Complex data analysis

1 Introduction

When data analysis is utilized for decision-making, it is critical to guarantee the traceability of the results. Data provenance in data analysis refers to all metadata describing its processing and is essential for traceability. In particular,

C. Strauss et al. (Eds.): DEXA 2021, LNCS 12923, pp. 65–77, 2021.
https://doi.org/10.1007/978-3-030-86472-9_6

data lineage refers to tracing the source data from which the analysis results are derived. This topic has been widely researched in the database domain [5,6,17].

Even with lineage, sufficient traceability remains a challenge. Modern data analysis has become more complex. Not only are the targets composed of diverse content data such as images, videos, and texts but processing often involves sophisticated technologies such as AI and machine learning (ML). Below is an example.

Example 1. Figure 1 shows an analysis of a financial institution summarizing the results of loan application examinations based on customer information using an ML model. The ML model takes (Income, Debt, LoanAmount) as inputs and returns the examination result (Accept, Marginal, Reject). The reason that an application was rejected is not obvious. In this case, the lineage of $\langle Reject, 1 \rangle$ is a set of tuples $\{\langle 0002, Joa, 25, 2200 \rangle, \langle 0002, 1350, 12/17 \rangle, \langle 0002, 20000, 12/19 \rangle\}$. This is insufficient to understand why $\langle Reject, 1 \rangle$ was derived from these source tuples. In other words, the source tuples do not fully explain "why this application was rejected." However, the reason for the decision in the ML model along with the source tuples may explain why an application was rejected. For example, "the model rejected this application because the debt and existing loan amount of the applicant are both large." This reason describes the basis of the result, leading to a higher traceability.

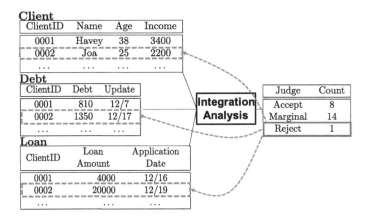

Fig. 1. Analysis summarizing the results for loan application examinations using an ML model. Tuples surrounded by dashed lines represent the source data of the result tuple $\langle Reject, 1 \rangle$.

In this paper, we propose the concept of *augmented lineage*, which is an extension of lineage that incorporates the basis for decisions (*reason*) in complex data analysis. Reason shows (1) which region of the content data (e.g., images or videos) the analytical model emphasizes and (2) the basis for the AI/ML model decision. The contributions of this paper are as follows: (i) proposal and formulation of augmented lineage, (ii) proposal of a basic algorithm to derive the augmented lineage, (iii) proposal of an enhanced method for its efficient derivation, and (iv) experimental evaluation of the proposed derivation method.

The rest of this paper is organized as follows. Section 2 overviews related work. Section 3 presents the data model used to represent data analysis. Section 4 proposes the augmented lineage, while Sect. 5 formulizes a basic algorithm to derive the augmented lineage. Section 6 details its enhanced method. Section 7 experimentally investigates the derivation method. Finally, Sect. 8 concludes this paper.

2 Related Work

Lineage has been researched widely in database and scientific workflow domains. Cui et al. [6] modeled queries in data warehouses using the relational model and proposed a method to find the source tuples from which the output tuples were derived by executing tracing queries. Bhagwat et al. [1] also targeted relational queries and showed how to obtain the lineage. They assigned identifiers to all values in the source tables, and annotated query results with the lineage tuple identifiers during query processing.

Modern data analysis includes content data analysis and AI/ML analysis (e.g., recommendation [18], anomaly detection [2,12], and medical diagnosis [7,13]). [5,17] derived the lineage for the analysis, which included user-defined functions (UDFs) and relational operators. Cui and Widom [5] proposed a method to derive a lineage for analysis, which included more generalized operators. Their method can deal with non-relational operators by focusing on the relationship between the input and output data (e.g., one-to-one, N-to-one, N-to-M relationships) of each operator and schema information. Wu et al. [17] presented a method to find lineage in SciDB, which handles multi-dimensional array data for an analysis involving user-defined operators. In their study, both the input cells from which an output cell was derived (backward lineage) and the output cells to which an input cell contributes (forward lineage) could be identified. The lineage was obtained by recording the input/output relationships of each operator in the workflow. They proposed a method to efficiently manage and later derive lineages by rerunning. In these studies, lineage means the correspondence between input/output tuples (cells). However, they did not show the important parts of the content data (images, texts, etc.). Moreover, they did not cover the basis for AI/ML decisions. Zheng et al. [19] proposed a method to track information about which parts of the content data were extracted by data extraction processing represented by UDFs as well as the traditional tuple-level lineage. Even this framework cannot provide the basis for decisions made by AI/ML processing.

Recently, many studies have investigated interpretability and explainability of AI/ML processing [8,9]. Using frameworks such as [14–16], the basis for the decisions in AI/ML processing can be shown. The augmented lineage presented in this paper is a general framework that integrates such reasoning functions into lineage and contributes to the improved traceability in complex data analysis, including AI/ML processing.

3   Data Model

This section describes the data model to represent data analysis. Complex data analysis is modeled as a *task* consisting of relation-like operators. First, we introduce the notations. Table $T(A_1, \ldots, A_n)$ has a set of tuples $\{t_1, \ldots, t_p\}$. $t.\boldsymbol{A}$ projects tuple t onto the tuple with an attribute set $\boldsymbol{A}$ ($\subseteq \{A_1, \ldots, A_n\}$). For a set of tuples $\{t_1, \ldots, t_q\}$ (including, in special cases, single values), $\langle t_1, \ldots, t_q \rangle$ denotes the tuple that concatenates them. Source data set D consists of the set of tables $\{T_1, \ldots, T_m\}$[1]. $\mathscr{T}$ denotes the task on the source data set D, and O denotes its output. That is, $O = \mathscr{T}(D) = \mathscr{T}(T_1, \ldots, T_m)$.

Next, we describe the operators that compose a task. To model data analysis, we assume seven set-relational operators: basic operators (Join $\bowtie$, Selection σ, Projection π, Aggregation α, Union $\cup$, Difference $-$) and Function operator (ϕ), which models complex data analysis. A task is represented as a tree consisting of these operators, while leaf nodes are source tables. For ease of exposition, this paper restricts the operators to relational operators. However, our framework can deal with more generalized processing and external programs as long as they have the same input/output relationships as relational operators. For example, when σ receives one tuple, it either (1) outputs the tuple or (2) outputs nothing. Processing by external programs with the same input/output relationships can be modeled in the same way as σ. Due to space limitations, we only introduce the function operator in detail. First, we define the *reason* for the function operator.

Definition 1 (Reason). *Reason is the information showing the basis for the result of complex data analysis.*

Example 2. Given a classifier that takes customer information and loan application information as inputs to examine the application, the classifier can show which source data contributes above a certain threshold in decision making as well as the decision results. We treat such information as reasons.

Definition 2 (Function operator). *Function operator ($\phi_{f(\boldsymbol{E})}$) applies UDF f, which models complex data analysis[2], to all tuples (more precisely, their values for attributes $\boldsymbol{E}$) in the input table. UDF f is the function with the following input and output, $f: Domain(\boldsymbol{E}) \rightarrow Value \times \boldsymbol{Reason}$[3], where Value*

---

[1] If a table is referred to more than once, each reference is regarded as access to a different table.

[2] UDF can be used to model target-specific simple computation (e.g., compute an area), too. For simplicity of discussion, we focus on the use of UDF for complex data analysis.

[3] If a reason is not needed (only the output value is needed), we can perform f in the non-reasoning mode, which produces no reason.

is the domain of the output values of the complex data analysis, and **Reason** *represents the domain of reasons. The UDF developers must determine what information is produced as reasons. The output of function operator* $\phi_{f(E)}$ *is* $\phi_{f(E)}(T) = \{\langle t, f(t.\boldsymbol{E}).Value, f(t.\boldsymbol{E}).Reason \rangle \mid t \in T\}.$

Example 3. Using our data model, the analysis in Fig. 1 can be represented as:

$$O = \mathscr{T}(D) = \alpha_{Judge,COUNT(*)}(\phi_{f(Income,Debt,LoanAmount)}(\\ \pi_{Income,Debt,LoanAmount}(Client \bowtie Debt \bowtie Loan)))$$

Figure 2(a) shows its operator tree. Note that $\alpha_{G,g(B)}(T)$ groups tuples in table T based on grouping key $\boldsymbol{G}$ and applies aggregate function g to attribute B for each group. Here, "Judge" refers to the UDF's Value attribute, and UDF f is assumed to show the set of attributes contributing above a certain threshold to the output (Value) as a reason. Table 1 shows the output of the function operator.

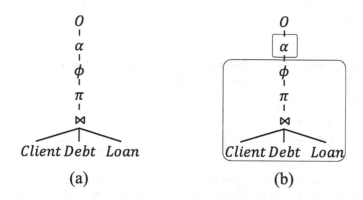

Fig. 2. (a) Operator tree. (b) Segmented operator tree.

Table 1. Output of the function operator in Example 3.

Income	Debt	LoanAmount	Judge	Reason
3400	810	4000	Accept	{Income, LoanAmount}
2200	1350	20000	Reject	{Debt, LoanAmount}
...	...	...	...	...

4   Augmented Lineage

This section defines *augmented lineage*. Augmented lineage is an extension of *lineage* in [4]. The set of the source tuples from which tuple $o(\in \mathscr{T}(D))$ of task $\mathscr{T}$ is derived is called the *source lineage* of tuple o in task $\mathscr{T}$. It is denoted by $SL(o) = \{T_1^*, \ldots, T_m^*\}$ $(T_i^* \subseteq T_i)$. The source lineage is called lineage in [4]. For a detail description, please refer to the paper. Re-executing task $\mathscr{T}$ on $SL(o)$ results in tuple o, and all tuples in $SL(o)$ contribute to it. The source lineage of tuple set $\bar{O}(\subseteq \mathscr{T}(D))$ in task $\mathscr{T}$ is $SL(\bar{O}) = \hat{\bigcup}_{o \in \bar{O}} SL(o)$. Note that $\hat{\bigcup}$ is an operator that produces the union for each table: $\{T_1^{1*}, \ldots, T_m^{1*}\} \hat{\bigcup} \{T_1^{2*}, \ldots, T_m^{2*}\} = \{T_1^{1*} \cup T_1^{2*}, \ldots, T_m^{1*} \cup T_m^{2*}\}$ s.t. $T_i^{j*} \subseteq T_i$.

Before defining the augmented lineage, we define *intermediate lineage* of tuple set $\bar{O}(\subseteq \mathscr{T}(D))$ in an intermediate result.

Definition 3 (Intermediate Lineage). *Let O be the output tuples of task $\mathscr{T}$. That is, $O = \mathscr{T}(D) = \mathscr{T}(T_1, \cdots, T_m)$. Furthermore, let task $\mathscr{T}$ be divided into two tasks, $\mathscr{T}'$ and $\mathscr{T}''$. Namely, $\mathscr{T}(D) = \mathscr{T}'(T_{O'}, T_{l_{i+1}}, \ldots, T_{l_m})$, $T_{O'} = \mathscr{T}''(T_{l_1}, \ldots, T_{l_i})$. The intermediate lineage $IL(\bar{O}, T_{O'})$ of tuple set $\bar{O}$ $(\subseteq \mathscr{T}(D))$ in intermediate result $T_{O'}$ is the source lineage $T_{O'}^* \in SL(\bar{O})$ of tuple set $\bar{O}$ in task $\mathscr{T}'$.*

Definition 4 (Augmented Lineage). *Augmented lineage $AL(\bar{O})$ of tuple set $\bar{O}(\subseteq \mathscr{T}(D))$ in task $\mathscr{T}$ consists of the following pair, where $T_{O_i'}$ denotes the intermediate result generated by each function operator $\phi_{f_i(\mathbf{E}_i)}$.*

- *Source Lineage (SL): $SL(\bar{O})$*
- *Reasoning Lineage (RL): $RL(\bar{O}) = \{\langle o.\mathbf{E}_i, o.Value, o.Reason \rangle \mid \forall i, o \in IL(\bar{O}, T_{O_i'})\}$*

Referring to the outputs of the function operator shown in Table 1, the augmented lineage of the tuple $\langle Reject, 1 \rangle$ in the task shown in Example 1 is as follows (Note that $[A, B]$ means a pair consisting of A and B.):

$$\left[\left\{ \begin{array}{l} \text{Client: } \{ \langle 0002, \text{Joa}, 25, 2200 \rangle \} \\ \text{Debt: } \{ \langle 0002, 1350, 12/17 \rangle \} \\ \text{Loan: } \{ \langle 0002, 20000, 12/19 \rangle \} \end{array} \right\}, \left\{ \begin{array}{l} \langle \langle 2200, 1350, 20000 \rangle, \\ \text{Reject}, \{\text{Debt}, \text{LoanAmount}\} \rangle \end{array} \right\} \right]$$

5   Augmented Lineage Derivation

There are two approaches to obtain the lineage [3,10]: the eager approach and lazy approach. The former produces lineage for all analysis results during execution. The latter derives the lineage for the chosen analysis results after execution. This paper uses the lazy approach.

In recent data analysis, analysts often develop analysis flows using trial and error to configure parameters, change source data, etc. In this context, the lazy approach is more desirable than the eager approach, which has a larger overhead for every analytical execution. This is why we focus on the lazy approach.

Our proposed method to derive the augmented lineage is based on [6], which proposed a method to obtain lineage of database queries via the lazy approach. Our extension enables use of UDFs for complex data analysis, a more practical treatment of task operators, and inclusion of reasoning lineage.

Given a task, we divide its operator tree into one or more operator sub-trees (*segments*) and then derive the augmented lineage for each segment. In Sect. 5.1, we explain segments in more details and present an algorithm to divide a task. In Sect. 5.2, we introduce a *tracing query*, which finds the source lineage of a single segment. Finally, we propose a basic algorithm to derive the augmented lineage of a task in Sect. 5.3.

5.1 Segment

There are two types of segments:

– Non-D-segment: A segment of operators except the difference in the order of ϕ-α-$\cup$-π-σ-$\bowtie$. The leftmost operator is located at the top of the operator tree, which is called the top of the segment. Namely, the operators are executed from right to left in a bottom-up manner. Note that all the operators do not need to actually appear in a Non-D-segment.
– D-segment: A segment consisting of a single difference operator.

An operator tree can be transformed by exchanging the order of commutative operators to be divided into fewer segments. We call such an operator tree its *canonical form*. Algorithm 1 transforms the operator tree of task $\mathscr{T}$ into the canonical form and divides it into segments. After canonicalizing, each Non-D-segment will satisfy one of the following conditions.

i. It is at the top of the operator tree.
ii. Otherwise,
    A. it has a function operator at the top,
    B. it has an aggregation operator at the top,
    C. it has a union operator at the top and is directly below a join operator or a D-segment, or
    D. it has a projection, selection, or join operator at the top and is directly below a D-segment.

Example 4. The task in Example 3 is divided into two Non-D-segments by Algorithm 1: (1) Non-D-segment consisting of α alone and (2) Non-D-segment ϕ-π-$\bowtie$ (Fig. 2(b)).

In the following explanation, it is assumed that the operator tree of task $\mathscr{T}$ is in a canonical form unless specified otherwise.

Algorithm 1. Canonicalize

Input: Operator tree of task $\mathscr{T}$
Output: Canonical form of $\mathscr{T}$
 1: Pull unions above projections and selections in $\mathscr{T}$;
 2: Pull projections above selections and joins in $\mathscr{T}$;
 3: Pull selections above joins in $\mathscr{T}$;
 4: Merge adjacent same operators;
 5: Split $\mathscr{T}$ into segments;
 6: **return** $\mathscr{T}$;

5.2 Tracing Query

First, we consider the case where task $\mathscr{T}$ consists of a single segment, and explain a tracing query to find the source lineage $SL(\bar{O})$ of a tuple set $\bar{O}(\subseteq \mathscr{T}(T_1, \ldots, T_m))$ in task $\mathscr{T}$. For this, we introduce the split operator.

Definition 5 (Split operator). *Let A denote the set of attributes in table T. The split operator produces a set of tables with attribute set $A_i \subseteq A$ using the projection as:*

$$Split_{A_1, \ldots, A_n}(T) = \{\pi_{A_1}(T), \ldots, \pi_{A_n}(T)\}$$

The tracing queries for Non-D-segments and D-segments are shown below. If a task consists of a single segment, its source lineage can be obtained using the following tracing queries.

Tracing Query for a Non-D-Segment: Given a non-D-segment $\mathscr{T}(D) = \phi_{f(E)}(\; \alpha_{G,g(B)}(\cup_i(\pi_{A_i}(\sigma_{C_i}(T_1^i \bowtie \cdots \bowtie T_{m_i}^i)))))$, the source lineage of tuple set $\bar{O}(\subseteq \mathscr{T}(D))$ in task $\mathscr{T}$ can be obtained by executing the following tracing query:

$$TQ_{\bar{O}, \mathscr{T}}(D) = \bigcup_i Split_{T_1^i, \ldots, T_{m_i}^i}(\sigma_{C_i}(T_1^i \bowtie \cdots \bowtie T_{m_i}^i) \ltimes \bar{O})$$

Note that $\ltimes$ denotes semi-join. If some operators are missing in the Non-D-segment, their counterparts are omitted in the tracing query. If tuple set $\bar{O}$ is the whole output $\mathscr{T}(D)$, the tracing query is denoted as follows using the notation **ALL**:

$$TQ_{\mathbf{ALL}, \mathscr{T}}(D) = \bigcup_i Split_{T_1^i, \ldots, T_{m_i}^i}(\sigma_{C_i}(T_1^i \bowtie \cdots \bowtie T_{m_i}^i))$$

Tracing Query for a D-Segment: Given a D-segment $\mathscr{T}(D) = T_1 - T_2$, the source lineage of tuple set $\bar{O}(\subseteq \mathscr{T}(D))$ in task $\mathscr{T}$ is represented as follows:

$$TQ_{\bar{O}, \mathscr{T}}(T_1, T_2) = \{\bar{O}, T_2\}$$

5.3 Augmented Lineage Derivation Procedure

We introduce an algorithm to derive the augmented lineage of tuple set $\bar{O}(\subseteq \mathscr{T}(D))$ in task $\mathscr{T}$. To obtain the augmented lineage, we (1) transform the operator tree of the task into the canonical form by executing Algorithm 1 and

---

Algorithm 2. Augmented Lineage Derivation Procedure

Input: Source data set D, Task $\mathscr{T}$, Tuple set $\bar{O}(\subseteq \mathscr{T}(D))$
Output: Augmented lineage of tuple set $\bar{O}$ in task $\mathscr{T}$
  1: $SL = RL = \varnothing$;
  2: Initialize $TQqueue$; // $TQqueue$ is a queue.
  3: Enqueue a pair $[\bar{O}, \mathscr{T}]$ into $TQqueue$;
  4: **while** $TQqueue$ is not empty **do**
  5:     Dequeue a pair $[\bar{O}, \mathscr{T}]$ from $TQqueue$;
  6:     **if** $\mathscr{T}'s$ top segment $\bar{\mathscr{T}}$ is a D-segment **then**
  7:         // $O = \bar{\mathscr{T}}(O_1, O_2) = O_1 - O_2, O_i = \mathscr{T}_i(D_i)$ s.t. $D_i \subseteq D$
  8:         **if** $\bar{O} = $ **ALL then**
  9:             $\bar{O} \leftarrow O$;
 10:         $\{O_1^*, O_2^*\} \leftarrow \{\bar{O}, $ **ALL**$\}$;
 11:     **else if** $\mathscr{T}'s$ top segment $\bar{\mathscr{T}}$ is a Non-D-segment **then**
 12:         // $O = \bar{\mathscr{T}}(O_1, \ldots, O_k), O_i = \mathscr{T}_i(D_i)$ s.t. $D_i \subseteq D$
 13:         **if** $\bar{\mathscr{T}}'s$ top operator is $\phi_{f(E)}$ **then**
 14:             **if** $\bar{O} = $ **ALL then**
 15:                 $\bar{O} \leftarrow O$;
 16:             **for** $o \in \bar{O}$ **do**
 17:                 $RL = RL \cup \{\langle o.E, o.Value, o.Reason \rangle\}$;
 18:         $\{O_1^*, \cdots, O_k^*\} \leftarrow TQ(\bar{O}, \bar{\mathscr{T}}, \{O_1, \cdots, O_k\})$;
 19:     **for** Each O_i^* **do**
 20:         **if** O_i^* corresponds to a source table T_j **then**
 21:             **if** $O_i^* = $ **ALL then**
 22:                 $O_i^* \leftarrow T_j$;
 23:             $SL = SL \cup \{O_i^*\}$;
 24:         **else if** O_i^* does not correspond to any source table **then**
 25:             Enqueue $[O_i^*, \mathscr{T}_i]$ into $TQqueue$;
 26: **return** $[SL, RL]$;

---

(2) apply Algorithm 2 to the task. Algorithm 2 recursively derives the source lineage of each segment and records the reasoning lineage if the tuples in the intermediate lineages contain reasons.

6   Implementation of Augmented Lineage Derivation

The procedure to derive the augmented lineage introduced in Sect. 5.3 assumes that when a task consists of multiple segments, the intermediate result of the lower segment is available to derive the source lineage of the upper segment. The following two approaches are commonly applied.

(1) **Full Materialization (Full):** This approach materializes (generates and stores) the intermediate results when the analysis runs. This approach somewhat affects the analytical execution, and the storage cost for managing the intermediate results becomes overhead.

(2) **Rerun:** This approach recovers all necessary intermediate results by rerunning the analysis before executing tracing queries. Although there is no runtime overhead and storage cost in the analytical execution, rerunning the analysis before deriving the augmented lineage is time consuming.

In the preliminary experiments, we evaluated the performance of the above two approaches on complex data analyses. As a result, when a task includes a function operator with an expensive UDF, the operator's re-execution is time consuming compared to other relational operators. Hence, we propose the following enhanced approach to derive the augmented lineage efficiently.

Function Materialization (FM): This approach materializes only intermediate results, which are directly generated by the function operators when running the analysis. The other intermediate results are recovered by rerunning the analysis before deriving the augmented lineage.

To improve the efficiency of Rerun and FM, it is possible to semi-join the source tables with the output tuples before rerunning the analysis. This optimization can reduce the source table size before the re-execution of the analysis. In addition, we can execute each segment after optimizing it in all approaches. We experimentally evaluate the cost of the above approaches.

7   Experiment

To evaluate the runtime and storage cost of our proposed method when deriving the augmented lineage, we implemented it on a relational database. The analysis run in our experiments used person recognition for images to determine the number of times that celebrities appeared on stage at large event places. We used [11] dataset for our experiments. The analysis employed two tables: $Image(\underline{ImageID}, PlaceID, Img)$ and $Event(\underline{EventID}, PlaceID, Visitors)$ as inputs, where the underlined attributes are keys. In our experiment, both source lineage and reasoning lineage were also stored in relational tables.

```
SELECT Value, COUNT(*)
FROM
(SELECT ImageID, PlaceID, avg, ret[1] AS Img, ret[2] AS Value, ret[3] AS Reason
 FROM
 (SELECT ImageID, PlaceID, avg, recognition(Img) AS ret
  FROM Image NATURAL JOIN
  (SELECT PlaceID, AVG(Visitors) AS avg FROM Event GROUP BY PlaceID) Seg2
  WHERE avg >= 50000) Selected ) Seg1
GROUP BY Value;
```

Fig. 3. SQL statement executed in our experiment.

Attribute Img contained the path (URI) of an external image file as a string. The path involved the name of the celebrity in the image. Figure 3 shows the SQL used for this experiment, and the analysis is modeled as follows:

$$O = \mathcal{T}(D)$$
$$= \alpha_{Value,COUNT(*)}(\phi_{recognition(E)}(\sigma_C(R_1 \bowtie (\alpha_{PlaceID,AVG(Visitors)}(R_2)))))$$
$$s.t.\ \boldsymbol{E} : Img,\ C : avg >= 50000,\ R_1 : Image,\ R_2 : Event$$

This operator tree consists of three Non-D-segments: (1) α, (2) ϕ-σ-$\bowtie$, and (3) α. *Recognition* function is a UDF that performed person recognition. To investigate the effect of UDF processing costs on the effectiveness of Function Materialization, we prepared two implementations of the *recognition* function.

Face Recognition: A person is identified by an ML-based face recognition model in the given image. It outputs the person's name as the value and the position of the bounding box around the face as the reason. Its processing cost is expensive.

String Processing: A person is recognized by extracting the name from the URI string. It outputs the person's name as the value and the position of the name in the URI as the reason. Its processing cost is cheap.

Each output value of the two *recognition* function implementations is the same.

We also prepared three different source table sizes (Small, Medium, Large). Table 2 shows the number of tuples in each case. Our implementation used PostgreSQL 9.6.21 and Python 3.7.8. All experiments were run on a machine with an Intel(R) Xeon(R) CPU E5-2620 v2 @ 2.10 GHz and 128 GB memory.

Table 2. Number of tuples in each source table.

	Image	Event
Small	4.5×10^4	6.075×10^5
Medium	4.5×10^5	6.075×10^6
Large	4.5×10^6	6.075×10^7

Table 3. Number of materialized tuples.

	Full	FM
Small	4.95×10^4	4.5×10^3
Medium	4.95×10^5	4.5×10^4
Large	4.95×10^6	4.5×10^5

Analysis with Face Recognition UDF (Expensive UDF): Figure 4 summarizes the processing time to derive the augmented lineage. Table 3 shows the number of materialized tuples as the intermediate results. Both Function Materialization (FM) and Full Materialization (Full) outperform Rerun with respect to processing time. Full is up to about 1578.3 times faster than Rerun for the medium-sized table and FM is up to about 516.8 times faster than Rerun for the small-sized table. FM drastically reduces the number of materialized tuples (by about 91%) compared to Full (Table 3). Hence, FM is an appropriate approach to realize a fair trade-off between processing time and storage cost.

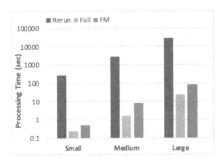

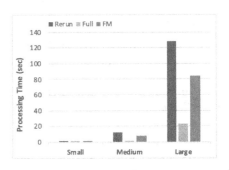

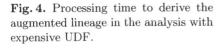

Fig. 4. Processing time to derive the augmented lineage in the analysis with expensive UDF.

Fig. 5. Processing time to derive the augmented lineage in the analysis with cheap UDF.

Analysis with String Processing UDF (Cheap UDF): Figure 5 summarizes the processing time to derive the augmented lineage. In this case, Full Materialization (Full) and Function Materialization (FM) exceed Rerun with respect to processing time. Full is about 6.6 times faster in medium table size and FM is about 2.2 times faster than Rerun in small table size. Furthermore, since the size of the intermediate results is the same as that in Table 3, FM had a smaller storage cost than Full. However, the advantage of materializing the intermediate results on the processing time is smaller because the processing cost of a cheap function operator is not very high compared with other relational operators, and its re-execution has a low overhead.

8   Conclusions and Future Work

In this paper, we have proposed the augmented lineage, which is an extended lineage combining reasons for complex data analysis. This augmented lineage ensures traceability of complex data analysis, including AI/ML processing. Additionally, we formulated an algorithm to derive the augmented lineage using the lazy approach. We also introduced Function Materialization (FM), which allows for a fair tradeoff between runtime cost and storage cost in deriving the augmented lineage. Experiments on a relational database showed that FM is effective, especially when expensive UDFs involving sophisticated AI/ML processing are included in the data analysis flows. Interesting future research topics include comparison with other existing approaches, extending the proposed data model for more generalized analysis contexts, and applying the proposed framework to big data processing systems and stream processing environments.

Acknowledgment. This work was partly supported by JSPS KAKENHI Grant Number JP19H04114 and the Project Commissioned by New Energy and Industrial Technology Development Organization (JPNP20006).

References

1. Bhagwat, D., Chiticariu, L., Tan, W.C., Vijayvargiya, G.: An annotation management system for relational databases. VLDB J. **14**(4), 373–396 (2005)
2. Chalapathy, R., Chawla, S.: Deep learning for anomaly detection: a survey. arXiv preprint arXiv:1901.03407 (2019)
3. Cheney, J., Chiticariu, L., Tan, W.C.: Provenance in Databases: Why, How, and Where. Now Publishers Inc, Hanover (2009)
4. Cui, Y., Widom, J.: Practical lineage tracing in data warehouses. In: Proceedings of 16th International Conference on Data Engineering, pp. 367–378 (2000)
5. Cui, Y., Widom, J.: Lineage tracing for general data warehouse transformations. VLDB J. **12**(1), 41–58 (2003)
6. Cui, Y., Widom, J., Wiener, J.L.: Tracing the lineage of view data in a warehousing environment. ACM Trans. Database Syst. **25**(2), 179–227 (2000)
7. Kermany, D.S., et al.: Identifying medical diagnoses and treatable diseases by image-based deep learning. Cell **172**(5), 1122–1131.e9 (2018)
8. Du, M., Liu, N., Hu, X.: Techniques for interpretable machine learning. Commun. ACM **63**(1), 68–77 (2019)
9. Gunning, D.: Explainable artificial intelligence (XAI). Defense Advanced Research Projects Agency (DARPA), nd Web 2, 2 (2017)
10. Herschel, M., Diestelkämper, R., Ben Lahmar, H.: A survey on provenance: what for? What form? What from? VLDB J. **26**(6), 881–906 (2017)
11. Huang, G.B., Ramesh, M., Berg, T., Learned-Miller, E.: Labeled faces in the wild: a database for studying face recognition in unconstrained environments. Technical report 07–49, University of Massachusetts, Amherst, October 2007
12. Kwon, D., Kim, H., Kim, J., Suh, S.C., Kim, I., Kim, K.J.: A survey of deep learning-based network anomaly detection. Clust. Comput. **22**(1), 949–961 (2017). https://doi.org/10.1007/s10586-017-1117-8
13. Litjens, G., et al.: Deep learning as a tool for increased accuracy and efficiency of histopathological diagnosis. Sci. Rep. **6**(1), 26286 (2016)
14. Lundberg, S.M., Lee, S.I.: A unified approach to interpreting model predictions. In: Advances in Neural Information Processing Systems, vol. 30, pp. 4765–4774. Curran Associates, Inc. (2017)
15. Ribeiro, M.T., Singh, S., Guestrin, C.: "Why should i trust you?" Explaining the predictions of any classifier. In: Proceedings of the 22nd ACM SIGKDD International Conference on Knowledge Discovery and Data Mining, pp. 1135–1144 (2016)
16. Selvaraju, R.R., et al.: Grad-CAM: visual explanations from deep networks via gradient-based localization. In: Proceedings of the IEEE International Conference on Computer Vision, pp. 618–626 (2017)
17. Wu, E., Madden, S., Stonebraker, M.: Subzero: a fine-grained lineage system for scientific databases. In: 2013 IEEE 29th International Conference on Data Engineering (ICDE), pp. 865–876 (2013)
18. Zhang, S., Yao, L., Sun, A., Tay, Y.: Deep learning based recommender system: a survey and new perspectives. ACM Comput. Surv. (CSUR) **52**(1), 1–38 (2019)
19. Zheng, N., Alawini, A., Ives, Z.G.: Fine-grained provenance for matching & ETL. In: 2019 IEEE 35th International Conference on Data Engineering (ICDE), pp. 184–195 (2019)

Neural Ordinary Differential Equations for the Regression of Macroeconomics Data Under the Green Solow Model

Zi-Yu Khoo[1]([✉]), Kang Hao Lee[2], Zhibo Huang[2], and Stéphane Bressan[1]

[1] National University of Singapore, 21 Lower Kent Ridge Road,
Singapore 119077, Singapore
{khoozy,steph}@comp.nus.edu.sg
[2] Singapore University of Technology and Design, 8 Somapah Road,
Singapore 487372, Singapore
{kanghao_lee,zhibo_huang}@mymail.sutd.edu.sg

Abstract. We are interested in the regression of data to a parameterised system of differential equations formalising a dynamical system. We study the case of the Green Solow model, a neoclassical economics model for sustainable growth. Faced with the challenges posed by the coupling of the equations, the scarcity of data, and their auto-correlation, we devise several solutions. We present a baseline model and propose three models leveraging neural ordinary differential equations, a recently proposed machine learning model. We empirically and comparatively evaluate the performance of the four models. The results demonstrate the advantages of the proposed approach using neural ordinary differential equations. We conclude by discussing the generality of this knowledge- and data-driven approach to the analysis of dynamical systems.

Keywords: Data analysis · Dynamical systems · Machine learning

1 Introduction

Dynamical systems [10] are physical systems characterised by the evolution of their state over time. The modelling of a dynamical system generally stems from a first principles approach where a general relationship is derived using fundamental theoretical concepts and empirical observations. It is a hand and brain iterative process of empirical observations, hypotheses formulation, and testing. The resulting model is generally a parameterised system of differential equations defining the time evolution rules of the physical systems.

Recent advances in machine learning, such as the identification of the governing equations from data [4,24] and the integration of knowledge into physics-informed machine learning models [17,18], create new opportunities for the data-driven modelling of dynamical systems.

Regression analysis [9] is a data-driven computational statistical analysis that models the relationship between variables. Given a parameterised model, regression finds values of the parameters that best fit the model to the observed data

© Springer Nature Switzerland AG 2021
C. Strauss et al. (Eds.): DEXA 2021, LNCS 12923, pp. 78–90, 2021.
https://doi.org/10.1007/978-3-030-86472-9_7

of the variables. However, these parameters may be inconsistent and not interpretable by domain experts if the coupled equations of the dynamical system are not solved simultaneously. The model may also assume ceteris paribus while the parameters themselves vary over time. Finally, the data for such a dynamical system is often scarce and auto-correlated. These complicate the search for parameters. We are interested in the regression of observed data to a parameterised system of differential equations formalising a dynamical system, to find values of the parameters that best fit the model to the observed data of the variables.

We therefore propose leveraging neural ordinary differential equations [5], a family of deep neural network models that combine neural networks with differential equation solvers. Neural ordinary differential equations have the ability to find consistent time-varying parameters based on coupled equations despite scarce and auto-correlated data.

The dynamical system that we consider is the Green Solow model [3]. It describes the dynamics of sustainable growth. Country-level macro-economics data for the variables of the model, such as capital, income, and emissions, are generally openly available from national and international organisations.

We present and empirically evaluate the performance of four approaches to the regression problem estimating the parameters of the dynamical system. We devise a conventional first approach to serve as a baseline reference. The three other original approaches illustrate how neural ordinary differential equations can be leveraged for the problem at hand.

The remainder of this paper is structured as follows. Section 2 presents the Green Solow model and describes the data sets. Section 3 synthesises related machine learning approaches to regression analysis, physics informed neural networks, and neural ordinary differential equations. Section 4 formulates the problem and presents the baseline model and the three proposed neural ordinary differential equation models. Section 5 is a comparative empirical performance evaluation and analysis of the four models. Section 6 summarises the findings and discusses research questions stemming from this work.

2   The Green Solow Model

The Green Solow model [3] is an augmentation of the Solow–Swan model [1, 21], a neoclassical economics model describing the dynamics of economic growth. The Solow–Swan model is a dynamical system that proposes to model the evolution of a country's economy over time due to factors such as labour force, capital, and savings rate, based on fundamental economics concepts. The Green Solow model augments the Solow–Swan model to describe the dynamics of economic sustainable growth considering pollution emissions and the rate of pollution abatement from technological progress within the Solow model.

The Green Solow model is represented by the coupled three-dimensional ordinary differential equation, Eqs. 1 to 3, defining the time evolution of the system.

It is a function of variables k_t, y_t, and e_t measuring the state of the physical system and parameters n, s, δ, α, g_B, and g_A quantifying the interaction between the variables and their time evolution.

$$\frac{dk_t}{dt} = (s \times y_t) - ((\delta + n + g_B) \times k_t) \tag{1}$$

$$\frac{dy_t}{dt} = y_t \times (g_B + (\alpha \times \frac{dk_t}{dt} \times \frac{1}{k_t})) \tag{2}$$

$$\frac{de_t}{dt} = e_t \times (-g_A + (\frac{dy_t}{dt} \times \frac{1}{y_t})) \tag{3}$$

The variables k_t and y_t represent capital per capita and income per capita at time t respectively, both in millions of United States dollars. The variable e_t represents emissions per capita at time t in metric tons. The parameters in Eqs. 1 to 3 are the fixed savings rate, s, growth rate of population, n, rate of capital depreciation, δ, rate of labour-augmenting technological progress, g_B, capital share of output, α, and rate of technological progress in emissions abatement, g_A. The parameters characterise the evolution of the dynamical system. However, the four parameters δ, α, g_B, and g_A are neither observable nor measurable. The values of these four unobserved parameters are to be estimated, to allow the best fit of the variables k_t, y_t, and e_t to the time evolution rules.

The Green Solow model was designed to show that the emissions per capita for a given country generally converge to a sustainable growth path described by the Environmental Kuznets Curve [22]. By modifying the differential equations describing the Green Solow model into a form that can be solved analytically and making use of aggregated variables for each country, a multivariate linear regression can be specified to estimate the unobserved variables α, g_B, and g_A (δ is substituted out). The results of this regression also allow making of cross-country comparisons regarding savings rates and population growth rates in emissions per capita and indirectly predicting future emissions per capita.

Data for the variable k_t is sourced from the Penn World Tables series named Capital stock at Current Purchasing Power Parities [6]. The data for the variable e_t is sourced from the World Development Indicators [23]. The data for the variable y_t is sourced from Angus Maddison and runs until 2008 [13], as is the parameter n_t. Lastly, data for the variable s_t is sourced from the World Development Indicators series named Adjusted savings, gross savings (as a percentage of the gross national income) [23]. The remaining parameters of the dynamical system are not measured and no data is available for them.

3   Related Work

Physics-informed neural networks [18] in general and neural ordinary differential equations [5,17] in particular attempt to reconcile observation data with the knowledge of systems of differential equations characterising a dynamical system.

Physics-informed neural networks refer to neural networks integrating prior knowledge about the physical systems being studied, by incorporating differential equations in the loss function of a neural network. They work on the basis of neural networks being universal function approximators [11, 17], but regularize the neural network based on constraints of the differential equation. These constraints can train neural networks to avoid physically inconsistent or nonsensible outputs and can supplement data scarcity issues faced by supervised learning tasks that typically require larger datasets [18].

Neural ordinary differential equations [17] constitute a noticeable breakthrough for the integration of dynamical systems and neural networks, allowing the design and implementation of effective and efficient physics-informed neural networks. They are made practical and effective by two properties of neural networks being differentiable objects [5] and universal approximators [11, 17]. Unlike other physics-informed neural networks that make use of existing knowledge to constrain a neural network, neural ordinary differential equations directly embed a neural network into differential equations, allowing the neural network to approximate all or part of the non-linear dynamics of a system while incorporating a variety of interactions and known mechanisms of the dynamical system [17]. Furthermore, instead of discretizing the states of a dynamical system, neural ordinary differential equations make use of an ordinary differential equation solver and compute gradients using the adjoint sensitivity method [15] which scales linearly with the dimensions of independent variables and explicitly controls numerical error [5].

Finding the parameters of the Green Solow model is complicated by the scarcity and auto-correlation of data. Neural ordinary differential equations have proven to perform well with scarce data [12, 19]: backpropagation reduces to calculating gradients of the differential equation solution with respect to their unobserved parameters via the adjoint method. The neural network hence only learns information about parameters that are unobserved, instead of all parameters in the dynamical system, and compensates for less data with more knowledge. Furthermore, because the neural network approximates fewer parameters, it has fewer trainable weights than one approximating the entire dynamical system. These two factors account for the ability of neural ordinary differential equations to reconstruct models based on scarce data and more knowledge to effectively and efficiently analyse and predict the dynamics of physical systems [17].

Auto-correlated data can also bias the parameters [8] of the Green Solow model found by regression. Ordinary differential equations solvers eliminate the problem of auto-correlation by computing the first difference of the dependent variable. Neural ordinary differential equations perform better with solvers using the adjoint method [2]. The adjoint method freely selects Lagrangian multipliers when the constraints of the ordinary differential equation are met. The Lagrangian multipliers are selected to prevent expensive differentiation of independent variables. They control for effects of the change in independent variables over the entire time period for which the ordinary differential equation is solved. This eliminates the problem of auto-correlation.

4 Methodology

Our problem statement is as follows: to find parameters of the Green Solow model, we formulate a regression problem that estimates coefficients by minimising the residual of the dependent variable, based on independent and control variables of the regression. It is possible and easier to ignore that the system is coupled and to consider and solve the three equations independently. However, this yields different parameter values for each equation. For instance, solving Eqs. 1–2 yield two different values of g_B. These values are mathematically inaccurate- they are not consistent with the coupling and are not possible to interpret consistently. Instead we propose four solutions that solve the coupled system with consistent parameter values. The estimated coefficients of each of the four regressions either directly find the parameters δ, α, g_A, and g_B or adjust the Green Solow model by allowing time-varying parameters δ_t, α_t, $g_{A,t}$, and $g_{B,t}$, or introducing multiplicative adjustment terms $M_{k,t}$, $M_{y,t}$, $M_{e,t}$, to fit the coupled equations. Model 0 is a baseline coupled regression of the three equations of the Green Solow model. It does not make use of neural ordinary differential equations and highlights the weakness of the Green Solow model in assuming constant parameters across time. Following it, three models using neural ordinary differential equations are introduced. We leverage the ability of neural ordinary differential equations to find consistent time-varying parameters based on coupled equations despite scarce and auto-correlated data.

In the following four models, the subscript t indicates time-dependent variables and coefficients while the subscript $data$ indicates observed data. The symmetric mean absolute percentage error [7] of each regression is minimised instead of the mean squared error. It allows the residual of each regression subproblem to be scaled into a dimensionless percentage value. The residuals of all sub-problems are then summed and minimised at the same time. Estimating coefficients of three regression sub-problems of different scales at the same time using the mean squared error would bias the estimation towards coefficients in regression sub-problems of a larger scale, but the squared residual can still be used for regression sub-problems that are estimated independently.

4.1 Baseline Method

Model 0 is a baseline non-linear regression of the three-dimensional coupled Green Solow model of Eqs. 2, 3 and 4. Its dependent variables are $\frac{dk_t}{dt}$, $\frac{dy_t}{dt}$, and $\frac{de_t}{dt}$ computed using the adjoint sensitivity analysis method. Its independent variables are y_t, k_t, and e_t. The estimation of s_t and n_t is not of interest therefore macroeconomics data for the variables $n_{t,data}$ and $s_{t,data}$ are used to control and remove their effects.

$$\frac{dk_t}{dt} = (s_{t,data} \times y_t) - ((\delta + n_{t,data} + g_B) \times k_t) \tag{4}$$

Model 0 simultaneously estimates parameters δ, α, g_B, and g_A as regression coefficients by minimising the symmetric mean absolute percentage error of the residual between solutions, y_t, k_t, and e_t and their corresponding macroeconomics data.

To accelerate the training process, the coefficients δ, α, g_B, and g_A are initialised using values obtained using the Nelder-Mead method [14] to solve each equation independently.

4.2   Proposed Models

Model 0 assumes constant parameters δ, α, g_B and g_A in the dynamical system of the Green Solow model. We propose two mutually exclusive methods that use neural ordinary differential equations to improve the fit of parameters to the dynamical system by either introducing time-dependent parameters or introducing a multiplicative time-dependent adjustment term to the Green Solow model.

Models 1 and 2 estimate time-dependent coefficients in the regression problem formulated from the Green Solow model. Neural networks are embedded within the dynamical system of the Green Solow model. The neural network in Model 1 takes k_t, y_t or e_t as inputs while that in Model 2 takes t. Both neural networks output δ, α, g_B and g_A as individual time-dependent parameters. The dynamical system is rewritten to reflect the time-varying parameters:

$$\frac{dk_t}{dt} = (s_{t,data} \times y_t) - ((\delta_t + n_{t,data} + g_{B,t}) \times k_t) \tag{5}$$

$$\frac{dy_t}{dt} = y_t \times (g_{B,t} + (\alpha_t \times \frac{dk_t}{dt} \times \frac{1}{k_t})) \tag{6}$$

$$\frac{de_t}{dt} = e_t \times (-g_{A,t} + (\frac{dy_t}{dt} \times \frac{1}{y_t})) \tag{7}$$

Model 1 is a non-linear regression of the three-dimensional coupled Green Solow model rewritten as Eqs. 5, 6, and 7. Its dependent variables are $\frac{dk_t}{dt}$, $\frac{dy_t}{dt}$, and $\frac{de_t}{dt}$ computed via the adjoint sensitivity analysis method. Its independent variables are k_t, y_t, and e_t, its control variables are n and s. The coefficients estimated by the regression are the parameters δ, α, g_B, and g_A, represented using a neural network $NN(k_t, y_t, e_t) = [\delta_t, \alpha_t, g_{B,t}, g_{A,t}]^\mathsf{T}$ that takes in the independent variables.

Model 2 is a non-linear regression of the three-dimensional coupled Green Solow model rewritten as Eqs. 5, 6 and 7. Its dependent variables are $\frac{dk_t}{dt}$, $\frac{dy_t}{dt}$, and $\frac{de_t}{dt}$ computed via the adjoint sensitivity analysis method. Its independent variables are k_t, y_t, e_t, and t, its control variables are n and s. The coefficients estimated by the regression are the parameters δ, α, g_B, and g_A, represented using a neural network that takes in the independent variable, a timestep t, according to $NN(t) = [\delta_t, \alpha_t, g_{B,t}, g_{A,t}]^\mathsf{T}$.

Both Models 1 and 2 are neural ordinary differential equations comprising a neural network embedded within the three ordinary differential equations of the Green Solow model. The optimal weights and biases of the neural network approximate the coefficients of the regression, by minimising the symmetric mean absolute percentage error loss of the residual. The neural network is initialised using δ, α, g_B, and g_A from the Nelder-Mead method [14].

Model 3 introduces a multiplicative time-dependent adjustment term to the Green Solow model. It is a non-linear regression of the three-dimensional coupled Eqs. 8, 9, and 10. Its dependent variables are $\frac{dk_t}{dt}$, $\frac{dy_t}{dt}$, and $\frac{de_t}{dt}$ computed via the adjoint sensitivity analysis method. Its independent variables are k_t, y_t, e_t, and t. Its control variables are δ, α, g_B, and g_A estimated using the Nelder-Mead method, and n and s. The coefficients estimated by the regression are the adjustment terms. Model 3 represents the multiplicative time-dependent adjustment terms each as a neural network $NN_x(t) = M_{x,t}$, for $x \in [k, y, e]$, respectively. The parameters δ, α, g_B, and g_A estimated using the Nelder-Mead method and used in the Green Solow model are sub-optimal, hence Model 3 multiplies them by an adjustment term, to adapt them for optimal time evolution of the dynamical system.

The neural ordinary differential equation comprises three neural networks, each embedded within one of the three ordinary differential equations of the Green Solow model. Each neural network is distinct, takes in the time-step t, and returns the respective multiplicative time-dependent term. As they are individually trained, $k_{t,data}$, $y_{t,data}$, and $e_{t,data}$ are substituted into Eqs. 8, 9, and 10 to ease the training. The optimal weights and biases of the neural network approximate the multiplicative time-dependent term as the regression coefficient by minimising the residual.

$$\frac{dk_t}{dt} = (s \times y_t) - ((\delta + n + g_B) \times k_t \times M_{k,t}) \tag{8}$$

$$\frac{dy_t}{dt} = y_t \times (g_B + (\alpha \times \frac{dk^c}{dt} \times \frac{1}{k_t})) \times M_{y,t} \tag{9}$$

$$\frac{de_t}{dt} = e_t \times (-g_A + (\frac{dy^c}{dt} \times \frac{1}{y_t})) \times M_{e,t} \tag{10}$$

After training, the optimal coefficients of each regression are substituted back into the dynamical system and a solver is used to obtain k_t, y_t, and e_t from initial conditions as a three-dimensional coupled ordinary differential equation. k_t, y_t, and e_t obtained from each model are compared in Sect. 5.2.

5   Performance Evaluation

5.1   Experimental Setup

We evaluate, compare, and analyse the performance of the four models for 73 countries, for which at least 20 years of data are available for each of the variables e_t, y_t, k_t, s_t, and n_t for years 1970 to 2008. We further choose the United States of America (USA) and Botswana as representative of developed and developing countries, respectively, for a detailed analysis. The former is a developed country with peak emissions before 1970 and a relatively linear trend of k_t, y_t, and e_t. The latter is a developing country with significant variation in k_t, y_t, and e_t.

The neural network of Model 1 has 3 hidden layers and 12 densely connected nodes in each layer. The neural network of Model 2 has 2 hidden layers and 12 densely connected nodes in each layer. The neural networks of Model 3 have 2 hidden layers and 20 densely connected nodes in each layer. The activation function is the hyperbolic tangent. The architecture and hyper-parameters of the neural networks are tuned to optimise their fit to data and minimise convergence time. The Nelder-Mead method is used to find numerically stable δ, α, g_B, and g_A for the intialisation of all models. The models are trained by combining the optimisation algorithms ADAM and Broyden-Fletcher-Goldfarb-Shanno and progressively reducing their learning rates [16] until the error plateaus. These training choices constrain the search space to accelerate model convergence.

5.2   Results and Discussion

This section compares the effectiveness of models in Sect. 4 in fitting data from USA and Botswana to the Green Solow model. Table 1 reports the mean squared error and mean absolute percentage error for k_t, y_t, and e_t, for each model. Figure 1 compares k_t in red, y_t in blue, and e_t in green against $k_{t,data}$, $y_{t,data}$ and $e_{t,data}$ for USA over 39 years. Figure 2 compares the same for Botswana over 34 years. All models follow the overall upward or downward trend of the data.

For USA, k_t and e_t for Model 0 are too linear to fit $k_{t,data}$ and $e_{t,data}$ well. However, Model 0 fits $y_{t,data}$ best among all models, because $y_{t,data}$ is highly linear. The poor fit of k_t and e_t to the Green Solow model comes from the constant and consistent δ, α, g_A, and g_B between Eqs. 2, 3 and 4. Values of δ, α, g_A, and g_B are compromised to fit multiple equations. Among the models, Model 1 has the lowest mean absolute percentage error in fitting k_t and y_t but not the lowest mean square error, likely because k_t deviates from $k_{t,data}$ in years 35–40, resulting in a heavier penalty under the mean squared error. Model 2 has the lowest mean squared error in fitting k_t among all models. Although k_t in Model 2 is more linear than that in Model 3, it constructs a best-fit line of $k_{t,data}$

Table 1. Metrics for USA and Botswana for Models 0 to 3.

	USA				Botswana			
	Model 0	Model 1	Model 2	Model 3	Model 0	Model 1	Model 2	Model 3
Ref.	Fig. 1a	Fig. 1b	Fig. 1c	Fig. 1d	Fig. 2a	Fig. 2b	Fig. 2c	Fig. 2d
Mean squared error								
k_t	3.53E−5	3.5E−5	**2.46E−5**	4.51E−5	4.47E−6	**3.13E−7**	4.01E−7	3.29E−6
y_t	**2.33E−7**	2.46E−7	3.37E−7	1.16E−6	2.19E−7	**1.56E−8**	3.99E−8	1.83E−7
e_t	6.66E−1	4.54E−1	5.93E−1	**3.72E−1**	5.25E−1	7.55E−2	8.44E−2	**6.69E−2**
Mean absolute percentage error								
k_t	3.81	**3.53**	3.98	5.64	1.34E+1	**6.00**	7.64	15.10
y_t	1.68	**1.63**	2.13	3.74	1.12E+1	**2.72**	4.12	11.71
e_t	3.16	2.37	3.04	**2.33**	3.98E+1	1.75E+1	1.82E+1	**1.64E+1**

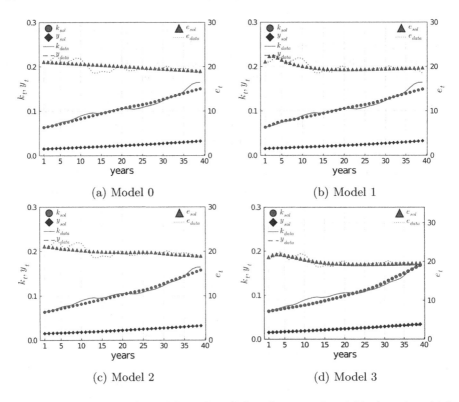

Fig. 1. Regression and data of k_t and y_t (left ordinate, red and blue), and e_t (right ordinate, green) between 1970 and 2008 inclusive for USA for Models 0–3. (Color figure online)

and minimises the residuals above and below k_t. It follows the global non-linear increasing trend in $k_{t,data}$ best. For Model 3, e_t matches the maximum point of $e_{t,data}$ and has the lowest mean absolute percentage error and mean squared error. This can be attributed to each neural network being trained independently and being less constrained compared to the neural networks in Models 1 and 2.

For Botswana, the results mirror those of USA. We highlight additional insights gained. For Model 0, the exponential trend observed in e_t highlights another difficulty in fitting coupled differential equations with constant parameters. The closed form solution of e_t has to fit the exponent (from solving Eq. 3) of an exponential term (from solving Eq. 4). For Model 1, k_t can fit the increasing gradient of $k_{t,data}$. Model 1 has the lowest mean squared error and mean absolute percentage error in fitting k_t and y_t. Model 2 slightly underperforms Model 1 in fitting k_t and y_t. Model 3 has the lowest errors in estimating e_t.

Figure 3 shows that on average across all 59 countries that converged for the training of all four models, Models 1–3 with neural ordinary differential equations outperform the baseline Model 0. Similar to previous results, Model 1 fits k_t and y_t best, while Model 3 fits e_t best. The complete code in Julia and results for the models discussed are available at github.com/zykhoo/nODE-Solow.

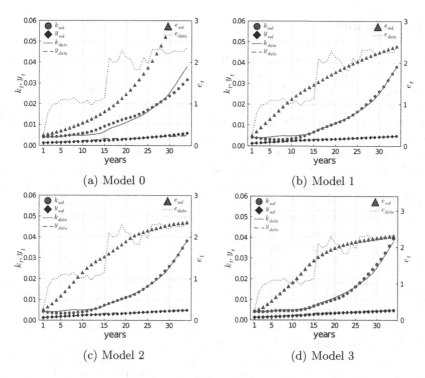

Fig. 2. Regression and data of k_t and y_t (left ordinate, red and blue), and e_t (right ordinate, green) between 1975 and 2008 inclusive for Botswana for Models 0–3. (Color figure online)

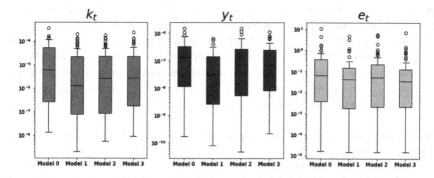

Fig. 3. Boxplots of the mean squared error loss for Models 0–3 (left to right) compared on a logarithm scale for k_t (in red), y_t (in blue), and e_t (in green). (Color figure online)

In terms of efficiency, Model 0 is simpler and therefore faster than Models 1 to 3. Models 1 and 2 require a training time of the order of 10^2 s per country. Model 3 trains sequentially a neural network for each equation. Each neural network requires training time of the order of 10^3 s per country.

6 Conclusion

We have presented four models to estimate the parameters of the Green Solow model: Model 0, a baseline model, and Models 1 to 3, three neural ordinary differential equation models. Models 1 and 2 consider time-dependent parameters while Model 3 introduces a multiplicative time-dependent adjustment term to the Green Solow model. The results indicate that the three proposed models are able to couple the differential equations and capture the time dependence of the parameters to achieve a better fit of the observed data to the model. We also verify that there are two different values and interpretations of α and δ in regression when ignoring that the system is coupled. We attribute the success of Models 1–3 to the ability of neural ordinary differential equations to solve coupled differential equations despite scarce and auto-correlated data.

The simplicity of design and quality of performance of the proposed models suggest neural ordinary differential equations potentially constitute a good generic tool for model discovery and improvement for a wide range of dynamical systems. This hybrid approach combines the interpretability of knowledge-driven models with the flexibility of a data-driven neural network training approach.

We are now investigating the application of the approach to other data analytics and machine learning tasks such as prediction and pattern recognition.

We have preliminarily evaluated the performance in prediction of the four models against a long short-term memory artificial recurrent neural network. The four models are competitive but do not perform as well. However, while the long short-term memory model makes three independent predictions for k_t, y_t, and e_t respectively, the four models optimise a more constrained problem and produce predictions for k_t, y_t, and e_t consistent with the coupling of the equations. This suggests further investigation of the integration of neural differential equations with long short-term memory models for consistency with coupled equations.

We have also qualitatively evaluated the performance of the four methods by using Grammarviz [20] to cluster the times series of the parameters. The clusters correspond to economies facing similar sustainable development challenges. This coincides with findings by Brock and Taylor [3] regarding emissions convergence and the shape of the Environmental Kuznets curve [22].

Acknowledgement. The first author is supported by a scholarship from the Agency of Science, Technology and Research (A*STAR). This research is partially supported by the National University of Singapore (NUS) under its NUS COVID-19 Research Seed Grant (Award No: NUSCOVID19RG-40). Any opinions, findings and conclusions or recommendations expressed in this material are those of the author(s) and do not reflect the views of NUS.

References

1. Abel, A.B., Bernanke, B.S., Croushore, D.: Macroeconomics. 7th edn., Person Publishing (2011)
2. Bradley, A.M.: PDE-constrained optimization and the adjoint method (2019). http://www.cs.stanford.edu/~ambrad/adjoint_tutorial.pdf. Accessed 3 Apr 2021
3. Brock, W.A., Taylor, M.S.: The green Solow model. J. Econ. Growth **15**, 127–153 (2010)
4. Brunton, S.L., Proctor, J.L., Kutz, J.N.: Discovering governing equations from data by sparse identification of nonlinear dynamical systems. Proc. Nat. Acad. Sci. **113**(15), 3932–3937 (2016)
5. Chen, R.T.Q., Rubanova, Y., Bettencourt, J., Duvenaud, D.K.: Neural ordinary differential equations. Adv. Neural Inf. Process. Syst. **31**, 6571–6583 (2018)
6. Feenstra, R.C., Inklaar, R., Timmer, M.P.: The next generation of the Penn World Table. Am. Econ. Rev. **105** (2015). ggdc.net/pwt. Accessed 28 Oct 2020
7. Flores, B.E.: A pragmatic view of accuracy measurement in forecasting. Omega **14**(2), 93–98 (1986)
8. Granger, C., Newbold, P.: Spurious regressions in econometrics. J. Econ. **2**(2), 111–120 (1974)
9. Hastie, T., Tibshirani, R., Friedman, J.: The Elements of Statistical Learning. Springer New York (2001). https://doi.org/10.1007/978-0-387-21606-5
10. Hirsch, M.W., Smale, S., Devaney, R.L.: Differential Equations, Dynamical Systems & An Introduction to Chaos. 2nd edn., Elsevier Academic Press (2004)
11. Hornik, K., Stinchcombe, M., White, H.: Multilayer feedforward networks are universal approximators. Neural Netw. **2**(5), 359–366 (1989)
12. Li, Z., et al.: Fourier neural operator for parametric partial differential equations (2020)
13. Maddison, A.: Statistics on world population, GDP and per capita GDP, 1–2008 AD (2008). ggdc.net/maddison/Maddison.htm. Accessed 28 Oct 2020
14. Nelder, J.A., Mead, R.: A simplex method for function minimization. Comput. J. **7**, 308–313 (1965)
15. Pontryagin, L.S., Boltyanskii, V.G., Gamkrelidze, R.V., Mishchenko, E.F.: The Mathematical Theory of Optimal Processes. Wiley, New York (1962)
16. Rackauckas, C.: Doing scientific machine learning with Julia's SciML ecosystem (2020). http://www.figshare.com/articles/presentation/Doing_Scientific_Machine_Learning_with_Julia_s_SciML_Ecosystem/12751949. Accessed 28 Oct 2020
17. Rackauckas, C., et al.: Universal differential equations for scientific machine learning (2020)
18. Raissi, M., Perdikaris, P., Karniadakis, G.E.: Physics-informed neural networks: a deep learning framework for solving forward and inverse problems involving nonlinear partial differential equations. J. Comput. Phys. **378**, 686–707 (2019)
19. Rubanova, Y., Chen, R.T.Q., Duvenaud, D.K.: Latent ordinary differential equations for irregularly-sampled time series. In: Advances in Neural Information Processing Systems, vol. 32 (2019)
20. Senin, P., et al.: GrammarViz 2.0: a tool for grammar-based pattern discovery in time series (2014)
21. Solow, R.: A contribution to the theory of economic growth. Q. J. Econ. **70**, 65–94 (1956)
22. Stern, D.I.: The rise and fall of the environmental Kuznets curve. World Dev. **32**(8), 1419–1439 (2004)

23. The World Bank: World Development Indicators (2019). databank.worldbank.org/source/world-development-indicators. Accessed 28 Oct 2020

24. Udrescu, S.M., Tegmark, M.: AI Feynman: a physics-inspired method for symbolic regression. Sci. Adv. **6**(16) (2020)

A Quantum-Inspired Neural Network Model for Predictive BPaaS Management

Ameni Hedhli[1](✉), Haithem Mezni[1,2], and Lamjed Ben Said[1]

[1] SMART Lab, Institut Supérieur de Gestion, Université de Tunis,
Bardo 2000, Tunisie
`ameni.hedhli@isg.u-tunis.tn, lamjed.bensaid@isg.rnu.tn`
[2] Taibah University, Al-Madinah, Saudi Arabia
`hmezni@taibahu.edu.sa`

Abstract. Nowadays, companies are more and more adopting cloud technologies in the management of their business processes rising, then, the Business Process as a Service (BPaaS) model. In order to guarantee the consistency of the provisioned BPaaS, cloud providers should ensure a strategical management (e.g., allocation, migration, etc.) of their available resources (e.g., computation, storage, etc.) according to services requirements. Existing researches do not prevent resource provision problems before they occur. Rather, they conduct a real-time allocation of cloud resources. This paper makes use of historical resource usage information for providing enterprises and BPaaS providers with predictions of cloud availability zones' states. For that, we first propose a Neural Network-based prediction model that exploits the superposition power of Quantum Computing and the evolutionary nature of the Genetic Algorithm, in order to optimize the accuracy of the predicted resource utilization. Second, we define a placement algorithm that, based on the prediction results, chooses the optimal cloud availability zones for each BPaaS fragment, i.e. under-loaded servers. We evaluated our approach using real cloud workload data-sets. The obtained results confirmed the effectiveness and the performance of our NNQGA approach, compared to traditional techniques.

Keywords: Cloud computing · BPaaS management · Resource prediction · Quantum Computing · Neural Network · Genetic Algorithm

1 Introduction

The success of cloud technologies has risen various service models like the proliferation of BPaaS as a new cloud service supporting companies' business processes [1]. Providing and managing BPaaS services over cloud environments deprive companies from owning and scaling the required resources [1]. Rather that is ensured by cloud providers through different strategies.

BPaaS management has been widely and diversely treated. Some researchers have proposed generic frameworks allowing the BPaaS design [5], modeling [4],

© Springer Nature Switzerland AG 2021
C. Strauss et al. (Eds.): DEXA 2021, LNCS 12923, pp. 91–103, 2021.
https://doi.org/10.1007/978-3-030-86472-9_8

allocation, execution evaluation [2] and reuse [1]. Other researches have concentrated on the trust and the security of the BPaaS management systems. However, existing BPaaS management approaches suffer from the following limits: (1) the specificity related to the multi-cloud environments has not been taken into account, neither as a BPaaS management supporting factor allowing to benefit from separated data centers, nor as a constraint against the resource sharing between the different cloud zones; (2) the majority of the approaches have not covered the fact of predicting future states of each cloud zone, as they are following an adaptive mode. In fact, that allows to inhibit potential Service Level Agreement (SLA) violations. It also allows the optimization of the future usage of cloud resources on the basis of historical data and BPaaS behaviour; (3) the existing frameworks have not proposed resource usage-aware management operations. Rather, they performed independent tasks allowing securing or reconfiguring the placement of BPaaS fragments. The need for a predictive management of BPaaS services comes from the fact that on-the-fly resources' provision may not be efficient. BPaaS management over cloud environments is a highly dynamic and complex task that, whenever it is managed in real time, it could lead to blocking situations like the lack of resources. Moreover, an unbalanced workload over cloud data centers brings to eventual execution delays or, even, to time wasting. Hence, an efficient management of cloud resources should not be limited on a real-time one. Rather, a predictive management mode could prevent several problems in the execution context of BPaaS services.

Our main contributions consist on a *prediction model* allowing to estimate the CAZs' future available resources in order to ensure a reliable BPaaS management. A *BPaaS management algorithm* is then defined to perform the placement of BPaaS fragments based on the predicted available resources.

The reminder of the paper is organized as follows. Section 2 presents the previous works in relation with BPaaS management. Section 3 details the proposed prediction model. Section 4 presents our experimental results. The last section discusses our future directions and concludes the paper.

2   Related Work

To deal with BPaaS management, some researchers have tackled one management operation like BPaaS placement in [2], BPaaS decomposition in [8], and BPaaS adaptation in [3]. Others have considered various objectives together, like the BPaaS design, allocation, execution and evaluation in the Cloud-Socket project [1]. Several BPaaS management techniques have been adopted, including fuzzy clustering, affinity rules, multi-modeling, decision models, etc. We cite, as example, the mapping of the BPaaS placement task to the DNA fragment assembly problem and the use of memetic genetic algorithm in [2]. Fuzzy logic has also been applied to deal with the changing and volatile cloud environment. In [3], a fuzzy c-means based QoS clustering method was proposed to facilitate the assessment of the QoS parameters' preservation. Other researchers, like [9], have proposed rule-based approaches for the deployment and the modeling of reusable business processes. That is based on the mapping of business needs to a set of organizational rules, which facilitates the management operations. However, the established rules should be updated periodically in order to still being relevant.

To trigger BPaaS management operations, researchers have mainly adopted the dynamic mode, allowing their approaches to pursue changes in terms of cloud resources or business requirements. We cite as examples, adaptive business process composition [11], adaptive deployment [9] and adaptive modeling [4]. This is unlike the static mode that does not guarantee cloud elasticity. An adaptive mode robustly follows parameters' variations (e.g., user requirements, available resources, etc.). Add to that, it allows to overpass blocking situations in an on-the-fly manner. However, that is consuming in terms of processing, whenever it is compared to the static mode. In fact, decision makers should survey the overall management context in real time. Despite the ability of this mode to readjust the offered tasks following environmental changes, that burdens the administrative practices with additional activities. The adaptive mode can lead to an imbalanced workload. These problems can not be over-passed by the static mode that, in addition to its inability to flexibly meet new requirements, it is unable to cover all of the possible BPaaS scenarios in advance. Moreover, in complicated business sectors, neither the static nor the adaptive mode can anticipate future challenging circumstances. Hence, that is specifically necessitating a predictive mode.

As for BPaaS management constraints, the security as a critical challenge for all cloud services, and especially for BPaaS, has been treated in [10,11], by addressing the privacy and the know-how preservation problems. Other constraints, like BPaaS execution time, have been covered only in [2]. The BPaaS cost has been optimized in [3]. In addition, the collaboration in the context of BPaaS, as stated in [10], seems to be highly recommended in order to deal with sophisticated scenarios. However, it has not been taken into account by existing works.

Although the majority of BPaaS management approaches propose dynamic solutions that automatically adapt to the cloud conditions, none of them has conducted predictions of BPaaS future changes. Moreover, the state of the cloud zones has not been neither taken into account, nor predicted. Despite the fact that BPaaS fragments can be dispersed over multiple CAZ, the selection of the target environments has not been done. In fact, that has a direct impact on the BPaaS management operations (e.g., placement of BPaaS fragments) and the cloud resource usage. Also, the dynamicity and uncertainty of cloud conditions, as well as the incomplete data of the managed BPaaS involve nearly the totality of the BPaaS management tasks. Their treatment bring several benefits like accuracy enhancement. Moreover, that strengthens providers' relationships with their customers. However, that has been considered only in [3]. Finally, most approaches have dealt with BPaaS management in traditional single cloud environments, while dispersing BPaaS fragments was proven to be an effective solution to some organization concerns (e.g., privacy policies) [10].

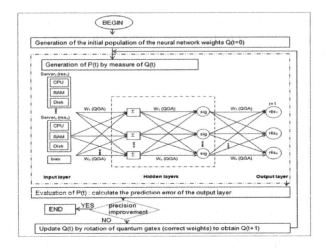

Fig. 1. Flowchart of the CAZ states' prediction process.

3   Resource Usage Prediction Model

Our prediction model consists on the hybridization of a neural network with a quantum genetic algorithm as a training tool. Like it is depicted in Fig. 1, we pursue the following steps:

1. An initial solution of equiprobable weights is, first, created. Then, it will be optimized following the proposed quantum genetic algorithm.
2. On the basis of the initially generated weights, we create the first neural network prediction P(t). That's by multiplying the historical resource usage values by the current weights.
3. The efficiency of these weights is, then, evaluated in terms of prediction accuracy. This step aims at training the neural network with weight values that are optimized by the proposed QGA.
4. If there is no improvement of the prediction's accuracy, we readjust the neural network's weights through the quantum gate rotation.
5. Finally, NN is used to evaluate the prediction of cloud zones' resource usage with QGA-optimized weights.

3.1   Multi-layer Perceptron for Resource Usage Prediction

The input of our prediction model consists of n neurons $(x_1, x_2, ..., x_n)$. Each neuron x_i $(1 \leq i \leq n)$ represents an observation of previous resource consumption within each cloud server (see Fig. 1). The weights binding the input layer to the hidden layer are first randomly initialized. Then, they are optimized through a proposed quantum genetic algorithm (see Sect. 3.2).

At the neurons of each hidden layer, the chosen transfer and activation functions are applied. For the activation, we use in this work, the sigmoid function [6] that is mostly employed to overcome linearity in the model.

3.2   Quantum Genetic Algorithm for the Neural Network's Training

In this section, we show how QGA is employed to optimize the training phase of our NN-based prediction model. The quantum chromosome is encoded as a two-dimensional vector that represents the superposition of two candidate solutions. For that, qubits allow the exploration of the search space in order to determine optimal weights of our NN model. The genes of the quantum chromosome, as it has been described, are a series of probabilities denoting respectively the bias and the weight values of the proposed network. The number of genes is equal to three (see Fig. 2): (1) a double-chain gene for the bias values ($b_1..b_n$ and $bb_1..bb_n$) (2) a double-chain gene for the weights of the input layer's neurons to those of the hidden layer, and (3) a double-chain gene for the weights linking the hidden layer's neurons to the output layer. After a measurement step, the most probable values are selected as shown in Fig. 3.

Quantum chromosome	b_1	...	b_n	w_{11}	...	w_{nn}	k_1	...	k_m
For NN training	bb_1	...	bb_o	ww_{11}	...	ww_{nn}	kk_1	...	kk_m

Fig. 2. Quantum chromosome for resource usage prediction with NN training

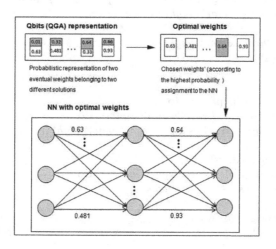

Fig. 3. Representation of the qubit for NN weights' optimization

As a quantum operator, we use the quantum gates to move from one generation of chromosomes to another much optimal one. For that, we adopt the rotation strategy [6]. Using the following formula, we first determine the rotation angle θ based on the evolution of the fitness function and the initial rotation angle θ_0:

$$\theta = \frac{(F_p - F_c)}{|F_p - F_c|} * \theta_0 \tag{1}$$

where F_c and F_p are, respectively, the current and previous fitness values. Then, we calculate the rotation gate through the following unitary operator $U(\theta)$:

$$U(\theta) = \begin{pmatrix} \cos(\theta) & -\sin(\theta) \\ \sin(\theta) & \cos(\theta) \end{pmatrix} \tag{2}$$

Finally, we apply the obtained rotation gate on the chromosomes to update their qubits as follows:

$$\begin{pmatrix} \alpha_i' \\ \beta_j' \end{pmatrix} = U_i * \begin{pmatrix} \alpha_i \\ \beta_i \end{pmatrix} \tag{3}$$

The QGA-based prediction steps are summarized in Algorithm 1.

Algorithm 1. NNQGA-based resource usage prediction

1: **Input:** Historical resource usage (HistRes).
2: **Output:** Predicted resource usage.
3: p ← 1   ▷ Population number.
4: Initialize (θ_0 , $ChromosomeWeight$) ▷ Generate the quantum rotation angle, and the initial NN weights
5: **while** ($e \leq e_{min}$) **do**
6:     **for** i in range (1, popSize) **do**
7:         **for** j in range (1, $Len(ChromWeights)$) **do**
8:             **if** (pMeasure $\leq ChromosomeWeight$ [i, j, 0]) **then**
9:                 $ChromosomeWeight$ [i] ← 0
10:            **else**
11:                $ChromosomeWeight$ [i] ← 1
12:            **end if**
13:        **end for**
14:    **end for**
15:    **for** i in range (1, popSize) **do**
16:        **for** j in range (1, $Len(ChromWeights)$) **do**
17:            $ws = ws + $ HistRes[i] $* ChromosomeWeight$ [i]   ▷ NN weighted sum
18:        **end for**
19:        H[i] ← sigmoid(ws)   ▷ NN hidden layer
20:        **for** j in range (1, $Len(ChromWeights)$) **do**
21:            $ws1$ ← $ws1$ + H[i] $*$ weights[k]
22:        **end for**
23:        $Output$=sigmoid ($ws1$)   ▷ NN output
24:        $e_c = O - Output$ ▷ Prediction error
25:    **end for**
26:    θ ← $(e_p - e_c)$ / $|e_p - e_c|$
27:    Calculate ($U_i(\theta)$)
28:    $ChromosomeWeight$ ← $U_i * ChromosomeWeight$
29:    p ← $p + 1$
30:    e_p ← e_c
31: **end while**

The algorithm's complexity mainly depends on the population size ($popSize$) and the number of generations (G), which is determined in function of the

solution evolution. Hence the global time complexity of our QGANN is $O(G *$
$popSize * Len(ChromWeights))$.

4   Experiments

We implemented our NNQGA-based approach using Python under Spider IDE.
Then based on an implementation of our previous work on BPaaS placement
[2], we tested our NNQGA model using resource usage historical data from two
datasets. The first one is from Google traces, whereas the second one[1] contains
8334 real world observations of CPU usage values with a timestamp of 300. 80%
of these data are devoted to the training of our NN model, whereas the remaining
data are used for the test.

4.1   Evaluation of the Prediction Model

Evaluation of the Prediction's Optimization Method. We established different tests for three NN training techniques (QGA, GA and back-propagation).
Figure 4 shows the impact of the quantum aspect with the genetic algorithm on
the accuracy of the cloud resource usage prediction. The QGA training method
outperforms the conventional GA and the gradient descent back-propagation
algorithm. The resource usage prediction error (MSE) considering the back-
propagation ranges from 0.11 to 0.38 as a maximal error value for both 500 and
1000 samples (see Fig. 4.a).

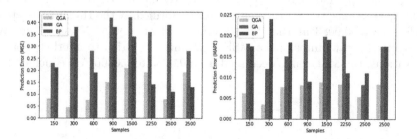

Fig. 4. Impact of the historical data size on the prediction's accuracy.

The major limit of back-propagation training, consists on the inability to
overcome local minimums. That is covered by the genetic algorithm through its
evolution operators (crossover and mutation), which allow the heritage of the
overall shape of the historical resource usage curve with a flexibility allowing the
prediction of sudden resource usage spikes. But that is done in a probabilistic
manner. The MSE values of the GA-training ranges from 0.23 to 0.42 for a fixed
number of generations. However, With QGA, where the prediction solutions

---

[1] http://tiny.cc/pnrg7y.

are encoded in a symbolic and non-binary manner, the evolution operators are replaced by the quantum gate. We tested in this work the Hadamard gate [7]. The qugate along with the qubit representation ensures a diversity, whenever exploring the search space. The error values of the QGA training range from 0.045 to 0.209 for 1250 samples, as it is depicted in Fig. 4.a. That shows a raise of 0.164 in the prediction error while varying the samples' number from 250 to 2000. This upgrading ability is justified by the potential of QGA to scale up and to treat a big number of input neurons. We also evaluated the resource usage prediction using the Mean Absolute Percentage Error (see Fig. 4.b). That clearly showed the outperforming of the QGA training technique. While increasing the samples' number from 250 to 500, we notice a decrease of the prediction error. It is the same for their variation from 1250 to 1750. That proves the scalability of our QGA algorithm. However, we note a slight increase (from 0.0034 to 0.0076) of the prediction error that is explained by sudden slights of the CPU usage.

Sensitivity to the Rotation Gates. QGA uses the quantum gate to explore the search space by determining the direction and the amplitude of each jump from one solution to another. In this series of tests, we varied the number of observations for the Rotation and Hadamard quantum gates. We assess their impact on the MSE prediction error. We notice that the rotation gate reaches the best solution at an early stage in comparison with the Hadamard gate. That is explained by the fact that the rotation gate determines its rotation amplitude in function of the best and worst solution. We acknowledge, based on Fig. 5, that the rotation angle converges before the Hadamard angle for a number of 1000 observations. In fact, in the 10th generation, its MSE achieves 0.0101, while, the Hadamard gate's error achieves 0.0162. However, for a number of 500 observations, the rotation gate is outperformed by the Hadamard gate by a difference of 0.0083 at the 10th generation of our QGANN algorithm. We, also, spotted great jumps of the MSE from one generation to another. For 250 observations, the MSE value decreased from 0.0733 to 0.0366. We conclude that

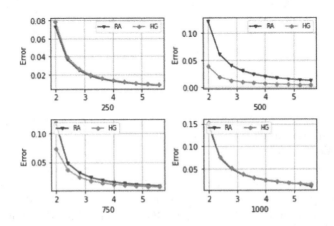

Fig. 5. Impact of the rotation gate on the resource usage prediction's accuracy

for a higher number of observations (e.g., 500), the Hadamard gate behaves better, in terms of prediction accuracy, than the rotation gate, and it is the inverse for a much higher observations' number (e.g., 1000).

Sensitivity to the Learning Rate. The learning rate allows to learn the model following the amplitude of this value. In our QGANN model, the learning is performed through both the quantum gate and the learning rate. In fact, the quantum gate, like it is depicted above, determines both the amplitude and the direction of better solutions' derivation. The learning rate is performed through mutation of the weights' chromosomes determining the step size of the training evolution. It also affects the convergence time of the fitness function, and it guides the propagation of the current solution to the best one. So to avoid either a premature or a late convergence, we trained the resource usage prediction model (see Fig. 6) using different learning rates (between 0.0001 and 10) for different numbers of observations (between 250 and 1000).

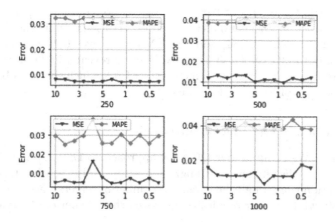

Fig. 6. Resource usage prediction's accuracy with different learning rates.

From the tests in Fig. 6, we notice slight fluctuations of the prediction accuracy in some learning rate intervals like [0.01, 0.001] in the case of 1000 observations. Moreover, there are some huge fluctuations of the prediction accuracy. For 1000 observations, the MSE decreases from 0.0159 to 0.0063 with a learning rate ranging from 10 to 0.05. When decreasing again this value into 0.0001 and keeping the number of training epochs, we notice a raise of the prediction error by 0.0049. Hence the optimal learning rate for these sub-series of tests is 0.05.

Comparison of the Predictions' Accuracy. We compared our approach with two workload prediction methods: the Functional Link Neural Network (FLNN) and the FLNN approach improved with Genetic Algorithm (FN-GANN) [12]. We assess the prediction error for a varying observation number from 250 to 2000. Using the first dataset, we notice in Fig. 7.a that both QGANN and FLGANN scale better than the FLNN algorithm. Moreover, for a number of observations varying from 250 to 1000 the QGANN scales much more better than the FLGANN with a difference of prediction error varying from 0.0086 to 0.0042.

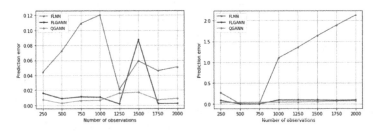

Fig. 7. Workload predictions' comparison between QGANN, FLNN and FLGANN

Using Google trace dataset (see Fig. 7.b), we notice the out-performance of the QGANN with an error of 0.036 for 250 observations, while the FLGANN's prediction error reaches 0.084, and the FNN's achieves 0.267. Then for 2000 observations, both QGANN and FLGANN's error reveal a little increase (0.071 and 0.0915 respectively). It is not same for the FLNN which showed a bad scaling by a huge increase of its prediction error.

4.2 Evaluation of the Placement Quality

In this section, We compare our QGANN approach to DFA [2] and CPSO [13] placement approaches in terms of migration rate, total execution time (TET), and computation time. In all the tests, the number of servers was varied between 150 and 2250, while for our predictive approach the number of observations was set to 200.

Quality of Predictions' Impact on the BPaaS Fragments' Migration Rate. The migration rate is calculated at an instant time t_1 after the establishment of a previous placement scheme. That is based on the resource usage prediction step. As shown in Fig. 8.a, the migration rate is highly decreased whenever compared to our non-predictive placement approach (DFA). We notice jumps from 0.21 to 0.0099 for DFA, and from 0.155 to 0.000879 for QGANN. That is justified by the fact that the preventive nature of the QGANN solution allows to limit the placement task to the underloaded and high-performance servers. That avoids blocking situations that cause useless migrations and increase the BPaaS fragments' migrating number.

Impact on the BPaaS Total Execution Time. We note, from Fig. 8.b, that the DFA placement which is based on fragments' reuse, outperforms the CPSO placement approach. We affirm also that when initially executed, the DFA and the QGANN have close values of TET. That is explained by the fact that they both consider the dependencies between BPaaS fragments. In fact, servers

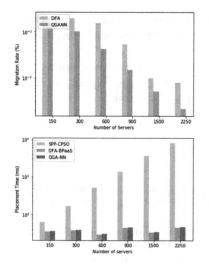
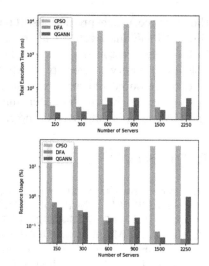

Fig. 8. Comparison of migration rates, TET, placement times, and resource usage

are selected based on their proximity expressed in terms of data transfer time (DTT). However, our predictive placement (QGANN) favors the performance and availability of cloud servers, as a prior selection criteria.

Impact on the Placement Time. By varying the number of servers, we notice in Fig. 8.c that the placement time of QGANN slightly exceeds the non-predictive DFA-based placement (e.g., 1.306 ms). That is due to the additional step of servers states' prediction (overloaded/underloaded) before triggering the placement process. However, this is considered being gainful because it is better for cloud providers to spend additional time in the prediction as a preventive process. In fact, that avoids, eventual SLA violations from one hand, and BPaaS fragments' useless migrations from another hand.

Impact on the Resource Usage. As for the TET, we note from Fig. 8.d that the resource usage rate is not necessarily reduced in comparison with the DFA placement. That is justified by the fact that the high-order objective of the QGANN placement consists on the continuous resource provision without any SLA violation. QGANN out-performed DFA and CPSO in most test cases. In fact, its prediction capabilities helped eliminate overloaded servers from the placement process, which is limited to the underloaded servers.

5   Conclusion and Future Work

In this paper, we addressed the issue of predictive BPaaS management. For that, we proposed a prediction model determining the future state of each cloud availability zone. We employed the quantum genetic algorithm in order to optimize the predictions supported by a neural network-based method. Second, we defined a BPaaS placement algorithm that is triggered depending on the resource usage predictions. The future work involves implementing additional BPaaS management operations and handling BPaaS/cloud uncertainty factors (e.g., customer-provider relationship, heterogeneity of BPaaS providers, policies' diversity), before proceeding to any BPaaS management operation. In fact, the resource allocation, as much as the provider-consumer relationships will be endangered, when the uncertainty of the captured data at both the BPaaS and CAZ level is not covered and handled.

References

1. Woitsch, R., Utz, W.: Business process as a service (BPaaS). In: Janssen, M., et al. (eds.) I3E 2015. LNCS, vol. 9373, pp. 435–440. Springer, Cham (2015). https://doi.org/10.1007/978-3-319-25013-7_35
2. Hedhli, A., Mezni, H.: A DFA-based approach for the deployment of BPaaS fragments in the cloud. Concurr. Comput. Pract. Exp. **32**, e5075 (2020)
3. Grati, R., Boukadi, K., Ben-Abdallah, H.: Business adaptation for BPaaS using fuzzy logic systems. In: 2017 IEEE/ACS 14th International Conference on Computer Systems and Applications, pp. 645–651 (2017)
4. Petcu, D., Stankovski, V.: Towards cloud-enabled business process management based on patterns, rules and multiple models. In: IEEE 10th International Symposium on Parallel and Distributed Processing with Applications, pp. 454–459. IEEE (2012)
5. Yu, D., Zhu, Q., Guo, D., Huang, B., Su, J.: jBPM4S: a multi-tenant extension of jBPM to support BPaaS. In: Bae, J., Suriadi, S., Wen, L. (eds.) AP-BPM 2015. LNBIP, vol. 219, pp. 43–56. Springer, Cham (2015). https://doi.org/10.1007/978-3-319-19509-4_4
6. Luo, W.: A quantum genetic algorithm based QoS routing protocol for wireless sensor networks. In: International Conference on Software Engineering and Service Sciences, pp. 37–40. IEEE (2010)
7. Tipsmark, A., et al.: Experimental demonstration of a Hadamard gate for coherent state qubits. Phys. Rev. A **84**(5), 050301 (2011)
8. Huang, Z., Huai, J., Liu, X., Zhu, J.: Business process decomposition based on service relevance mining. In: 2010 IEEE/WIC/ACM International Conference on Web Intelligence and Intelligent Agent Technology, vol. 1, pp. 573–580, August 2010
9. Moreno-Vozmediano, R., Montero, R.S., Huedo, E., Llorente, I.M.: Orchestrating the deployment of high availability services on multi-zone and multi-cloud scenarios. J. Grid Comput. **16**(1), 39–53 (2018)
10. Schwarzbach, B., Glöckner, M., Makarov, S., Franczyk, B., Ludwig, A.: Privacy preserving BPMS for collaborative BPaaS. Ann. Comput. Sci. Inf. Syst. **11**, 925–934 (2017)
11. Nacer, A.A., Goettelmann, E., Youcef, S., Tari, A., Godart, C.: Business process design by reusing business process fragments from the cloud. In: IEEE 8th International Conference on Service-Oriented Computing and Applications, pp. 193–200 (2015)

12. Nguyen, T., et al.: A resource usage prediction system using functional-link and genetic algorithm neural network for multivariate cloud metrics. In: IEEE 11th International Conference on Service-Oriented Computing and Applications, pp. 49–56 (2018)
13. Hajji, M.A., Mezni, H.: A composite particle swarm optimization approach for the composite SaaS placement in cloud environment. Soft Comput. **22**(12), 4025–4045 (2018)

Predicting Psychiatric Diseases Using AutoAI: A Performance Analysis Based on Health Insurance Billing Data

Markus Bertl[1]([✉]) [iD], Peeter Ross[1], and Dirk Draheim[2] [iD]

[1] Department of Health Technologies, Tallinn University of Technology,
Tallinn, Estonia
{mbertl,peeter.ross}@taltech.ee
[2] Information Systems Group, Tallinn University of Technology, Tallinn, Estonia
dirk.draheim@taltech.ee

Abstract. Digital transformation enables a vast growth of health data. Because of that, scholars and professionals considered AI to enhance quality of care significantly. Machine learning (ML) algorithms for improvement have been studied extensively, but automatic artificial intelligence (autoAI/autoML) has been widely neglected. AutoAI aims to automate the complete AI lifecycle to save data scientists from doing low-level coding tasks. Additionally, autoAI has the potential to democratize AI by empowering non-IT users to build AI algorithms. In this paper, we analyze the suitability of autoAI for mental health screening to detect psychiatric diseases. A sooner diagnosis can lead to cost savings for healthcare systems and decrease patients' suffering. We evaluate AutoAI using the open-source machine learning library auto-sklearn, as well as the commercial Watson Studio's AutoAI platform to predict depression, post-traumatic stress disorder, and psychiatric disorders in general. We use health insurance billing data from 83,986 patients with a total of 687,697 ICD-10 coded diseases. The results of our research are as follows: (i) on average, an accuracy of 0.6 (F_1–score 0.58) with a precision of 0.61 and recall of 0.56 was achieved using auto-sklearn. (ii) The evaluation metrics for Watson Studio's autoAI were 0.59 accuracy, 0.57 F_1–score, a precision of 0.6, and a recall of 0.55. We conclude that the prediction quality of autoAI in psychiatry still lacks behind traditional ML approaches by about 24% and is therefore not ready for production use yet.

Keywords: Artificial intelligence · AI · Machine learning · ML · AutoAI · AutoML · IBM Watson AutoAI · Auto-sklearn · Decision support systems · Psychiatry · Depression · Post-traumatic stress disorder (PTSD)

1 Introduction

Artificial Intelligence (AI) is nowadays a major driver for innovation in digital government and will play a significant role in tackling the challenges our society

© Springer Nature Switzerland AG 2021
C. Strauss et al. (Eds.): DEXA 2021, LNCS 12923, pp. 104–111, 2021.
https://doi.org/10.1007/978-3-030-86472-9_9

is currently facing. This especially applies to the healthcare sector. The creation of health data is rising year by year. Together with the drastic increase in computing power and the rising acceptance of AI by the general public, research about AI-driven digital decision support systems (DDSS) gets more and more popular. The Cambridge Dictionary defines AI as "the study of how to produce computers that have some of the qualities of the human mind, such as the ability to understand language, recognize pictures, solve problems, and learn" [14]. Sauter 1997 defines DDSSs as "computer-based systems that bring together information from a variety of sources, assist in the organization and analysis of information and facilitate the evaluation of assumptions underlying the use of specific models" [15]. Especially in mental health, this has enormous potential to optimize patient care. The prevalence of mental illnesses, suffering, and stigmatization are high – at the same time, diagnostic accuracy is low. 52.7% of people with depression are not correctly diagnosed in a primary care [11]. Mental illness has a severe impact on the society as a whole; In [8], Greenberg et al. estimate that major depressive disorder alone results in an economic burden of approximately \$210.5 billion annually.

These obstacles, together with the exponential data growth, create increasing opportunities for DDSS. Recent meta-reviews showed that current research of DDSS promises high accuracy scores using ML algorithms [1,2]. The used data was mostly transformed and algorithms were specifically selected and tuned. Building, maintaining, and operating AI algorithms that way not only requires advanced data science skills but is also time-intensive.

The relatively new research area of automatic AI (autoAI), sometimes also called automatic machine learning (autoML), tackles this problem by providing systems that automate the whole ML lifecycle end to end (e.g., data preparation, feature engineering, model selection, pipeline optimization, and hyperparameter optimization). There are both open-source products (such as auto-sklearn [6], or autokeras [10]), as well as commercial products (such as IBM Watson Studio's AutoAI [9]) available. AutoAI speeds up the process of developing ML models and will be inevitable in the future of data science [17]. Additionally, autoAI can democratize AI by enabling non-technical users to apply AI technology easily because they do not need to understand the statistical background for selecting and tuning the right AI algorithm. AutoAI takes data, automatically transforms it as needed, selects a proper algorithm, and automatically tunes the hyperparameter of the algorithm. Currently, autoAI has not found wide adoption in scientific literature, nor clinical practice [18].

The described benefits led us to our research question on how autoAI algorithms in the healthcare sector perform compared to traditional approaches, and if autoAI is an alternative for increasing DDSS adoption rates by enabling a faster implementation. Our main contributions in this paper are as follows:

- A novel performance evaluation of two popular autoAI frameworks (open-source, as well as commercial) against a large, real-world dataset in psychiatry.
- We put this evaluation into the context of traditional, manual, machine learning approaches to test the suitability of autoAI for AI-based DDSSs.

The paper is organized as follows. In Sect. 2, we provide an overview of the used data set and describe the investigated autoAI frameworks and the development environment. In Sect. 3, we present the results of our evaluation. In Sect. 4, we discuss our approach and put the results in context of other research. Finally, we finish the paper with a conclusion in Sect. 5.

2   Method

2.1   Data

The data used was collected from the Estonian Health Insurance Fund. Our dataset includes information on sex, birth year, diagnosis, and diagnoses year and month from 83,986 adults (18 years or above) with a total of 687,697 diagnoses in 2019. The data consists of all publicly insured people in Estonia with a depression diagnosis, either single episode (F32), or recurrent (F33), an equally-sized random sample of people with other psychiatric disorders and no depression diagnosis, and an also equally sized random sample of people without any psychiatric diagnosis. Records with only one diagnosis were excluded.

Patients with and without the disease to predict were equally sampled, and test and training data were divided in a 1:4 ratio. Since our goal is a benchmark of how well autoAI performs on its own, we did no further data preparation or feature engineering on the training data. The test data was selected using proportionate stratified random sampling with the two strata 'healthy' and 'the disease to predict' in a 1:1 ratio, to decrease the risk of sampling bias. Additionally, this solves the problem of misleading accuracy values because of the oversampled 'healthy' group.

The diagnoses were coded using the International Statistical Classification of Diseases and Related Health Problems, tenth revision (ICD-10) which is an international standardized coding system for reporting diseases [19]. Each ICD-10 code is structured based on an alpha character and two digits describing the category of the disease followed by a dot and further digits representing more details such as cause, location, severity, or other clinical information. As one example, F32.2 codes for major depressive disorder, single episode, severe without psychotic features. F stands for mental and behavioral disorders, F30–F39 code mood [affective] disorders where F32 is the sub-item for major depressive disorder, single episode. The '.2' in the end specifies the severity.

Our performance benchmark is based on the prediction of F32 (major depressive disorder, single episode), F33 (major depressive disorder, recurrent), F43 (reaction to severe stress, and adjustment disorders), and if any F-diagnosis present is present. For the prediction of each diagnosis we report:

$$precision = \frac{true\ positives}{true\ positives\ +\ false\ positives} \tag{1}$$

$$recall = \frac{true\ positives}{true\ positives\ +\ false\ negatives} \tag{2}$$

$$F_1\text{-score} = \frac{2 * precision * recall}{precision + recall} \qquad (3)$$

$$accuracy = \frac{true\ positives + true\ negatives}{true\ positives + false\ positives + true\ negatives + false\ negatives} \qquad (4)$$

Definitions of the measurements can be found in [16].

2.2 AutoAI Frameworks, Development Environment, and Configuration

Our development environment consisted of *JupyterLab 3.0.12* with *Python 3*, *Pandas 1.2.3*, and *auto-sklearn 0.12.4* with the Auto-Sklearn 2.0 classifier [5]. To analyze the models produced by auto-sklearn, we are using the Python package *PipelineProfiler* [13].

Auto-Sklearn was chosen because it is a state-of-the-art open-source autoAI framework claiming to outperform its competitors [6]. Auto-Sklearn works based on Bayesian optimization, meta-learning, and ensemble construction [6]. Auto-sklearn supports 15 classifiers, 14 feature pre-processing techniques, and 4 data pre-processing methods, with 110 hyperparameters [5,6]. We used a seed value of 7 for the Auto-Sklearn 2 classifier. Since auto-sklearn is non-deterministic, we execute each training five times to see how the different runs compare.

We also included the results from IBM Watson Studio's AutoAI, the autoAI platform on the IBM public cloud [9] to see if commercial platforms perform differently. Watson Studio's AutoAI supports 7 classifiers and 20 data transformations [9]. It uses a model-based, derivative-free global search algorithm, called RBfOpt [4] for hyperparameter optimization, in contrast to Auto-sklearn's Bayesian optimization. We configured accuracy as optimization metric for both frameworks. For both frameworks, no restrictions concerning time or memory were given. We ensured that enough RAM and disk space are available for each training run to finish without out-of-memory errors.

3 Results

3.1 Watson AutoAI

Table 1 gives an overview of the different classifiers' metrics resulting from Watson Studio's AutoAI on IBM Cloud. Since the F diagnoses data set was not supported by autoAI because of file size limitations, it is omitted in the results table. The average accuracy of all runs for all diseases is 0.59, with a F_1-score of 0.57, a precision of 0.6 and a recall of 0.55. In all cases, Watson Studio's AutoAI chose LGBM classifier (gradient boosting with leaf-wise tree-based learning) with first applying principal component analysis to the data.

Table 1. Watson AutoAI – performance.

Disease	Precision	Recall	F_1-score	Accuracy	Test data
F32	0.65	0.52	0.58	0.62	4331
F33	0.60	0.59	0.59	0.60	4325
F43	0.55	0.55	0.55	0.55	1195

3.2 Auto-Sklearn

Table 2 shows the aggregated metrics of the different classifier ensembles created by the five runs of auto-sklearn.

The average accuracy of all runs for all diseases is 0.6, 95% CI [0.596, 0.604], with a F_1–score of 0.58, a precision of 0.61 and a recall of 0.56. In the following subsections, we report more details of the construction and the metrics for each classifier ensemble. Since auto-sklearn is not deterministic, different runs can lead to different results. However, the average standard deviation of the accuracy values was with $\sigma = 0.0057$ small.

Table 2. Auto-sklearn classifiers – average performance.

Disease	Precision	Recall	F_1-score	Accuracy	Test data
F32	0.60	0.56	0.58	0.59	4331
F33	0.63	0.57	0.6	0.61	4325
F43	0.59	0.63	0.61	0.59	1195
F	0.63	0.47	0.53	0.60	17194

F32 - Major Depressive Disorder, Single Episode. During our five runs, auto-sklearn analyzed between 200 and 213 target algorithms for this classification task. On average, an accuracy of $\mu = 0.59$ with a standard deviation of $\sigma = 0.0055$ was achieved with five independent runs. The final ensemble consisted of 35 pipelines. Categorical and numerical transformers were used for pre-processing, Gradient Boosting [12], Random Forest [3], and Extra Trees [7] as classifiers.

F33 - Major Depressive Disorder, Recurrent. Auto-sklearn analyzed between 215 and 251 target algorithms for this classification task. On average, an accuracy of $\mu = 0.612$ with a standard deviation of $\sigma = 0.0084$ was achieved with five independent runs. The final ensemble consisted of 30 pipelines. Categorical and numerical transformers were used for pre-processing and Gradient Boosting [12] and Extra Trees [7] as classifiers.

F43 - Reaction to Severe Stress, and Adjustment Disorders. Auto-sklearn analyzed between 99 and 140 target algorithms for this classification task. On average, an accuracy of $\mu = 0.594$ with a standard deviation of $\sigma = 0.0089$ was achieved with five independent runs.

The final ensemble consisted of 18 pipelines. Categorical and numerical transformers were used for pre-processing, and Random Forest [3] and Gradient Boosting [12] as classifiers.

All F – Diagnoses. Auto-sklearn analyzed between 148 and 150 target algorithms for this classification task. On average, an accuracy of $\mu = 0.6$ with a standard deviation of $\sigma = 0.0$ was achieved with five independent runs. The final ensemble consisted of 27 pipelines. Categorical and numerical transformers were used for pre-processing and Gradient Boosting [12] as classifier.

4 Discussion

We demonstrated that auto-sklearn can be used to predict psychiatric diseases using health insurance billing data. However, the resulting evaluation metrics are behind the ones that are reported using traditional AI approaches. While we achieved an accuracy of 0.6 with auto-sklearn and 0.59 with Watson Studio's AutoAI Experiment, a recent literature survey reports an average accuracy of 0.84 for predicting psychiatric diseases [2]. Notable is, that the studies found by the named survey have a by far smaller sample size (mean of $\mu = 5569$ records with a standard deviation of $\sigma = 19194.28$ and a median of $\eta = 237$) than we used in our evaluation (687,697 records). While most research just has samples from one hospital or healthcare provider, we were able to use data from all patients with depressions in Estonia to test autoAI on a realistic, BigData sample from a whole country.

We observed that auto-sklearn is easy to use, even for non-IT professionals, and gives fast results without the need for data science skills. Watson Studio's AutoAI Experiment does not even require coding skills. All tasks can be carried out through a web interface on the IBM Cloud. Concerning the initial research question, our experiment showed that the tested autoAI frameworks still lag behind traditional approaches where data scientists develop and tune ML models. AutoAI can by no means replace data scientists. While autoAI offers functionality for feature engineering, model selection and tuning, data scientists still need to attend to the human side of AI model implementation like finding the right business problem to solve, analyze the requirements, and determining the superiority of an AI solution compared to currently used solutions.

Nevertheless, autoAI holds the potential to save data scientists time by automating basic, low-level tasks, enabling the professionals to focus on understanding the business problem and later on, the individual fine-tuning of the out-putted autoAI models. This can decrease the time for data scientists to design new AI models. Since autoAI is a relatively new field in computer science, it is not fully matured and we expect it to deliver better results in the near future.

One limitation of our research is that we compare the autoAI results to studies using other data for evaluation. In further research, we will use the applied dataset for traditional machine learning algorithms, as well as deep learning algorithms to see how they perform compared to autoAI. Additionally, our research is limited to the binary classification functionality of auto-sklearn and IBM Watson Studio's AutoAI. Other libraries might perform differently. Because of that, our results are not generalizable for the whole autoAI area. Neither auto-sklearn nor Watson Studio's AutoAI support deep learning. Considering the complexity and high dimensionality of the dataset, deep learning algorithms could lead to better results. Another limitation comes from the way auto-sklearns optimization works. The system is non-deterministic, so that results may change between different runs. To get a comprehensive picture, further research needs to be done to compare these libraries and traditional AI approaches based on a standard benchmark dataset.

5   Conclusion

AutoAI is an emerging research field in computer science with high potential. The ease of use enables a faster development of AI algorithms without extensive data science or programming knowledge. Therefore, it contributes to democratizing AI-based solutions. One possible area where autoAI could be applied in the healthcare sector are DDSSs. Despite the theoretical potential, there is currently no thorough evaluation of autoAI on healthcare datasets.

We presented a novel evaluation of autoAI libraries based on real-world data. In our setting, the accuracy of autoAI (namely auto-sklearn classifier 2.0 and Watson Studio's AutoAI Experiment on IBM Cloud) without any human intervention could not achieve as good evaluation metrics as traditional approaches. Our main finding is that AutoAI's accuracy was, on average, 24% behind conventional techniques. Because of that, we argue that autoAI is not yet ready to be used for the detection of psychiatric diseases based on health insurance billing data. Manual tuning and especially domain knowledge is still needed to create an accurate machine learning model. Because of the ease of use, autoAI can serve as a rapid prototyping tool for ML. It gives an initial direction and can be seen therefore as a smart assistant for data scientists, saving their time for the human side of machine learning projects and advanced fine-tuning tasks after the initial model creation.

References

1. Bertl, M., Metsallik, J., Ross, P.: Digital decision support systems for post-traumatic stress disorder - implementing a novel framework for decision support systems based on a technology-focused, systematic literature review (2021). https://doi.org/10.13140/RG.2.2.12571.28965/1
2. Bertl, M., Ross, P., Draheim, D.: A survey on AI and decision support systems in psychiatry - uncovering a dilemma (2021). https://doi.org/10.13140/RG.2.2.10810.82880/2

3. Breiman, L.: Random forests. Mach. Learn. **45**(1), 5–32 (2001)
4. Costa, A., Nannicini, G.: RBFOpt: an open-source library for black-box optimization with costly function evaluations. Math. Program. Comput. **10**(4), 597–629 (2018)
5. Feurer, M., Eggensperger, K., Falkner, S., Lindauer, M., Hutter, F.: Auto-sklearn 2.0: the next generation. arXiv:2007.04074 [cs, stat] (2020)
6. Feurer, M., Klein, A., Eggensperger, K., Springenberg, J., Blum, M., Hutter, F.: Efficient and robust automated machine learning. In: Cortes, C., Lawrence, N., Lee, D., Sugiyama, M., Garnett, R. (eds.) Proceedings of NIPS 2015 - The 28th Annual Conference on Neural Information Processing Systems, pp. 1–9 (2015)
7. Geurts, P., Ernst, D., Wehenkel, L.: Extremely randomized trees. Mach. Learn. **63**(1), 3–42 (2006)
8. Greenberg, P.E., Fournier, A.A., Sisitsky, T., Pike, C.T., Kessler, R.C.: The economic burden of adults with major depressive disorder in the United States (2005 and 2010). J. Clin. Psychiatry **76**(2), 155–162 (2015)
9. IBM: AutoAI-implementation details - IBM Watson studio (2021). https://dataplatform.cloud.ibm.com/docs/content/wsj/analyze-data/autoai-details.html?audience=wdp
10. Jin, H., Song, Q., Hu, X.: Auto-keras: an efficient neural architecture search system. In: Proceedings of the 25th ACM SIGKDD International Conference on Knowledge Discovery & Data Mining, pp. 1946–1956. ACM (2019)
11. Mitchell, A.J., Vaze, A., Rao, S.: Clinical diagnosis of depression in primary care: a meta-analysis. Lancet **374**(9690), 609–619 (2009)
12. Natekin, A., Knoll, A.: Gradient boosting machines, a tutorial. Front. Neurorobot. **7**, 21 (2013)
13. Ono, J.P., Castelo, S., Lopez, R., Bertini, E., Freire, J., Silva, C.: PipelineProfiler: a visual analytics tool for the exploration of AutoML pipelines. arXiv:2005.00160 [cs] (2020). http://arxiv.org/abs/2005.00160
14. Procter, P. (ed.): Cambridge International Dictionary of English. Cambridge University Press, Cambridge (1995)
15. Sauter, V.L.: Decision Support Systems for Business Intelligence. Wiley, Hoboken (1997)
16. Tohka, J., van Gils, M.: Evaluation of machine learning algorithms for health and wellness applications: a tutorial. Comput. Biol. Med. **132**(104324), 1–15 (2021)
17. Wang, D., et al.: Human-AI collaboration in data science: exploring data scientists' perceptions of Automated AI. In: Proceedings of the ACM on Human-Computer Interaction 3(CSCW), pp. 211:1–211:24 (2019)
18. Waring, J., Lindvall, C., Umeton, R.: Automated machine learning: review of the state-of-the-art and opportunities for healthcare. Artif. Intell. Med. **104**(101822), 1–12 (2020)
19. World Health Organization: ICD-10: International Statistical Classification of Diseases and Related Health Problems: Tenth Revision, 2nd edn. World Health Organization (2004)

Improving Billboard Advertising Revenue Using Transactional Modeling and Pattern Mining

P. Revanth Rathan[1(✉)], P. Krishna Reddy[1], and Anirban Mondal[2]

[1] IIIT Hyderabad, Hyderabad, India
revanth.parvathaneni@research.iiit.ac.in, pkreddy@iiit.ac.in
[2] Ashoka University, Sonipat, India
anirban.mondal@ashoka.edu.in

Abstract. Billboard advertisement is among the dominant modes of outdoor advertisements. The billboard operator has an opportunity to improve its revenue by satisfying the advertising demands of an increased number of clients by means of exploiting the user trajectory data. Hence, we introduce the problem of billboard advertisement allocation for improving the billboard operator revenue, and propose an efficient user trajectory-based transactional framework using coverage pattern mining. Our experiments validate the effectiveness of our framework.

Keywords: Billboard advertisement · Pattern mining · Transactional modeling · User trajectory · Ad revenue

1 Introduction

Billboard advertisement is among the dominant modes of traditional outdoor advertisements. Notably, outdoor ads constitute a $500 billion market around the world [1]. In fact, billboards are the most widely used medium for outdoor ads with a market share of about 65%, and 80% people notice them while driving [2]. A given billboard operator (BO) rents/owns and manages a set of billboards by allocating the ads of *clients* on the billboards in lieu of a cost, while *users* view the billboard ads during transit. Observe that budget is a critical factor for the client. For example, the average cost of renting a billboard ad space in New York can run into thousands of dollars per month.

The key goal of BO is to maximize its revenue by allocating ads of clients on billboards. This paper introduces the problem of *efficient* billboard allocation to clients for improving the revenue of BO by assigning ads to billboards such that a typical user is exposed to *distinct* ads in her trajectory as far as possible. We refer to this problem as the **B**illboard **AL**location (**BAL**) problem.

Existing works have focused on the allocation of billboard ad space to clients based on user trajectory-driven data [10] and crowd-sensed vehicular trajectory data [9]. Selection of billboard locations by visually analyzing large-scale taxi

© Springer Nature Switzerland AG 2021
C. Strauss et al. (Eds.): DEXA 2021, LNCS 12923, pp. 112–118, 2021.
https://doi.org/10.1007/978-3-030-86472-9_10

trajectories has been studied in [6]. Greedy heuristics to compute the k-billboard location set with most influence have been examined in [5]. Our work differs from existing works in that we approach the problem as a pattern mining problem as opposed to an optimization problem. Further, we focus on improving the revenue of *BO* instead of improving billboard advertising from the client perspective.

We propose a framework, designated as Billboard Allocation Framework (BAF), to improve the revenue of *BO* by extending the notions of transactional modeling and *coverage*. In BAF, each user trajectory is modeled as a transaction. Given a transactional dataset of user trajectories, the issue is to determine the potential combinations of billboards such that each combination *covers* a required number of unique trajectories (users) by minimizing *overlap*. Such combinations could be allocated to different clients based on the requirement. However, for determining potential combinations of billboards from m billboards, we need to examine $2^m - 1$ combinations, which leads to combinatorial explosion problem. Hence, there is an opportunity to extend the framework of coverage patterns (CPs) [8], which can be extracted from transactional databases by exploiting apriori based pruning property.

Approaches to extract coverage patterns (CPs) were first proposed in [8] for banner advertisements. The notion of coverage patterns has been extended to improve display advertising performance [4], and search engine advertising [3].

The concept of CPs can be extended to extract the potential combinations of billboards from user trajectory data in an efficient manner. BAF comprises the formation of transactions from user trajectory data and a CP-based billboard views allocation approach. In BAF, we extract CPs from transactions, which are formed by processing user trajectory data. Each CP is assigned with the values of coverage support (CS) and the overlap ratio (OR). Notably, CS of a CP indicates the number of unique transactions and the value of OR represents the extent of overlap, which is equal to the number of repetitions among the transactions covered by CP. We propose the mapping function to convert the CS of the extracted CPs to views and a ranking function to order the CPs. Finally, we allocate the generated CPs (i.e., a set of billboards) to the clients subject to her requirements. Our key contributions are two-fold:

- We introduce the billboard allocation problem (BAL) for improving the billboard operator revenue.
- We propose an efficient user trajectory-based transactional framework using coverage pattern mining for addressing the billboard allocation problem.

We conducted a performance study with a real dataset to demonstrate the effectiveness of the proposed framework.

2   Proposed Framework of the Problem

Billboards are visible road-side ad hoardings. Consistent with real-world scenarios, we assume that only *one* advertisement is displayed in a given billboard at a specific point in time. A given *billboard operator (BO)* rents/owns the billboards in a particular region. *Clients* wish to display their ads on the billboards. *BO* assigns the sets of billboards to the clients for a particular cost, namely *billboard cost*. *User trajectory* is a sequence of GPS locations traversed by a given user. If a user stops at any location for more than a threshold amount of time, we consider her trajectory after the stoppage as a new user trajectory.

Views of a billboard are defined as the total number of user trajectories, which are influenced by the billboard. We consider that a user trajectory is influenced by the billboard if it passes through the sector of radius λ and angle θ in the vicinity of the billboard. Furthermore, we assume the existence of conversion schemes, which convert the budget of the client to the views of the billboard.

Consider a set of m billboards B $\{b_1, b_2, .., b_m\}$ and a set $C = \{c_1, c_2, .., c_n\}$ of n clients, where c_i represents a client's unique identifier. Let $t(c_i)$ be the budget of client c_i, while $v(c_i)$ is the threshold number of views corresponding to $t(c_i)$. Moreover, $v(b_i)$ is the number of user trajectories, which viewed billboard b_i.

The **Billboard ALlocation (BAL)** problem aims to improve the revenue of *BO* by allocating different sets of billboards to the maximum number of clients in set C by satisfying clients' budget constraint. The sum of all views of the billboards in the set s_i i.e., $\{b_1, b_2, ..., b_{|s_i|}\}$ is equal to $\sum_{k=1}^{|s_i|} v(b_k)$. Mathematically, we can allocate the billboards set s_i to client c_i if $\sum_{k=1}^{|s_i|} v(b_k) \geq v(c_i)$.

The billboard set s_i assigned to the client c_i should have *overlap* less than or equal to the maximum overlap value $(maxOV)$, i.e., $Overlap(s_i, c_i) \leq maxOV$. By using this *overlap constraint*, *BO* ensures that the client's advertisement in the billboard is guaranteed to be viewed by at least a minimum number of *distinct* user trajectories. Now we define the BAL problem as follows:

BAL Problem: Consider a billboard operator BO, a set UT of user trajectories, a set of m billboards B $\{b_1, b_2, \ldots, b_m\}$ and a set C of n clients such that $C = \{c_1, c_2, \ldots, c_n\}$. BO has to assign a set of billboards from $S = \{s_1, s_2, \ldots, s_o\}$, where each candidate set s_i comprises a set of billboards, to maximum number of clients in the set C such that $\sum_{k=1}^{|s_j|} v(b_k) \geq v(c_i)$ and $Overlap(s_j, c_i) \leq maxOV$, $\forall s_j \in S$ and $\forall c_i \in C$ to improve the revenue.

3 Proposed Billboard Allocation Framework

BAF comprises three steps, as depicted in Fig. 1.

1. Formation of user trajectory transactions $(UTTs)$: Here, the input is a set of user trajectories UT and billboards B and output is UTTs. We convert each UT into UTT by including billboard locations covered by UT.

2. Extraction of CPs: Based on minRF, minCS and maxOR, we employ the existing CP mining algorithm proposed in [8] and extract S. Note that we use the notion of maximum overlap ratio (maxOR) defined in [8] to capture the notion of $maxOV$ in BAL problem.

3. Computing views of CPs and allocation: It consists of three sub-steps.

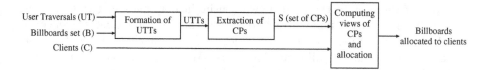

Fig. 1. Proposed Billboard Allocation Framework (BAF)

(i) Computing views and uniqueness rank of a CP: We compute the number of user views for each CP in S. CS of a CP is the effective number of unique users influenced by the billboards in the pattern. Let s_i be a CP, $|T_{b_i}|$ be the number of $UTTs$ influenced by the billboard b_i and $views(s_i)$ denote the views of s_i. Then $views(s_i) = \sum_{\forall b_i \in s_i} |T_{b_i}|$.

The number of views computed in Step (i) includes the repeated views. Since a large number of CPs are extracted from the $UTTs$, when the views of several patterns are equal, it is beneficial to allocate the pattern with the maximum number of distinct views. Given two patterns having the same number of views, a pattern is considered to have more unique views if it has more CS and less OR. For a given s_i, we capture the uniqueness of a pattern with uniqueness rank heuristic $U(s_i)$, which equals $(CS(s_i) * (1 - OR(s_i)))$.

(ii) Computing views required by clients: We employ marketing conversion schemes to convert the budget to views as follows: $v_j = convert(t_j)$, where v_j is the views demanded by client for budget t_j and $convert()$ is the conversion function decided by BO.

(iii) Allocation: Several allocation methods are possible. We now present one of the methods. We sort S in the decreasing order of both number of views and uniqueness. We sort the list of the clients C in the decreasing order of the required views. Each client c_j is allocated with the eligible candidate set s_i, which satisfies the requirement of the client c_j, i.e., $views(s_i) \geq v_j$. Multiple candidate sets may satisfy the above requirement. Hence, we allocate the s_i to c_j, which satisfies the above condition and has the minimum difference of views required by c_j and views of s_i. After allocation, we need to remove all the sets where the allocated billboards are present. The above procedure continues until all the clients are allocated with the potential candidate sets, or remaining candidate sets are unable to satisfy the clients.

4   Performance Evaluation

Our experiment used Microsoft's *Geolife* [11] real dataset. The dataset contains 17,621 trajectories with a total distance of 1,292,951 kilometers out of which we considered 13,971 trajectories within Beijing's Fifth Ring Road. We used OpenStreetMap[1] for Beijing to obtain the road network. First, we mapped each

---

[1] https://www.openstreetmap.org.

trajectory of the user to the road network using a map-matching tool Graph-hopper[2]. Then we used nominatim[3] to obtain the corresponding sequence of edge IDs of road segments as an user trajectory. The Geolife dataset does not contain information about billboard locations. We placed billboard locations in the Geolife dataset only in road segments, where the number of user trajectories exceeds a pre-specified threshold. We assume that a billboard influences a user trajectory if the edge ID corresponding to the billboard exists in the sequence of edge IDs of the user trajectory. We set the billboard's cost to be proportional to the frequency of user trajectories in the road segment containing the billboard.

We set the number of billboards available (N_B) to 50. The default values of maxOR and minCS are set to 0.5 and 0.1 respectively. Performance metrics are execution time (ET), the number of billboards allocated (BA) and the number of views assigned (VA). ET is the time required for the allocation of the billboard set to the clients. Recall that we compute the knowledge of coverage patterns in an *offline* manner and use it to allocate billboards at run-time. ET does not include the time required to generate coverage patterns since these patterns are generated offline. BA computes the number of billboards utilized in the process of allocation. VA is the number of *unique views* assigned in the process of allocation.

Recall that given a set of billboards with their respective costs and user trajectory data, the approach proposed in [10] determines the maximum influence set within a pre-specified budget. However, it satisfies only a single client. Hence, as reference, we adapted the approach proposed in [10] as follows. First, we represent the views of the billboard in terms of budget. We represent the cost of a billboard b as $views(b)/10$. We use the notion of overlap ratio discussed in [8] as the metric to divide the billboards into clusters such that the overlap between the clusters is always less than a pre-specified threshold. We sort the clients in descending order of budget (in terms of views). Second, using the approach proposed in [10], we compute the billboard set with maximum influence within the budget of the first client (after sorting). We allocate the computed billboard set to the client if the aggregate views of the set of billboards are greater than 90% of the views specified by the client since this approach maximizes the views within the budget. We refer to this approach as PartSel Framework (**PSF**).

In contrast, our proposed BAF scheme divides the user trajectories into multiple user trajectory transactions ($UTTs$) based on billboard locations. BAF deploys the *parallel C-Mine* algorithm [7] to obtain all candidate coverage patterns from $UTTs$ with the value of minRF = 0 for the Geolife dataset in an *offline* manner. Notably, for all the approaches, we generated the budget of clients randomly in the range between the values of minCS and $|UTT|$.

Figure 2 depicts the results for varying the number N_C of clients requested. As N_C increases, ET increases for both schemes because the allocation step should run for every client. ET for BAF is much lower than that of PSF because most of the computations in case of BAF (i.e., computation of coverage patterns) are done *offline*. In contrast, for PSF, there is no scope for performing any offline

---

[2] https://graphhopper.com/api/1/docs/map-matching/.
[3] https://nominatim.openstreetmap.org/.

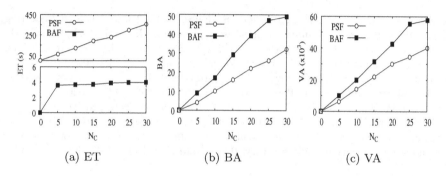

Fig. 2. Effect of varying the number of clients requested for Geolife dataset

computations in advance, hence all required computations for every client are done during run-time. From the results in Fig. 2b, observe that BAF always allocates more billboards than PSF. BAF utilizes available billboards better than PSF since it uses almost all the billboards for allocation, thereby increasing revenue for *BO*. The value of VA is proportional to the revenue of *BO* (since we convert budget to views). From the results in Fig. 2c, observe that BAF provides more views than PSF. Hence, BAF provides more revenue to *BO*.

5 Conclusion

We have introduced the problem of determining an efficient billboard allocation approach to improve the revenue of the billboard operator. We have proposed an efficient user trajectory-based transactional framework using coverage pattern mining for addressing the aforementioned problem. Our performance study using a real dataset has demonstrated the effectiveness of the proposed framework.

References

1. https://www.statista.com/topics/979/advertising-in-the-us/
2. http://www.runningboards.com.au/outdoor/relocatable-billboards
3. Budhiraja, A., Ralla, A., Reddy, P.K.: Coverage pattern based framework to improve search engine advertising. Int. J. Data Sci. Anal. **8**(2), 199–211 (2019)
4. Kavya, V.N.S., Reddy, P.K.: Coverage patterns-based approach to allocate advertisement slots for display advertising. In: Bozzon, A., Cudre-Maroux, P., Pautasso, C. (eds.) ICWE 2016. LNCS, vol. 9671, pp. 152–169. Springer, Cham (2016). https://doi.org/10.1007/978-3-319-38791-8_9
5. Li, Y., et al.: Mining the most influential k-location set from massive trajectories. IEEE Trans. Big Data **4**(4), 556–570 (2017)
6. Liu, D., et al.: SmartAdP: visual analytics of large-scale taxi trajectories for selecting billboard locations. IEEE Trans. Vis. Comput. Graph. **23**(1), 1–10 (2016)
7. Ralla, A., Siddiqie, S., Reddy, P.K., Mondal, A.: Coverage pattern mining based on MapReduce. In: Proceedings ACM IKDD CoDS and COMAD, pp. 209–213 (2020)

8. Srinivas, P.G., Reddy, P.K., Bhargav, S., Kiran, R.U., Kumar, D.S.: Discovering coverage patterns for banner advertisement placement. In: Tan, P.-N., Chawla, S., Ho, C.K., Bailey, J. (eds.) PAKDD 2012. LNCS (LNAI), vol. 7302, pp. 133–144. Springer, Heidelberg (2012). https://doi.org/10.1007/978-3-642-30220-6_12
9. Wang, L., et al.: Efficiently targeted billboard advertising using crowdsensing vehicle trajectory data. IEEE Trans. Ind. Inform. **16**(2), 1058–1066 (2020)
10. Zhang, P., Bao, Z., Li, Y., Li, G., Zhang, Y., Peng, Z.: Trajectory-driven influential billboard placement. In: Proceedings ACM SIGKDD, pp. 2748–2757 (2018)
11. Zheng, Y., Xie, X., Ma, W.Y.: Geolife: a collaborative social networking service among user, location and trajectory. IEEE Data Eng. Bull. **33**(2), 32–39 (2010)

Sarcasm Detection for Japanese Text Using BERT and Emoji

Yoshio Okimoto[1], Kosuke Suwa[2], Jianwei Zhang[2(✉)], and Lin Li[3]

[1] Cosmo Computing System, Inc., Tokyo, Japan
y.okimoto@cosmocomputing.co.jp
[2] Iwate University, Morioka, Japan
zhang@iwate-u.ac.jp
[3] Wuhan University of Technology, Wuhan, China
cathylilin@whut.edu.cn

Abstract. In this paper, we propose methods to detect sarcasm from Japanese text on Twitter by using the BERT language model and analyzing emoji as well as text. After constructing a Japanese Twitter dataset, we extract feature vector for both text and emoji. Our experimental results show that the usage of BERT and emoji can help improve the performance of sarcasm detection.

Keywords: Sarcasm detection · BERT · Emoji · Twitter

1 Introduction

In recent years, with the spread of SNS, people can easily deliver their opinions. Sentiment analysis that analyzes people's opinions or emotions towards entities, events, individuals or organizations, etc. is a challenging task. One factor that makes this task difficult is the existence of sarcasm that conveys negative sentiment using positive expressions or vice versa. Although machine learning is applied for sentiment analysis, sarcasm detection is especially difficult since there is a difference between the meaning of the words used and the intent indicated.

Although there are some researches on sarcasm detection, most of them analyze English data and few works focus on Japanese text. Japanese grammar is usually complex and the context is difficult to understand. Moreover, except for conventional words, there are the proliferating net neologisms such as emoji, emoticon, etc. To the best of our knowledge, there is no previous work targeting Japanese that analyzes expressions other than text. Therefore, in this paper, we propose a method to detect sarcasm by adopting a language model called BERT that can provide different contextualized embedding for a same word appearing in different sentences, and extracting features including emoji in the sentence. The experimental results show that the usage of BERT and emoji can help improve the performance of sarcasm detection, outperforming the baseline method.

This work was supported by JSPS KAKENHI Grant Number 19K12230.

C. Strauss et al. (Eds.): DEXA 2021, LNCS 12923, pp. 119–124, 2021.
https://doi.org/10.1007/978-3-030-86472-9_11

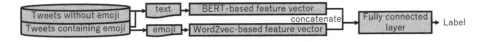

Fig. 1. Flow of the proposed method

2   Related Works

There are some approaches for sarcasm detection [1]. In recent years, new language models such as Transformer [2] and BERT [3] are used. Moreover, there are some researches that consider other types of expressions except for conventional text. Chaudhary et al. [4] showed using emoji in English texts is effective.

The above researches target English texts. There are few works on analyzing Japanese texts. Hiai et al. [5] proposed a LSTM and attention-based method. However, these researches only focus on text data, not considering net neologisms. In this paper, we aim to improve the performance by applying BERT to understand Japanese context and using features extracted from emoji.

3   Proposed Method

3.1   Definition of Sarcasm

Hiai et al. [6] define sarcasm as "positive expression that conveys negative sentiment". Referring to their definition, we extend the definition like "the text that seems positive (or neutral) intuitively but the author actually intends to express negative sentiment." or "the text that seems negative (or neutral) intuitively but the author actually intends to express positive sentiment.". Most of the texts collected from Twitter are the former.

3.2   Overview of Proposed Method

We describe the flow of the proposed method shown in Fig. 1. First, we extract emojis from the dataset and split the data into text and emoji[1]. Next, we create feature vectors for the text and emoji separately. We use BERT for text feature vector and Word2Vec for emoji feature vector. We describe the detail of BERT in Subsect. 3.3 and Word2Vec in Subsect. 3.4. Finally, we concatenate these vectors and conduct supervised learning with the fully connected neural network.

3.3   BERT: Bidirectional Encoder Representations from Transformers

BERT is a NLP model proposed by Devlin et al. [3]. BERT takes into account the context and has achieved state-of-the-art performance on a number of NLP tasks such as translation, document classification, Q&A, etc.

---

[1] Emoji contains 2,811 characters defined by Unicode 13.1.

BERT's learning has two steps: pre-training and fine-tuning. The step of pre-training uses only an unlabeled plain text corpus and creates a generative model. The step of fine-tuning uses the weights obtained from pre-training as initial values and performs supervised learning for tasks with labelled learning data. Except for fine-tuning tasks, the weights of hidden layers in the pre-trained model can be used as feature vector. We use the feature vector with 1024 dimensions.

3.4 Word2Vec

Word2vec is another NLP model proposed by Mikolov et al. [7]. This represents a word with a vector in which each dimension is a real number. Since Word2Vec produces a distributed representation of words, semantic similarity between two words can be calculated based on the similarity between the vectors.

We utilize the idea of Word2Vec to obtain distributed representation for emojis. The dimension of feature vector for each emoji is set to 300. When there are multiple emoji pictograms in one text, those emojis' values on each dimension are averaged to construct the feature vector for that text.

4 Experiment

4.1 Dataset and Preprocessing

For experimental evaluation, we first construct a Japanese dataset. The data is collected by retrieving Twitter with the query "#皮肉" and "(皮肉)"[2] and the period is from Jan. 2010 to Oct. 2020. As the result, 12,070 tweets are collected.

Next, we perform preprocessing for collected data. We remove the keywords "#皮肉" and "(皮肉)" from the collected tweets since we intend to construct the model that are independent of this obvious feature word. In this work, we focus on sarcasm text that can be judged individually without considering context. Thus, we also remove the tweets including URLs, "@(mention)" from the dataset.

After preprocessing, remaining 6,020 tweets are sent to Yahoo! crowdsourcing[3]. Each tweet is checked by three crowd workers and the label of sarcasm is assigned to the tweets that more than two crowd workers judged as sarcasm. Finally, 3,025 sarcasm tweets are decided for constructing the model.

We also randomly select the same number of tweets from Twitter as non-sarcasm.

4.2 Experimental Setup

For the Japanese pretrained BERT model, we use the Kurohashi Lab's model[4]. For the pretrained Word2Vec model for emoji, we use emoji2vec[5] [8].

---

[2] "皮肉" means "sarcasm" in Japanese.

[3] Yahoo! crowdsourcing: https://crowdsourcing.yahoo.co.jp/.

[4] BERT model: https://nlp.ist.i.kyoto-u.ac.jp/index.php?ku_bert_japanese.

[5] emoji2vec: https://github.com/uclnlp/emoji2vec.

Table 1. Experiment contents

ID	Method	Target data (Sarcasm/Non-sarcasm/Total)
1-1	Baseline (Bi-LSTM and Attention)	All data (with or without emoji)
1-2	BERT	(3,025/3,025/6,050)
2-1	Proposed method (text feature)	Data containing emoji
2-2	Proposed method (text & emoji feature)	(705/705/1,410)

We conduct four experiments with the setting shown in Table 1. Experiment 1-1 and 1-2 are conducted on all the 6,050 labeled data, and the purpose is to verify the effectiveness of BERT. Experiment 2-1 and 2-2 are done on 1,410 data containing emoji, and the purpose is to verify the effectiveness of emoji.

In experiment 1-1, a model with Bi-LSTM and attention adopted in many previous works is used as the baseline. The batch size is 16, the number of epochs is 10, and the number of units in hidden layers is 512. In experiment 1-2, BERT is used for the task. The percentage of dropout is 0.1, the learning rate is 1.0×10^{-5}, the batch size is 16, and the number of epochs is 10.

In experiment 2-1, the weights of the hidden layer before the output layer in the pre-trained BERT model is used as the input to a FNN. The percentage of dropout is 0.1, the number of units in a 4-layer neutral network is 1024, 512, 256, 128, the batch size is 8, and the number of epochs is 30. In experiment 2-2, BERT-based text feature is concatenated with Word2Vec-based emoji feature to construct the FNN. The parameters are the same as experiment 2-1.

We perform stratified 5-fold cross-validation.

4.3 Experimental Results

As mentioned in the last section, we performed two kinds of experiments for Japanese sarcasm detection. The results are shown in Table 2. The number of tweets whose labels change in each experiment are shown in Table 3.

The purpose of experiment 1 is to verify the effectiveness of BERT. From the confusion matrix and scores, we can see BERT has higher TP, TN, and scores than the baseline. Comparing the baseline to BERT, the number of tweets whose labels change from false based on the baseline method to true using BERT is larger than the reverse case. This also indicates BERT is useful.

The purpose of experiment 2 is to verify the effectiveness of emoji. From the confusion matrix and scores, we can see the method using text and emoji has higher TP, TN, and scores than only texts. Comparing proposed methods using emoji or not, the number of the tweets whose classification results change from false to true by using emoji is higher than the reverse case. This also indicates emoji is useful. However, since the increase in the number of correct classification is only 10, emoji's effect should be further investigated on a larger dataset.

At last, we show some examples whose labels change in the following. In the examples, "JA" means original Japanese text and "EN" means translated English one. Texts with multiple emojis tend to be classified wrongly.

Table 2. Result of experiments

ID	Method	Confusion matrix			Score		
		Ground truth	Sarcasm	Non-sarcasm	Prec.	Recall	F_1
1-1	Baseline	Sarcasm	623	152	0.78	0.79	0.79
		Non-sarcasm	226	549			
1-2	BERT	Sarcasm	648	127	0.80	0.84	0.82
		Non-sarcasm	159	616			
2-1	Proposed method (text feature)	Sarcasm	103	38	0.73	0.75	0.74
		Non-sarcasm	35	106			
2-2	Proposed method (text & emoji feature)	Sarcasm	111	30	0.79	0.80	0.80
		Non-sarcasm	33	108			

Table 3. Classification label changed by using BERT or emoji

Compared methods	How changed	Ground truth		
		Sarcasm	Non-sarcasm	Total
Baseline to BERT	False to true	88	128	216
	True to false	63	61	124
Proposed method (text feature) to	False to true	15	12	27
Proposed method (text & emoji feature)	True to false	7	10	17

Examples 1 Comparing the baseline to BERT

Example 1-1 False to true/Sarcasm

JA 某ショッピングモールのフードコートに来ていますが，民度の低さが肌に合っていて心地よいです.

EN I'm coming to a food court in a shopping mall. the low cultural level of the people fits my skin and makes me feel comfortable.

Example 1-2 False to true/Non-sarcasm

JA 昔は卒アルに連絡先書いてありましたが今は書いてないのかもですね.

EN In the old days, contact information was written on the graduation album. Maybe it isn't written now.

Example 1-3 True to false/Sarcasm

JA 毎年毎年ハズレを連れてくるスカウトを表彰したい.

EN I want to commend the scout for bringing poor players every year.

Example 1-4 True to false/Non-sarcasm

JA こんにちわ！日本語が下手ですからなにか間違いがあったら気軽に訂正してね！

EN Hello! I'm bad at Japanese. So if you notice a mistake, please feel free to correct it.

Examples 2 Comparing the only text method to the text & emoji method

Example 2-1 False to true/Sarcasm

JA 元検（犬🐕）が当選したんだって www 自民党の皆さんおめでとう！

EN The former prosecutor (lap dog 🐕) won! LOL Congratulations to the liberal democratic party!

Example 2-2 False to true/Non-sarcasm

JA 月って無駄にお金使ってる気がするしめんどいことがよく起こる🐷.

EN In June, I feel I've wasted money and often get into some trouble 🐷.

Example 2-3 True to false/Sarcasm

JA 他人を晒して机上の理論並べてる人自分を持ってて好きですよ．🌀🐷♂

EN I like the persons who expose others to academic gossip. Because they are confident. Because they are confident in themselves. 🌀🐷♂

Example 2-4 True to false/Non-sarcasm

JA マジックレベル 10 の豆アズめちゃくちゃカッコいいから見てほしい 😝😝🙏🙏

EN Mameazu who became magic level 10 is extremely cool, so I want everyone to see him. 😝😝🙏🙏

5   Conclusion

In this paper, we proposed a method for Japanese sarcasm detection by using BERT and emoji. We constructed a dataset from Japanese Twitter and the experiments on this dataset showed that both BERT and emoji can contribute.

Since the volume of our dataset is not large enough and the labeling is conducted by crowd workers, one of our future works is to construct a larger and more trustworthy dataset. Another work is to verify the classification performance with other models such as XLNet [9], ALBERT [10].

References

1. Joshi, A., Bhattacharyya, P., Carman, M.J.: Automatic sarcasm detection: a survey. ACM Comput. Surv. (CSUR) **50**(5), 73:1-73:22 (2017)
2. Vaswani, A., et al.: Attention is all you need. In: Advances in Neural Information Processing Systems, pp. 5998–6008 (2017)
3. Devlin, J., Chang, M.W., Lee, K., Toutanova, K.: BERT: pre-training of deep bidirectional transformers for language understanding. In: CL-HLT, pp. 4171–4186 (2019)
4. Chaudhary, A., Hayati, S.A., Otani, N., Black, A.W.: What A Sunny Day: toward emoji sensitive irony detection. In: W-NUT, pp. 212–216 (2019)
5. Hiai, S., Shimada, K.: Sarcasm detection using RNN with relation vector. In: IJDWM 2019, pp. 66–78 (2019)
6. Hiai, S., Shimada, K.: A sarcasm extraction method based on patterns of evaluation expressions. In: IIAI-AAI, pp. 31–36 (2016)
7. Mikolov, T., Chen, K., Corrado, G., Dean, J.: Efficient estimation of word representations in vector space. In: ICLR Workshop (2013)
8. Eisner, B., Rocktaschel, T., Augenstein, I., Bosnjak, M., Riedel, S.: emoji2vec: learning emoji representations from their description. In: SocialNLP (2016)
9. Yang, Z., Dai, Z., Yang, Y., Carbonell, J., Salakhutdinov, R., Quoc, V.L.: XLNet: generalized autoregressive pretraining for language understanding. In: NeurIPS 2019, pp. 5754–5764 (2019)
10. Lan, Z., Chen, M., Goodman, S., Gimpel, K., Sharma, P., Soricut, R.: ALBERT: a lite BERT for self-supervised learning of language representations. In: 8th ICLR (2020)

Sigmalaw PBSA - A Deep Learning Model for Aspect-Based Sentiment Analysis for the Legal Domain

Isanka Rajapaksha[✉], Chanika Ruchini Mudalige[✉], Dilini Karunarathna[✉],
Nisansa de Silva[✉], Amal Shehan Perera[✉], and Gathika Ratnayaka[✉]

Department of Computer Science and Engineering, University of Moratuwa,
Moratuwa, Sri Lanka
{israjapaksha.16,chanikaruchini.16,dilinirasanjana.16,NisansaDdS,
shehan,gathika.14}@cse.mrt.ac.lk

Abstract. Legal information retrieval holds a significant importance to lawyers and legal professionals. Its significance has grown as a result of the vast and rapidly increasing amount of legal documents available via electronic means. Legal documents, which can be considered flat file databases, contain information that can be used in a variety of ways, including arguments, counter-arguments, justifications, and evidence. As a result, developing automated mechanisms for extracting important information from legal opinion texts can be regarded as an important step toward introducing artificial intelligence into the legal domain. Identifying advantageous or disadvantageous statements within these texts in relation to legal parties can be considered as a critical and time consuming task. This task is further complicated by the relevance of context in automatic legal information extraction. In this paper, we introduce a solution to predict sentiment value of sentences in legal documents in relation to its legal parties. The Proposed approach employs a fine-grained sentiment analysis (Aspect-Based Sentiment Analysis) technique to achieve this task. Sigmalaw PBSA is a novel deep learning-based model for ABSA which is specifically designed for legal opinion texts. We evaluate the Sigmalaw PBSA model and existing ABSA models on the SigmaLaw-ABSA dataset which consists of 2000 legal opinion texts fetched from a public online data base. Experiments show that our model outperforms the state-of-the-art models. We also conduct an ablation study to identify which methods are most effective for legal texts.

Keywords: Legal information extraction · Legal domain · Aspect-based sentiment analysis · Deep learning · NLP

1 Introduction

Factual scenario analysis of previous court cases holds a significant importance to lawyers and legal officers whenever they are handling a new legal court case.

© Springer Nature Switzerland AG 2021
C. Strauss et al. (Eds.): DEXA 2021, LNCS 12923, pp. 125–137, 2021.
https://doi.org/10.1007/978-3-030-86472-9_12

Legal officials are expected to analyse previous court cases and statutes to find supporting arguments before they represent a client at a trial. As the number of legal cases increases, legal professionals typically endure heavy workloads on a daily basis, and they may become overwhelmed and as a result of that, be unable to obtain quality analysis. In this analysis process, identifying advantageous and disadvantageous statements relevant to legal parties [1–4] can be considered a critical and time consuming task. By automating this task, legal officers will be able to reduce their workload significantly. In this paper, we introduce a solution to predict sentiment value of sentences in legal documents in relation to its legal parties. The proposed approach employs a fine-grained sentiment analysis technique to achieve this task.

Sentiment analysis (SA) is identifying opinions and then classifying them into several polarity levels (*Positive, Neutral*, or *Negative*) using computational linguistics and information retrieval [5]. Sentiment analysis can be divided into 4 levels; document level SA, sentence level SA, phrase level SA, and aspect level SA. Sentences in a legal case usually contain two or more members/entities which belong to main legal parties (*plaintiff, petitioner, defendant*, and *respondent*). Extracting opinions with respect to each legal party cannot be performed only by using document-level, sentence-level, or phrase-level sentiment analysis. Aspect-based sentiment analysis (ABSA) is the most appropriate and fine-grained solution to perform Party-Based Sentiment Analysis (PBSA) in the legal domain [1]. In aspect-based sentiment analysis, we can identify there processing steps such as "identification, classification, and aggregation" [6]. Generally, in ABSA aspects are extracted from a given text and then each aspect is allocated a sentiment level (*positive, negative,* or *neutral*) [7]. The members of legal parties in a court case are considered as aspects and therefore performing ABSA in the legal opinion texts can also be termed as Party-Based Sentiment Analysis (PBSA) [1].

A number of studies have addressed Aspect-based Sentiment Analysis in different domains such as restaurants, hotels, movies, products reviews, government services, mobile phones and telecommunication [8]. When it comes to the legal domain, sentiment analysis becomes a challenging area because of the domain-specific meanings and behaviour of words in the legal opinion texts [9]. Languages being used are sometimes mixed (i.e., English, Latin, etc.) and in some situations, the meaning of the words and context varies from that of domain interpretations. The complexity structure and the length of the sentences also increase the difficulty. As a result of the above factors, it is difficult to obtain a comparative accuracy to other areas such as customer feedback, movie or product reviews, and political comments.

Example 1

– Sentence 1.1: *After obtaining a warrant, the officials searched Lee's house, where they found drugs, cash, and a loaded rifle.*

Example 1 contains a sentence extracted from Lee v. United States [10] which mentions two legal party members: *Lee* and *officials*. As the illegal materials were found at *Lee's house*, this sentence clearly shows a negative sentiment towards *Lee* and Positive sentiment towards *officials*.

The rule-based approach proposed by Rajapaksha et al. [1] can be identified as the first and only attempt to perform ABSA in the legal domain to the best of our knowledge. However, that approach has two weaknesses: (1) it significantly depends on the phrase-level sentiment annotator, (2) manually created rules may not cover all the sentence patterns. There are many existing deep-learning models with different architectures trained for different domains to fulfil a wide array of tasks. Despite that, to the best of our knowledge, there is no existing deep learning-based approach for ABSA in the legal domain. In this paper we show that, as the sentences in legal documents are often long and have a complex semantic structure, the existing model architectures, created for short sentences in general use, do not perform well for the legal domain. The main objective of this study is to propose a novel deep learning-based model (SigmaLaw-PBSA) for ABSA, designed specifically for the legal domain.

2   Related Work

2.1   Legal Information Extraction

When referring to the past literature, it shows that within the legal domain, there exist very few studies related to sentiment analysis. The study by Gamage et al. [11] introduced a sentence-level sentiment annotator using transfer learning for the legal domain. In this proposed approach, the sentiment of a given sentence is classified into one of the two classes; *negative* and *non-negative*. But it does not take into consideration any party mentioned in the sentence when detecting the sentiment of the sentence. Moreover, the study by Ratnayaka et al. [12] have proposed methodologies to identify relationships among sentences in the legal documents. They have demonstrated that sentiment analysis can be used to identify sentences that provide different opinions on the same topic (contradictory opinions) within a legal opinion text. The study of Rajapaksha et al. [1] developed a rule-based approach which is built around a phrase level sentiment annotator [11] and manually created rules for sentiment detection of legal sentences with respect to legal parties. This can be identified as the first attempt to use ABSA in the legal domain.

2.2   Existing Aspect-Based Sentiment Analysis Models

Lexicon-based approaches, machine learning-based approaches, and hybrids of machine learning and lexicon-based approaches are the main types of methods to perform Aspect-Based Sentiment Analysis (ABSA) [13]. Recently, deep neural network approaches have shown better results on aspect-based sentiment classification tasks due to its ability to generate the dense vectors of sentences without handcrafted features. Tang et al. [14] proposed TD-LSTM which uses two Long Short-term Memory (LSTM) networks in order to extract important information from the left and right sides of the target. Although it improves the LSTM architecture, it is often impossible to distinguish between various sentiment polarities at a fine-grained level. A number of subsequent studies employed attention mechanisms to learn the key parts of sentences that should be given special focus in order to enhance the sentence representation. In that perspective, Wang et al. [15] proposed AT-LSTM and ATAE-LSTM, incorporating attention mechanisms to model relationships between aspects and context. In order to better understand target information, Cheng et al. [16] introduced the HiErarchical ATtention (HEAT) network with sentiment attention and aspect attention. Chen et al. [17] designed the RAM model by adopting multiple attentions to extract important information from memory. IAN, which was proposed by Ma et al. [18], utilizes a bidirectional attention mechanism and learns the attention for the contexts and the targets separately via interactive learning.

Although attention-based models have shown promising results over many ABSA tasks, they are not adequate to catch syntactic dependencies between aspect and the context words within the sentence. The important feature of Graph Convolutional Network (GCN) is that, it has the ability to draw syntactically related terms to the target aspect and then manipulate, multi-word associations and syntactical knowledge in long-range, utilizing GCN layers [19]. Zhao et al. [20] proposed ASGCN, adopting GCN for ABSA. Zhao et al. concluded that GCN improves overall efficiency by exploiting both syntactic knowledge and long-range word dependency. Zhao et al. [20] introduced the SDGCN model with the aim of modeling sentiment dependencies within a sentence among different target aspects.

3   Methodology

The ultimate goal of our proposed approach is to detect the sentiment polarity of sentences in legal texts with respect to each legal party mentioned in the sentence. Legal texts usually consist of multiple legal parties having different interdependencies among them. Hence, the sentiment classifier should be developed in order to classify sentiment polarity values of multiple legal parties. In our approach, *positive*, *negative* and *neutral* are considered as sentiment polarities. The overall architecture of our proposed model is illustrated in Fig. 1. To perform the aspect sentiment classification, our model architecture is designed with the following layers; word embedding layer, Recurrent Neural Network (RNN) layer, position aware attention mechanism, GCN layer, and sentiment classification layer.

3.1   Word Embedding Layer

Word embedding layer maps each word to a high dimensional vector space. It is widely known that a strong word embedding is extremely important for composing a strong and efficient text representation for use at later stages. We used

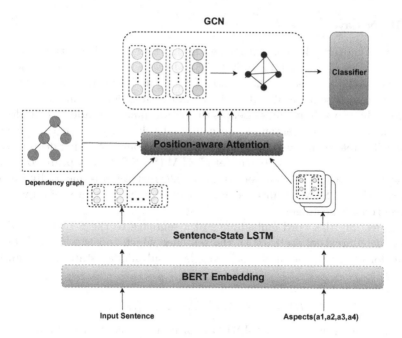

Fig. 1. Overall architecture

a pre-trained BERT (Bidirectional Encoder Representations from Transformers) model[1] [9] post-trained using the criminal court case legal opinion texts available in the SigmaLaw dataset to obtain the word embedding.

An input sentence (S), of N number of words is represented as $S = \{w_{s_1}, w_{s_2}, ..w_{s_N}\}$. A given sentence S, would include a set of aspect terms (S_a) of cardinality K where the ith aspect term is represented by A_i, $S_a = \{A_1, A_2, ..A_K\}$. Further, the ith aspect term, A_i, contains M_i number of words such that $M_i \in [1; N)$ represented by $A_i = \{w_{A_{i1}}, w_{A_{i2}}, ..w_{A_{iM_i}}\}$. By the virtue of aspects not overlapping each other, $\sum_{i=1}^{K} M_i \le N$ holds.

We use the above BERT model to get word embedding of the input sentence and all the aspect terms in the sentence. First, we construct the input as "[CLS] + input + [SEP]" and feed it to the BERT tokenizer. The special token [CLS] is added at the beginning of our text and the special token [SEP] is added to mark the end of a sentence. The BERT tokenizer then outputs tokens which correspond to BERT vocabulary. After mapping the token strings to their vocabulary indices, indexed tokens are next fed into the BERT model. Each word of the context and aspects are represented by a 768 dimensional embedding vector. The BERT model is used only for the word embedding purpose.

---

[1] Legal-BERT model - https://osf.io/s8dj6/.

3.2   RNN Layer

In order to capture the contextual details for every word, on the top of the embedding layer we use Sentence-State LSTM (S-LSTM) [21]. Most of the existing model architectures use LSTM, Bi-LSTM, and Bi-GRU as the encoder. LSTM processes sequential data while maintaining long-term dependencies. However, when encoding long sentences the performance degrades. In our domain (legal documents), the sentences are comparatively longer than that of other domains. Therefore, aiming to address these limitations of existing deep-learning approaches, we leverage a sentence state LSTM (S-LSTM) to capture contextual information due to its proven performance [22]. Instead of sequentially processing words, the S-LSTM simultaneously models the hidden states of all words in each recurrent time stage.

After feeding the word embeddings of a sentence to the S-LSTM model, it returns the contextual state H^t of the sentence which consists of a sub hidden state h_i^t for each word w_i and a sentence-level sub hidden state s^t as shown in the Eq. 1.

$$H^t = <h_0^t, h_1^t, h_1^t, h_2^t, ..., h_{n-1}^t, s^t> \tag{1}$$

In our architecture, we use S-LSTM in order to get contextual hidden output of the sentence and contextual hidden outputs of aspects.

3.3   Position Aware Attention Mechanism

In a sentence, the sentiment polarity is heavily associated with the aspect-words and opinion terms of the sentence. Hence, the method that we adopt to rely on these aspect-terms is quite important in the process of sentiment analysis. The main weakness of RNN models is the inability to understand the most critical parts of the sentence for sentiment analysis. As a solution to this, we employ an attention mechanism which can grab the most important parts in a sentence. However, every word in a sentence is not equally important for determining sentiment polarity. Words which are closer to the target or having modifier relation to the target word should be given higher weights [23]. To ease this problem, we used an attention mechanism incorporating position information of each word in the sentence based on the current aspect term. We use position information here to incorporate the claim by He et al. [23] that the aspect sentiment polarity is mainly influenced by the context words that are situated very close to the target aspect.

Lee	was	found	guilty	because	the	attorney	had	provided	constitutionally	ineffective	assistance
[0,6]	[1,5]	[2,4]	[3,3]	[4,2]	[5,1]	[6,0]	[7,1]	[8,2]	[9,3]	[10,4]	[11,5]

Fig. 2. Basic relative distances to the aspects *Lee* and *attorney*

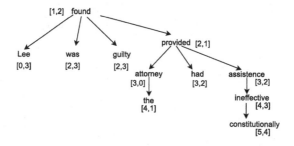

Fig. 3. Distances along the dependency tree to the aspects *Lee* and *attorney*

Here we used the bidirectional attention mechanism introduced by Zhao et al. [20] with two attention modules as context-to-aspect attention module and aspect-to-context attention module. We followed the same methodology for the calculation of attention weights. However, for position-aware representation, we used the distances along the dependency tree instead of the basic relative distances used in their approach. In our approach, as the distance, the length of the path from the specific word to the aspect in the dependency tree is used to encode the syntactic structure of the legal text. Figure 2 illustrates the example sentence with basic relative distances to aspects and Fig. 3 shows distances along the dependency tree. When considering the two types of distances, we can see that the vital opinion words such as *guilty* and *ineffective* are closer to the relevant aspects in the Fig. 3 than in Fig. 2. The sentences in the court cases are comparatively much longer than other domains. Hence, opinion words are sometimes not close to the target. Therefore it is not suitable to get the basic relative distance between each word and the current aspect for position representation. The final output of the attention mechanism is the Aspect-specific representation between the target aspect (party) and context words given as $X = [x_1, x_2, .., x_K]$ where K denotes the number of aspects.

3.4 Graph Convolution Network

In order to capture the inter-dependencies between multiple aspects/parties in a sentence, we used GCN in our study following the observations reported by Zhao et al. [20] in their study. GCNs can be identified as a basic and efficient convolution neural network running on graphs which has the ability to collect interdependent knowledge from rich relational data. As the first stage of implementing the GCN layer, it is needed to construct a graph which we name as *Sentiment Graph*, where a node is a party (aspect) mentioned in the sentence and an edge is the inter-dependency relation between two nodes. If there is a dependency relationship between two parties in the sentence, we denote that by marking an edge between the corresponding two nodes. As shown in the Fig. 4, when creating the *Sentiment Graph*, we initially defined a fully-connected graph assuming that each aspect has a relationship with every other aspect of the sentence.

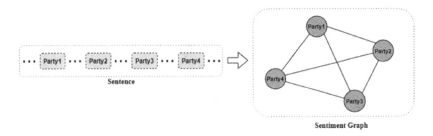

Fig. 4. Sentiment graph

GCN generates a new vector representation for each node by discovering all relevant information about the neighboring nodes of the selected node. Moreover, when generating the new vector representation, it is needed to put attention on the information of the node itself. For that, we assume that each node has a self-loop. The new representation for a node can be defined as shown in the Eq. 2 where given v node, $N(v)$ defines the all neighbors of v, $W_{cross} \in \mathbb{R}^{d_m \times d_n}$, $W_{self} \in \mathbb{R}^{d_m \times d_n}$, $b_{cross} \in \mathbb{R}^{d_m \times 1}$, $b_{self} \in \mathbb{R}^{d_m \times 1}$, x_u is the uth aspect-specific representation taken from the output of attention layer.

$$x_v^1 = ReLU(\sum_{u \in N(v)} W_{cross} x_u + b_{cross}) + ReLU(W_{self} x_v + b_{self}) \quad (2)$$

We can expand the neighborhood for each node by stacking multiple GCN layers. As the input, each GCN layer gets the output form the previous layer and returns the new node representation. From the experiments, we identified that using more than two GCN layers reduces the accuracy. Therefore in our case, we use two GCN layers (see Eq. 3).

$$x_v^2 = ReLU(\sum_{u \in N(v)} W_{cross}^1 x_u^1 + b_{cross}^1) + ReLU(W_{self}^1 x_v^1 + b_{self}^1) \quad (3)$$

3.5   Sentiment Classification

Once the output of the GCN layer(x) is obtained, it is fed to a *Softmax* layer to obtain a probability distribution over polarity decision space of C classes (where W and B are the learned weights and bias):

$$z = Softmax(Wx + b) \quad (4)$$

3.6   Model Training

The model is trained by the gradient descent algorithm with cross entropy loss and L2 regularization.

$$Loss = -\sum_{c=1}^{C} y \log \hat{y} + \lambda ||\theta||^2 \quad (5)$$

C denotes the number of classes (3 in our case), y is the true label, $\hat{y}$ is the predicted label, θ denotes all the parameters that need to regularized, and λ is the coefficient of L2-regularization.

4 Experiments

4.1 Dataset

Experiments and evaluations were carried out on the SigmaLaw-ABSA [2] data set which consists of 2000 human-annotated legal sentences taken from previous court cases. The said court cases were originally fetched from the *SigmaLaw - Large Legal Text Corpus and Word Embedding data set* [9]. To the best of our knowledge SigmaLaw-ABSA is the only existing dataset for the Aspect-Based Sentiment Analysis in the legal domain. The dataset has been annotated by legal experts and it contains entities of different parties, their polarities, aspect category (*Petitioner* or *defendant*), and the category polarities. The data set has been designed to perform various research tasks in the legal domain including aspect extraction, polarity detection, aspect category identification, aspect category polarity detection. It is the only existing dataset for the aspect based sentiment analysis in the legal domain.

Legal sentence, members of legal parties in sentence, their polarities are the fields used for this study from the SigmaLaw-ABSA dataset. We feed the legal sentence as the input sentence and the legal party members as aspects into the BERT model. Polarity of the legal party members are used to evaluate the model.

4.2 Parameter Setting

For experiments, word embeddings for both context and targets are initialized by using 300-dimensional pretrained Glove word vectors and 760-dimensional Bert embeddings. Dimension of hidden state vectors of RNN is set to 300 and weights of the model are randomly initialized with uniform distribution. 600 is set as the output dimension of the GCN layer. We used Spacy[2] to calculate the distance through the dependency tree for attention mechanism and hidden states of the attention layer are set to 300. During the training, we set the batch size to 16, dropout to 0.1, coefficient of L2 is 10^{-5}, and used Adam optimizer with a learning rate of 0.001.

4.3 Word Embedding Models Comparison

In our experiments, we tried two word embedding methods: 300-dimensional GloVe [24] and BERT [25]. In BERT, two different BERT models were tried for the embedding layer: the base uncased English model and the pre-trained BERT model specially fine-tuned for the legal corpus. Table 1 shows the comparison of the results of above models. The legal-BERT model outperformed the other

---

[2] Spacy Toolkit - https://spacy.io/.

models. The BERT models use 12 layers of transformer encoders, and each output per token from each layer of these and initial input embedding can be used as a word embedding. We tried various vector combinations of hidden layers to get state-of-art results. Table 2 illustrates the result of various word-embedding strategies using the BERT model for legal domain.

4.4 RNN Models Comparison

LSTM, Bi-LSTM, Bi-GRU and Sentence-State LSTM (S-LSTM) models were tested as the encoder for our approach as shown in Table 3. As S-LSTM offers richer contextual information exchange with more parallelism compared to BiLSTMs, it outperformed the other models. This is because it has strong representation power compared to the other RNNs [21]. This feature became relevant given that the sentences of Legal documents are often long and have a complex semantic structure.

Table 1. Word embedding models comparison

Model	Accuracy	F1 score
GloVe [24]	0.6615	0.5798
BERT (base) [25]	0.6997	0.6193
BERT (legal domain) [26]	**0.7068**	**0.6281**

Table 2. Different word embedding strategies comparison of BERT model

Strategy	Accuracy	F1 score
Initial embedding	0.6670	0.5705
Last hidden layer	0.6921	0.6193
Concat last 4 layers	0.6987	0.6105
Sum all layers	0.6954	0.6098
Sum last 4 layers	**0.7086**	**0.6281**

Table 3. RNN models comparison

Model	Accuracy	F1 score
LSTM	0.6721	0.5964
Bi-LSTM	0.6987	0.6204
Bi-GRU	0.6854	0.6045
S-LSTM	**0.7086**	**0.6281**

Table 4. Performances of different models on SigmaLaw-ABSA

Model	Accuracy	F1 score
TD-LSTM [14]	0.6512	0.5647
TC-LSTM [14]	0.6182	0.5438
AE-LSTM [15]	0.6228	0.5588
AT-LSTM [15]	0.6272	0.5592
ATAE-LSTM [15]	0.6542	0.5802
IAN [18]	0.6332	0.5650
PBAN [27]	0.6332	0.5650
Cabasc [28]	0.6123	0.5643
RAM [17]	0.6639	0.6022
MemNet [29]	0.5389	0.4361
SDGCN [20]	0.6781	0.6121
ASDGCN [19]	0.6699	0.6001
SigmaLaw-PBSA	**0.7086**	**0.6281**

4.5   Overall Performance

The experimental results generated on different existing models using the SigmaLaw-ABSA dataset [2] are shown in Table 4. By analysing the obtained results, we can conclude that, in the legal domain our proposed model outperforms every other existing model. We claim that it is mainly due to the complexity and the length of the sentences in the legal domain as it makes it difficult to those models to understand the sentence to an adequate degree.

4.6   Ablation Study

In order to study the efficiency of the various modules in our proposed approach, we conducted an ablation study on the SigmaLaw-ABSA dataset as shown in Table 5. It is observed that removing both attention mechanisms and GCN drops the F1 score by 0.0617. Introducing the attention mechanism (with dependency tree distance) to the baseline increases the F1 score by 0.0439. This verifies the significance of the position-aware attention mechanism. The results gained from using the dependency tree distance to calculate position weights shows higher performance than the calculating position weights through basic relative distances. This verifies the impact of the syntactic information introduced by the dependency trees.

Further, we can see that the model shows higher results with the introduction of the GCN layer. Therefore we can conclude that the GCN layer contributes significantly to increase the results since it helps to capture the inter-dependencies among multiple aspects and relationships between words at long ranges.

Table 5. Results of ablation study

Setting	Accuracy	F1 score
Base	0.6287	0.5664
Base + Attention (relative distance)	0.6689	0.5989
Base + Attention (dependency tree distance)	0.6793	0.6103
Base + Attention (dependency tree distance) + 1_layer GCN	0.6938	0.6215
Base + Attention (dependency tree distance) + 2_layer GCN	0.7086	0.6281

5   Conclusion

We analysed the existing deep-learning based model architectures and pointed out suitable model components for tackling the challenges of the legal domain. Accordingly, we introduced a deep learning-based approach to perform party-based sentiment analysis in legal opinion texts. First, the model utilizes a pretrained BERT model (further fine tuned on a legal corpus) for a strong word embedding. Then, the model employs a position-aware attention mechanism, to capture the critical parts of the sentence relevant to aspects, with incorporating position information, using the dependency tree. Because multiple legal party members are involved in a single sentence, a GCN is employed over the attention mechanism to model the inter-dependencies between members. Experiments were carried out using the SigmaLaw-ABSA dataset and the experimental results demonstrate that our proposed approach outperforms all other existing state-of-art ABSA models.

References

1. Rajapaksha, I., Mudalige, C.R., Karunarathna, D., de Silva, N., Rathnayaka, G., Perera, A.S.: Rule-based approach for party-based sentiment analysis in legal opinion texts. In: 2020 20th International Conference on Advances in ICT for Emerging Regions (ICTer)
2. Mudalige, C.R., et al.: SigmaLaw-ABSA: dataset for aspect-based sentiment analysis in legal opinion texts. In: 2020 IEEE 15th International Conference on Industrial and Information Systems (ICIIS). IEEE (2020)
3. Samarawickrama, C., de Almeida, M., de Silva, N., Ratnayaka, G., Perera, A.S.: Party identification of legal documents using co-reference resolution and named entity recognition. In: 2020 IEEE 15th International Conference on Industrial and Information Systems (2020)
4. de Almeida, M., Samarawickrama, C., de Silva, N., Ratnayaka, G., Perera, A.S.: Legal party extraction from legal opinion text with sequence to sequence learning. In: 2020 20th International Conference on Advances in ICT for Emerging Regions (ICTer) (2020)
5. Moralwar, S., Deshmukh, S.: Different approaches of sentiment analysis. Int. J. Comput. Sci. Eng. **3**(3), 160–165 (2015)
6. Schouten, K., Frasincar, F.: Survey on aspect-level sentiment analysis. IEEE Trans. Knowl. Data Eng. **28**(3), 813–830 (2015)
7. Bhoi, A., Joshi, S.: Various approaches to aspect-based sentiment analysis. ArXiv, abs/1805.01984 (2018)
8. Pontiki, M., Galanis, D., Papageorgiou, H., et al.: Semeval-2016 task 5: aspect based sentiment analysis, pp. 19–30, January 2016
9. Sugathadasa, K., et al.: Synergistic union of word2vec and lexicon for domain specific semantic similarity. In: IEEE International Conference on Industrial and Information Systems (ICIIS), pp. 1–6 (2017)
10. Lee v. United States, in US, vol. 432, no. 76-5187, p. 23, Supreme Court (1977)
11. Gamage, V., Warushavithana, M., de Silva, N., Perera, A.S., Ratnayaka, G., Rupasinghe, T.: Fast Approach to build an automatic sentiment annotator for legal domain using transfer learning. In: Proceedings of the 9th Workshop on Computational Approaches to Subjectivity, Sentiment and Social Media Analysis (2018)
12. Ratnayaka, G., Rupasinghe, T., de Silva, N., Gamage, V., Warushavithana, M., Perera, A.S.: Shift-of-perspective identification within legal cases. In: Proceedings of the 3rd Workshop on Automated Detection, Extraction and Analysis of Semantic Information in Legal Texts (2019)
13. Piryani, R., Gupta, V., Singh, V.K., Ghose, U.: A linguistic rule-based approach for aspect-level sentiment analysis of movie reviews. In: Bhatia, S.K., Mishra, K.K., Tiwari, S., Singh, V.K. (eds.) Advances in Computer and Computational Sciences. AISC, vol. 553, pp. 201–209. Springer, Singapore (2017). https://doi.org/10.1007/978-981-10-3770-2_19

14. Tang, D., Qin, B., Feng, X., Liu, T.: Effective LSTMs for target-dependent senti-ment classification. arXiv preprint arXiv:1512.01100 (2015)
15. Wang, Y., Huang, M., Zhu, X., Zhao, L.: Attention-based LSTM for aspect-level sentiment classification. In: Proceedings of the 2016 Conference on Empirical Meth-ods in Natural Language Processing (2016)
16. Cheng, J., Zhao, S., Zhang, J., King, I., Zhang, X., Wang, H.: Aspect-level senti-ment classification with heat (hierarchical attention) network. In: Proceedings of the 2017 ACM on Conference on Information and Knowledge Management, pp. 97–106 (2017)
17. Chen, P., Sun, Z., Bing, L., Yang, W.: Recurrent attention network on memory for aspect sentiment analysis. In: Proceedings of the 2017 Conference on EMNLP (2017)
18. Ma, D., Li, S., Zhang, X., Wang, H.: Interactive attention networks for aspect-level sentiment classification. arXiv preprint arXiv:1709.00893 (2017)
19. Zhang, C., Li, Q., Song, D.: Aspect-based sentiment classification with aspect-specific graph convolutional networks. arXiv preprint arXiv:1909.03477 (2019)
20. Zhao, P., Hou, L., Wu, O.: Modeling sentiment dependencies with graph convolu-tional networks for aspect-level sentiment classification. Knowl. Based Syst. **193**, 105443 (2020)
21. Zhang, Y., Liu, Q., Song, L.: Sentence-state LSTM for text representation. In: ACL (2018)
22. Demotte, P., Senevirathne, L., Karunanayake, B., Munasinghe, U., Ranathunga, S.: Sentiment analysis of Sinhala news comments using sentence-state LSTM net-works. In: Moratuwa Engineering Research Conference (MERCon) 2020, pp. 283–288 (2020)
23. He, R., Lee, W.S., Ng, H.T., Dahlmeier, D.: Effective attention modeling for aspect-level sentiment classification. In: Proceedings of the 27th International Conference on Computational Linguistics, pp. 1121–1131 (2018)
24. Pennington, J., Socher, R., Manning, C.: GloVe: global vectors for word repre-sentation. In: Association for Computational Linguistics, pp. 1532–1543, October 2014
25. Devlin, J., Chang, M.-W., Lee, K., Toutanova, K.: BERT: pre-training of deep bidirectional transformers for language understanding. In: ACL, pp. 4171–4186, June 2019
26. Ratnayaka, G., de Silva, N., Perera, A.S., Pathirana, R.: Effective approach to develop a sentiment annotator for legal domain in a low resource setting. arXiv preprint arXiv:2011.00318 (2020)
27. Gu, S., Zhang, L., Hou, Y., Song, Y.: A position-aware bidirectional attention net-work for aspect-level sentiment analysis. In: Proceedings of the 27th International Conference on Computational Linguistics. ACL, August 2018
28. Liu, Q., Zhang, H., Zeng, Y., Huang, Z., Wu, Z.: Content attention model for aspect based sentiment analysis. In: Proceedings of the 2018 World Wide Web Conference (2018)
29. Tang, D., Qin, B., Liu, T.: Aspect level sentiment classification with deep memory network. arXiv preprint arXiv:1605.08900 (2016)

BERT-Based Sentiment Analysis: A Software Engineering Perspective

Himanshu Batra, Narinder Singh Punn$^{(\boxtimes)}$, Sanjay Kumar Sonbhadra, and Sonali Agarwal

Indian Institute of Information Technology Allahabad, Prayagraj, India
{mit2019119,pse2017002,rsi2017502,sonali}@iiit.ac.in

Abstract. Sentiment analysis can provide a suitable lead for the tools used in software engineering along with the API recommendation systems and relevant libraries to be used. In this context, the existing tools like SentiCR, SentiStrength-SE, etc. exhibited low f1-scores that completely defeats the purpose of deployment of such strategies, thereby there is enough scope for performance improvement. Recent advancements show that transformer based pre-trained models (e.g., BERT, RoBERTa, ALBERT, etc.) have displayed better results in the text classification task. Following this context, the present research explores different BERT-based models to analyze the sentences in GitHub comments, Jira comments, and Stack Overflow posts. The paper presents three different strategies to analyse BERT based model for sentiment analysis, where in the first strategy the BERT based pre-trained models are fine-tuned; in the second strategy an ensemble model is developed from BERT variants, and in the third strategy a compressed model (Distil BERT) is used. The experimental results show that the BERT based ensemble approach and the compressed BERT model attain improvements by 6–12% over prevailing tools for the F1 measure on all three datasets.

Keywords: Sentiment analysis · Transformer · Deep learning · Software engineering · BERT

1 Introduction

Sentiment analysis is a technique to classify sentences to positive, negative or neutral based upon the point of view expressed through that text. Studies reveal that there are several applications of sentiment analysis for software engineering (SE) like analysing developers' opinions through Stack Overflow posts, the scope of improvement in a library from the comments on GitHub issues section [26], tracking negative comments to inspect bugs in API [33].

The existing research shows that off-the-shelf sentiment analysis tools [18,20] when tested on SE domain-specific texts tend to disagree with each other. Lin et al. [15] reported results of five tools: SentiStrength, NLTK, Stanford

---

All authors have contributed equally.

C. Strauss et al. (Eds.): DEXA 2021, LNCS 12923, pp. 138–148, 2021.
https://doi.org/10.1007/978-3-030-86472-9_13

CoreNLP, SentiStrength-SE, Stanford CoreNLP SO on app reviews, Stack Overflow and JIRA issues dataset, with the f1-scores ranging from 0.15 to 0.5. SEntiMoji [5], based upon emoji labelled posts, reported an f1-score of 0.47 for the positive sentences and 0.64 for the negative sentences. But still, the utility of the prevailing tools remains quite low for software engineering applications.

The present paper focuses on the scope of improvement of the existing work for the software engineering domain through sentiment analysis. Specifically leveraging the current state-of-the-art performance of BERT (Bidirectional Encoder Representations from Transformers), which has been widely used for major NLP tasks like sentiment analysis, question answering, translation, etc.

1.1 BERT

BERT [6] was introduced in 2018 by Devlin et al., for language modelling, based on bidirectional training of transformer (attention model) [31]. It uses the attention mechanism that gets the actual context of the text and has two important parts: the encoder (to get the input text) and the decoder (that produces the output or the prediction of the task), through which the pre-training and fine-tuning is carried out as shown in Fig. 1. There are two major techniques to train the BERT model:

- *Masked language modelling*: This technique is used to mask 15% of the words randomly in a sentence and the model tries to predict the masked word based upon the context of the other (non-masked) words in the sentence.
- *Next sentence prediction*: In this, the model receives pairs of sentences as input and learns to predict if the second sentence in the pair is the subsequent sentence in the original document. During training, 50% of the inputs are a pair in which the second sentence is the subsequent sentence in the original document, while in the other 50% a random sentence from the corpus is chosen as the second sentence.

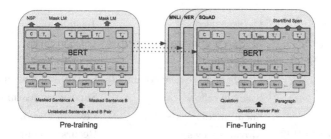

Fig. 1. Pre-training and fine-tuning procedures for BERT [6].

Insufficiency of task-specific labeled data can be handled by BERT through pre-training. This pre-training is done on a large corpus of text (∼2500 million words), which forms a common architecture for a generic task (sentiment analysis, question answering, translation, etc.). This model can then be fine-tuned for our downstream task, to analyse sentiments. On sentiment treebank dataset [28] BERT attained a GLUE score of 94.9, considered as a benchmark dataset, it has fine-grained sentiment labels for 215,154 phrases in the parse trees of 11,855 sentences. Following the state-of-the-art potential of BERT model, the present research proposes to exploit the features of BERT variants with the ensemble approach that uses three variants of BERT i.e., the base BERT model, RoBERTa and ALBERT. Furthermore, a compressed BERT model (Distil BERT) is used to establish more robust results for sentiment analysis from the perspective of software engineering.

With the objective and contributions highlighted above, the rest of the paper is organized into several sections. The literature review in Sect. 2 presents the recent developments in the deep learning classification task approaches followed by the methods in Sect. 3 to cover the background knowledge and proposed methodology. The later Sect. 4 covers the exhaustive experimental trials over multiple datasets followed by the results in Sect. 5. Finally, the concluding remarks are presented with future research directions in Sect. 6.

2   Related Work

With the advent of advancements in the deep learning algorithms, many state-of-the-art frameworks are developed across various domains such as healthcare, natural language processing, target feature learning, etc. [24,25,27,30]. In consideration of sentiment analysis for software engineering domain, [8,9,11], mainly focus on deriving developers' opinions/emotions, along with the context. It is observed that the existing tools make dataset driven predictions, where each prediction conflicts with one another [21]. Table 1 shows the techniques used for the tools along with the datasets over which the training was performed.

Table 1. Prevailing tools.

Tool	Dataset used	Technique
SentiStrength [29]	MySpace informal short texts	Lexicon based
SentiStrength-SE [12]	Jira comments	Lexicon based
SentiCR [1]	Oracle code review comments	Supervised learning
Senti4SD [3]	Stack overflow	Supervised learning
NLTK [18]	Social media texts	Lexicon based
Stanford CoreNLP [20]	Movie reviews	Recursive neural tensor network

The results of NLTK [18] and Stanford CoreNLP [20] were not very promising as they were not trained on SE-specific datasets and lead to diverging conclusions. SentiCR [1] leveraged boosting algorithms and was based on oracle code review comments but only had two classes i.e., negative and non-negative, unlike others classifying on three classes (positive, negative, and neutral). And it provided the highest f1-score among all the datasets. Senti4SD [3] being trained on the Stack Overflow dataset with a supervised learning approach had the highest f1-score. Studies have reflected that only to a certain extent these SE-specific tools for sentiment analysis agree with each other. To examine the performance of these tools, Novielli et al. [21] performed a benchmark study on standard datasets (Jira and Stack Overflow comments) and to identify potential weak points for improvement and recorded an f1-score of 0.85 on SentiCR and on SentiStrength-SE [12] it was 0.83 for the Stack Overflow dataset.

Along with the ones mentioned in the Table 1, there are two more related tools, Emotxt [4], based on classifying emotions (anger, sadness, love, joy, fear) and is trained on the Jira and Stack Overflow datasets; whereas, DEVA [10] is a dictionary-based approach to detect valence and arousal in the text that can capture individual emotional states (depression, relaxation, excitement, stress, and neutrality). However, in contrast, BERT with task-specific fine-tuning can provide much better results.

Several other convenient techniques help to derive the sentiment from a text. Data augmentation [14] is one of such techniques that have been used for various downstream tasks and some of the data augmentation techniques used widely are - Lexical substitution of words in the text, which is basically thesaurus based substitution (replacing a word with its synonym) and word-embedding substitution, this technique primarily takes the pre-trained word embeddings to consideration for the replacement of any word in a text through the nearest neighbor in the embedding space (Word2Vec [7], GloVe [23], FasText [2], Sent2Vec [22]). Alternatively, masking of a particular word in a sentence is also a great way to predict a word that has the relative context (BERT). Xie et al. [32] used a method called back translation for augmentation of unlabelled text on the IMDB dataset for a semi-supervised model which had only 20 labelled examples. Back translation follows translating an English sentence to some other language (say Spanish) and then translating it back to get the new sentence and that sentence is taken as an augmented sentence for training. However, the utility of the prevailing tools remains quite low for software engineering applications. Following this, the present research work exploits the state-of-the-art potential of the BERT variants for sentiment analysis in the software engineering domain.

3   Proposed Work

As per the above discussion, it is evident that software engineering sentiment analysis has always been a challenging task. In this context, the present research proposes a robust solution, where initially, the data augmentation is chosen as a pre-processing strategy to aid in training or fine-tuning the BERT variants for efficient results. In the fine-tuned model an additional untrained layer on top of the BERT model is provided for task-specific learning whereas an ensemble approach is proposed that combines the fine-tuned models with a weighted voting scheme. The sentiment analysis is further extended using a compressed model (Distil BERT) to establish robust results.

3.1 Data Augmentation

Data augmentation essentially helps to increase the diversity of the dataset that is available for training the model, without actually getting new data or rather scraping data from relevant websites like Stack Overflow, Jira, GitHub, etc. The available techniques used for text data augmentation are as follows:

Lexical Based Substitution. This technique is based on taking into account the synonym of a random word from a given sentence and substituting it using a thesaurus. Similarly, one basic approach is to provide substitution based on word-embeddings, i.e. to replace a word with its nearest neighboring word in the embedding space.

Back Translation. It is primarily the translation of a sentence to a different language and then translating it back to the original language, where the sentence translated back is likely to use similar words from the language and the context remains unchanged.

3.2 Fine-Tuning BERT Model

BERT can be used to get some high-quality language features and can be fine-tuned for the classification task performed on the software engineering datasets to achieve promising results. Fine-tuning provides a few advantages like much faster development, lesser epochs to train on, and limited data requirements. With the pre-trained model, an additional untrained layer of neurons is appended, to fine-tune the model with 2–4 training epochs as suggested in the original BERT paper. Following this three BERT variants are fine-tuned using multiple datasets for better comparative analysis of the proposed strategies for sentiment analysis in software engineering.

3.3 Ensemble BERT Models

Based on the fact that combining several models could produce a stronger model hence an ensemble technique is utilized to aggregate the performance of the BERT variants. The scope of prediction is much diverse in the case of the ensemble because there might be certain diversified conclusions on each of the models (that are part of the ensemble), as the core functioning is different for each model. Each model has been pre-trained on a different language modelling task like next sentence prediction used in BERT, sentence order prediction used in ALBERT and dynamic masking used in RoBERTa. Hence it is evident that a stronger model is obtained and the final prediction can be fetched through a voting scheme. For the ensemble technique, BERT base model [6], RoBERTa [17], and ALBERT [13] models are used, as shown in Fig. 2. This article used a weighted voting scheme, where the confidence score (aggregated) is considered, i.e. the last Softmax layer output, for each model in the ensemble to get the final weighted prediction.

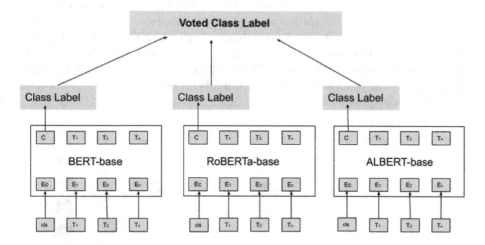

Fig. 2. BERT model variations used for the ensemble technique

3.4   Compressed Model

Apart from quantization and model pruning, a distilled model is really helpful to boost up the model's performance rather than to train the model from scratch which requires a large amount of data. The distillation framework mainly contains two models: a larger one and a smaller one. The smaller one mimics the working of the larger one by learning from the probabilities generated by the larger one before the final activation function and produces comparable results. These probabilities are helpful for the generalization capabilities of the compressed model as they are still higher than absolute zero values. Hence, this compressed model is different from the base BERT wherein the model suppresses the lower probability for a particular class and enhances the value for the higher probability class, in its final distribution. The Distil BERT model retains 97% of the language understanding capabilities and the size is reduced by 40% in contrast to BERT base model.

4   Experiments

The experiments are aligned to have an overview of the performance on different datasets [15,16] and to check: 1) how the model performs after fine-tuning, 2)

how well the BERT models perform when used as an ensemble, and 3) to test the performance of the Distil BERT model. The datasets used for the analysis are GitHub commit comments, Jira issue comments, and Stack Overflow posts. Furthermore, a publicly available synthetic dataset was used to check the performance of the model. For the training part, the number of unique sentences present is 150,000 and the test dataset has 30,000 sentences. It's an added advantage to train and test the model on such a huge dataset, as compared to the Stack Overflow, Jira and GitHub datasets with limited sentences. The complete training part is done in a cloud environment having high-end NVidia GPUs. Figure 3 shows the correlation among the three BERT models used for ensemble strategy. The correlation scores for RoBERTa-ALBERT and ALBERT-BERT is high, which signifies the models have high probability to generate similar predictions in the ensemble approach.

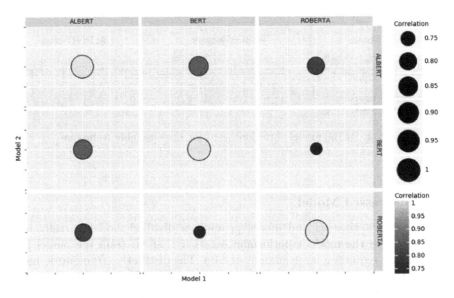

Fig. 3. Correlation plot for the three BERT variants.

Table 2. Train, validation and test split for the three datasets.

Dataset	Train data	Validation data	Test data
Github	4480	498	2137
Stack overflow	2787	310	1326
Jira	3700	410	1759

4.1 Training and Testing

The proposed models are trained and evaluated on the training and test set derived from three datasets as shown in Table 2. The training data is further split into train and validation set for better learning of the model. The training phase is assisted with the earlystopping technique (halt the training process as soon as the performance stops improving) to avoid the overfitting problem. The trained models are then evaluated on the test set using the standard evaluation metrics, precision, recall and f1-score, for multi-class classification (positive, negative, and neutral). These metrics can be computed using the following equations:

$$Precision = \frac{TP}{TP + FP} \tag{1}$$

$$Recall = \frac{TP}{TP + FN} \tag{2}$$

$$F1-\text{Score} = \frac{2 * Recall * Precision}{Recall + Precision} \tag{3}$$

5 Results and Discussion

Following from the first proposed strategy involving the fine-tuning of the BERT models, the classification performance concerning sentiment analysis is presented in Table 3. It is observed that from fine-tuning, the lowest f1-score of 0.67 was recorded for the ALBERT model on Stack Overflow dataset. While the RoBERTa model performed better on GitHub and Jira datasets with the f1-scores as 0.91 and 0.83 respectively. The ensemble approach on the other hand was a motivation to obtain a stronger model with slightly varied conclusions from RoBERTa, ALBERT, and BERT-base models, where the f1-score for the GitHub dataset is improved to 0.92, as shown in Table 4.

The trend with deep learning models is that the training time is directly proportional to the number of parameters. Therefore it is evident that the compressed model can improve the inference time of the model if deployed as a tool. For the Stack Overflow and the Jira datasets, the f1-score improved to 0.88 and 0.84 respectively, with the Distil BERT model, as shown in Table 5. Additionally, the f1-score recorded for the synthetic data when tested on the ensemble model was 0.85 which proved to be a major improvement.

Table 3. Fine-tuning results for the three available datasets.

Dataset	Model	Neutral			Negative			Positive		
		P	R	F1	P	R	F1	P	R	F1
GitHub	BERT	**0.92**	0.91	**0.91**	0.89	**0.91**	0.90	**0.93**	0.93	**0.93**
	ALBERT	0.85	**0.94**	0.89	0.92	0.80	0.85	0.90	0.93	0.91
	RoBERTa	**0.94**	0.91	**0.92**	**0.90**	0.91	**0.90**	0.91	**0.94**	**0.93**
Stack overflow	BERT	0.85	0.86	0.85	**0.85**	**0.87**	**0.86**	**0.94**	**0.92**	**0.93**
	ALBERT	0.92	**0.94**	0.93	0.62	0.62	0.62	0.71	0.34	0.48
	RoBERTa	**0.96**	0.92	**0.94**	0.78	0.82	0.80	0.57	0.76	0.65
Jira	BERT	**0.92**	0.89	**0.91**	0.85	0.68	0.76	0.76	**0.94**	0.84
	ALBERT	0.86	**0.94**	0.90	**0.86**	**0.70**	**0.78**	0.78	**0.94**	**0.86**
	RoBERTa	**0.92**	0.90	**0.91**	0.80	0.73	0.76	0.78	**0.88**	0.83

*Bold values indicate the highest metric value

Table 4. Ensemble model results over different datasets.

Dataset	Neutral			Negative			Positive		
	P	R	F1	P	R	F1	P	R	F1
GitHub	0.92	0.93	0.92	0.92	0.90	0.91	**0.93**	**0.93**	**0.93**
Stack overflow	**0.93**	**0.92**	**0.92**	0.91	0.91	0.91	0.77	0.79	0.78
Jira	0.91	0.90	0.90	0.86	0.65	0.74	0.76	0.92	0.83

*Bold values indicate the highest metric value

Table 5. Compressed model results over different datasets.

Dataset	Neutral			Negative			Positive		
	P	R	F1	P	R	F1	P	R	F1
GitHub	0.93	0.92	0.92	0.91	0.92	0.91	**0.93**	**0.93**	**0.93**
Stack overflow	0.87	0.84	0.86	0.85	0.88	0.86	**0.93**	**0.94**	**0.94**
Jira	0.92	0.89	0.91	0.79	0.77	0.78	0.77	0.89	0.83

*Bold values indicate the highest metric value

6 Conclusion

From the perspective of software engineering applications, BERT model shows significant improvement on SE-specific datasets for the classification task when compared to the performance of the prevailing tools. This article proposes that the fine-tuning of the pre-trained model is more effective than taking a large corpus of text and training the model from scratch, while also the ensemble approach can boost the overall performance. The compressed model on the other hand produced similar results and is overall advantageous when deployed as a tool with limited resources. The extensive trials show that the approaches used outperformed the prevailing tools with an improved f1-score of 0.88 and 0.84 for the Stack Overflow and the Jira datasets respectively. Similarly for the GitHub dataset, the f1-score enhanced for positive and negative classes as 0.93 and 0.91 respectively, with an overall f1-score recorded as 0.92. Furthermore, the study shows the results with synthetic data had an overall f1-score of 0.85 on the ensemble model. The paper highlights the improved results based on the BERT model and can further be extended on domain-specific research, while there is still scope to improve these results as the dataset availability is limited concerning the software domain. Bug prediction [19] is another important task that could be studied along with the API recommendation and mining developer's opinions on text/comments.

Acknowledgment. We thank our institute, Indian Institute of Information Technology Allahabad (IIITA), India and Big Data Analytics (BDA) lab for allocating the centralised computing facility and other necessary resources to perform this research. We extend our thanks to our colleagues for their valuable guidance and suggestions.

References

1. Ahmed, T., Bosu, A., Iqbal, A., Rahimi, S.: SentiCR: a customized sentiment analysis tool for code review interactions, October 2017
2. Bojanowski, P., Grave, E., Joulin, A., Mikolov, T.: Enriching word vectors with subword information (2016)
3. Calefato, F., Lanubile, F., Maiorano, F., Novielli, N.: Sentiment polarity detection for software development (2017)
4. Calefato, F., Lanubile, F., Novielli, N.: EmoTxt: a toolkit for emotion recognition from text, October 2017
5. Chen, Z., Cao, Y., Lu, X., Mei, Q., Liu, X.: SEntiMoji: an emoji-powered learning approach for sentiment analysis in software engineering, July 2019
6. Devlin, J., Chang, M.W., Lee, K., Toutanova, K.: BERT: pre-training of deep bidirectional transformers for language understanding, October 2018
7. Goldberg, Y., Levy, O.: word2vec explained: deriving Mikolov et al'.s negative-sampling word-embedding method, February 2014
8. Imtiaz, N., Middleton, J., Murphy-Hill, E., Girouard, P.: Sentiment and politeness analysis tools on developer discussions are unreliable, but so are people, June 2018
9. Islam, M., Zibran, M.: Leveraging automated sentiment analysis in software engineering, May 2017
10. Islam, M., Zibran, M.: DEVA: sensing emotions in the valence arousal space in software engineering text, April 2018
11. Islam, M., Zibran, M.: SentiStrength-SE: exploiting domain specificity for improved sentiment analysis in software engineering text. J. Syst. Softw. **145**, 125–146 (2018)
12. Islam, M., Zibran, M.: SentiStrength-SE: exploiting domain specificity for improved sentiment analysis in software engineering text, August 2018
13. Lan, Z., Chen, M., Goodman, S., Gimpel, K., Sharma, P., Soricut, R.: ALBERT: a lite BERT for self-supervised learning of language representations, September 2019
14. Liesting, T., Frasincar, F., Trusca, M.M.: Data augmentation in a hybrid approach for aspect-based sentiment analysis (2021)
15. Lin, B., Zampetti, F., Bavota, G., Di Penta, M., Lanza, M., Oliveto, R.: Sentiment analysis for software engineering: how far can we go? May 2018

16. Lin, B., Zampetti, F., Oliveto, R., Di Penta, M., Lanza, M., Bavota, G.: Two datasets for sentiment analysis in software engineering, September 2018
17. Liu, Y., et al.: RoBERTA: a robustly optimized BERT pretraining approach, July 2019
18. Loper, E., Bird, S.: NLTK: the natural language toolkit, July 2002
19. Mangnoesing, G.V.H., Trusca, M.M., Frasincar, F.: Pattern learning for detecting defect reports and improvement requests in app reviews (2020)
20. Manning, C., Surdeanu, M., Bauer, J., Finkel, J., Bethard, S., McClosky, D.: The Stanford coreNLP natural language processing toolkit, January 2014
21. Novielli, N., Girardi, D., Lanubile, F.: A benchmark study on sentiment analysis for software engineering research, March 2018
22. Pagliardini, M., Gupta, P., Jaggi, M.: Unsupervised learning of sentence embeddings using compositional n-gram features. In: NAACL 2018 - Conference of the North American Chapter of the Association for Computational Linguistics (2018)
23. Pennington, J., Socher, R., Manning, C.: GloVe: global vectors for word representation, January 2014
24. Punn, N.S., Agarwal, S.: CHS-Net: a deep learning approach for hierarchical segmentation of COVID-19 infected CT images. arXiv preprint arXiv:2012.07079 (2020)
25. Punn, N.S., Agarwal, S.: Multi-modality encoded fusion with 3d inception U-Net and decoder model for brain tumor segmentation. Multimedia Tools Appl., 1–16 (2020)
26. Rahman, M.M., Roy, C., Kievanloo, I.: Recommending insightful comments for source code using crowdsourced knowledge, September 2015
27. Rajora, H., Punn, N.S., Sonbhadra, S.K., Agarwal, S.: Web based disease prediction and recommender system (2021)
28. Socher, R., et al.: Recursive deep models for semantic compositionality over a sentiment treebank, January 2013
29. Thelwall, M., Buckley, K., Paltoglou, G., Cai, D., Kappas, A.: Sentiment strength detection in short informal text, December 2010
30. Torfi, A., Shirvani, R.A., Keneshloo, Y., Tavaf, N., Fox, E.A.: Natural language processing advancements by deep learning: a survey. arXiv preprint arXiv:2003.01200 (2020)
31. Vaswani, A., et al.: Attention is all you need, June 2017
32. Xie, Z., Genthial, G., Xie, S., Ng, A., Jurafsky, D.: Noising and denoising natural language: Diverse backtranslation for grammar correction, January 2018
33. Zhang, Y., Hou, D.: Extracting problematic API features from forum discussions, May 2013

A Stochastic Block Model Based Approach to Detect Outliers in Networks

Fabrizio Angiulli, Fabio Fassetti, and Cristina Serrao[✉]

DIMES, University of Calabria, 87036 Rende, CS, Italy
{f.angiulli,f.fassetti,c.serrao}@dimes.unical.it

Abstract. Finding outliers in networks is a central task in different application domains. Here, we exploit the stochastic block model framework to study the network from a generative point of view and design a score able to highlight those nodes whose connection with the rest of the network violates in some way the law according to which the rest of the nodes are interconnected. The peculiarity of our approach is that no pre-defined notion of outlier is employed; rather outliers emerge as deviations from the underlying network generating mechanism.

Keywords: Networks · Stochastic block model · Anomaly detection

1 Introduction

Nodes in real networks are often organized into modules or communities so that connections are relatively more abundant within modules than between modules. However, networks often come with nodes with a notable behaviour: they may be uniformly connected to nodes from different communities, thus violating the community structure, or they can fullfill a special role within the community as in case of hubs. The special or abnormal role of such nodes can be studied by designing a score that is able to evaluate the characteristics of each node's neighborhood with respect to the network structure. In order to do that, we consider the family of Stochastic Block Models (SBMs) and propose a mathematical and computational framework whose aim is to detect anomalies in unlabeled undirected graphs based on their structure.

Stochastic Block Models are particularly suitable to model those situations in which nodes belong to groups and interact with each other depending on their group membership. For this reason, they have been widely used to solve clustering and community detection problems [3] but, from the best of our knowledge, there is still little literature about the application of this framework in the field of outlier mining.

Much more literature exists about methods exploiting the graph structure to find patterns and highlight anomalies. These kind of approaches can be grouped into two main families [2]: structure-based and community-based techniques.

Structure-based methods exploit the graph structure to extract graph-centric features such as node degree and subgraph centrality or measure closeness of

© Springer Nature Switzerland AG 2021
C. Strauss et al. (Eds.): DEXA 2021, LNCS 12923, pp. 149–154, 2021.
https://doi.org/10.1007/978-3-030-86472-9_14

objects in the graph to identify associations. Among them a technique called ODDBall is proposed in [1] to discover outlier nodes in graphs with weighted edges. The method focuses on each node neighborhood, defined as a sphere around it, and consider the induced sub-graph of its neighboring nodes, which is referred to as the *egonet*, to deduce nodes with an anomalous behaviour.

On the other hand, methods that exploit communities or proximity of nodes in the graph to spot anomalous nodes include [10], which is based on random-walk, and [11] that relies on matrix factorization and some other methods that directly focus on network clustering and identify hubs and outliers as a by-product of the clustering process [9, 12].

Our contribution is discussed by dividing the paper into two main section: Sect. 2 introduces the main definitions and presents the proposed model while Sect. 3 is devoted to the discussion of some experimental results.

2   Model Definition

A *graph* $\mathcal{G}$ is a pair (V, E) where V is a finite set of *vertices* or *nodes*, $E \subseteq V \times V$ is a finite set of unordered pairs of vertices called *edges*. Given a node v, a node w is a *neighbor* of v if there is an edge (v, w) in E and $\mathcal{N}(v) = \{w \mid (v, w) \in E\}$ denotes the set of all neighbors of v, also called *neighborhood* of v.

Let $\mathcal{G} = (V, E)$ be a graph, a *partition* $\mathcal{P}$ of $\mathcal{G}$ is any partition $\mathcal{P} = \{\pi_1, \ldots, \pi_K\}$ of V into K non-empty groups. We will refer to the set of all possible *partitions* of V as $\mathbf{P}$ throughout the paper. The further notation employed through the paper is the following:

l_{π_i, π_j}: Given two groups π_i and π_j in $\mathcal{P}$, l_{π_i, π_j} is the number of edges (v, w) in E such that $v \in \pi_i$ and $w \in \pi_j$;

l_{v, π_j}: Given a node v and a group π_j in $\mathcal{P}$, l_{v, π_j} is the number of edges from v to any node in π_j;

r_{π_i, π_j}: Given two groups π_i and π_j in $\mathcal{P}$, r_{π_i, π_j} is the maximum number of edges that can exist between nodes in π_i and nodes in π_j, namely $|\pi_i| \cdot |\pi_j| \; \forall i \neq j$ and $|\pi_i| \cdot (|\pi_j| - 1)/2$ for $i = j$.

A stochastic block model [5] is a generative model $M = (\mathcal{P}, \mathbf{Q})$ that provides networks with a community structure. The model is determined by a partition $\mathcal{P}$ of the nodes into K groups and a symmetric matrix $\mathbf{Q} \in \{0 \ldots 1\}^{\mathbf{K} \times \mathbf{K}}$ of probabilities of linkage between groups. The model assigns each node to one of the K groups and the probability of the connection between two nodes $v \in \pi_i$ and $w \in \pi_j$ is $\mathbf{Q}_{i,j}$ thus, the probability of any two nodes being connected depends only on the groups to which they belong.

According to [5, 8], a Bernoulli approach can be used to decide whether an edge is created or not. Thus, assume you are provided with a graph $\mathcal{G}$ generated by a stochastic block model $M = (\mathcal{P}, \mathbf{Q})$, then the probability of observing $\mathcal{G}$ given the parameters $(\mathcal{P}, \mathbf{Q})$ is

$$\Pr(\mathcal{G}|\mathcal{P}, \mathbf{Q}) = \prod_{i \leq j} \mathbf{Q}_{i,j}^{l_{\pi_i, \pi_j}} (1 - \mathbf{Q}_{i,j})^{r_{\pi_i, \pi_j} - l_{\pi_i, \pi_j}} \tag{1}$$

The model that describe better the observed network $\mathcal{G}$ is the one maximizing the probability in Eq. (1) in two stages, first with respect to the unknown model parameters $\mathbf{Q}$, then with respect to the group assignments.

As l_{π_i,π_j} are independent, in [5] it has been proved that the maximum-likelihood estimate for parameter $\mathbf{Q}$ is obtained by maximizing each of the marginal likelihood functions $\mathbf{Q}_{i,j}^{l_{\pi_i,\pi_j}}(1-\mathbf{Q}_{i,j})^{r_{\pi_i,\pi_j}-l_{\pi_i,\pi_j}}$ with respect to $\mathbf{Q}$. The maximum occurs when $\mathbf{Q}_{i,j} = l_{\pi_i,\pi_j}/r_{\pi_i,\pi_j}$.

When $\mathbf{Q}$ is fixed, the observed network could be generated according to a set of models that differ from each other in the way nodes are partitioned. This framework can be used to estimate the probability of an arbitrary network property [4]. As we perceive a node to be an outlier if its neighborhood is unusual, we are interested in estimating the probability of each node's neighborhood: low probability values suggest that such a node may be anomalous.

Let $\mathcal{N}(v)$ be the neighborhood of node v. The probability of observing a certain neighborhood for a node v in $\mathcal{G}$ can be defined as follows:

$$\Pr(\mathcal{N}(v)|\mathcal{G}) = \sum_{\mathcal{P}\in\mathbf{P}} \Pr(\mathcal{N}(v)|\mathcal{P}) \Pr(\mathcal{P}|\mathcal{G}) \tag{2}$$

where $\Pr(\mathcal{N}(v)|\mathcal{P})$ is the probability of observing $\mathcal{N}(v)$ in a network generated by a model having parameter $\mathcal{P}$.

Using the Bayes Theorem you can rewrite $\Pr(\mathcal{P}|\mathcal{G})$ and $\Pr(\mathcal{N}(v)|\mathcal{G})$ results in the weighted mean of all contributions $Pr(\mathcal{N}(v)|\mathcal{P})$ in $\mathbf{P}$:

$$\Pr(\mathcal{N}(v)|\mathcal{G}) = \frac{\sum_{\mathbf{P}} \Pr(\mathcal{N}(v)|\mathcal{P}) \Pr(\mathcal{G}|\mathcal{P}) \Pr(\mathcal{P})}{\sum_{\mathbf{P}} \Pr(\mathcal{G}|\mathcal{P}) \Pr(\mathcal{P})}.$$

Since we are only interested in ranking nodes according to their outlierness and the denominator is just a normalization factor common to each node, we can safely ignore it. Moreover, since all the partitioning are equally probable, also the term $\Pr(\mathcal{P})$ is negligible. Thus, the score definition become the following:

$$outscore(v) = -\sum_{\mathbf{P}} \Pr(\mathcal{N}(v)|\mathcal{P}) \cdot \Pr(\mathcal{G}|\mathcal{P})$$

We have already discussed about the likelihood factor $\Pr(\mathcal{G}|\mathcal{P})$, now we need to derive the probability of observing $\mathcal{N}(v)$ given a partitioning $\mathcal{P}$. In the *Bernoulli* framework, the neighborhood probability can be computed as:

$$\Pr(\mathcal{N}(v) \mid \mathcal{P}) = \prod_{j} \mathbf{Q}_{\pi^v,j}^{l_{v,\pi_j}} (1 - \mathbf{Q}_{\pi^v,j})^{|\pi_j|-l_{v,\pi_j}}. \tag{3}$$

where is π^v is the group in $\mathcal{P}$ containing the node v. Note that, both Eqs. (1) and (3) can be handled more easily by moving to the logarithm.

Such a score is able to highlight those nodes violating in some way the law according to which edges are distributed. Figure 1 shows the main structures our method is able to discover. A bridge node (Fig. 1a) is one connecting two or more clusters without belonging to any; neighbors of these nodes are almost uniformly

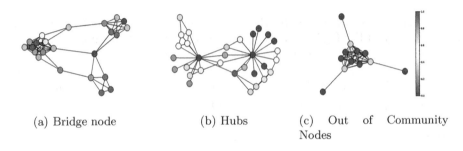

| (a) Bridge node | (b) Hubs | (c) Out of Community Nodes |

Fig. 1. Notable structures.

distributed among different clusters. More generally, the technique brings out nodes whose elimination cause two or more clusters to be disconnected. For this reason, hubs emerge as well (Fig. 1b). Lastly, the more intuitive concept of out of community node is highlighted (Fig. 1c).

Our mining procedure has the aim of ranking nodes according to their outlierness score which should be computed as a sum over all partitions. However, the cardinality of the partitions set rapidly grows with the number of nodes making this evaluation infeasible, even for small network. To identify those partition that significantly contribute to the sum, we exploit the main ideas discussed in [7] about the Metropolis-Hastings algorithm and design our own mining strategy which is performed in a parallel way: we start a certain number of concurrent processes, each of which navigates thorough the partition space starting from a random partition, samples the relevant ones and evaluates their contribution to the scores. Results coming from each process are then joined and the change in node-ranking is evaluated to decide whether to stop.

3   Experiments

The preliminary evaluation we discuss here has been done on synthetic datasets generated using the LFR Benchmark [6]. Such a framework has been proposed to test the performances of community detection techniques on graphs with heterogeneous distributions of node degree and heterogeneous community sizes, both approximated by a power laws. LFR networks represent a good approximation of real networks, thus we perturb them by injecting some outliers and test the ability of our method to detect these perturbations.

We build different 200-nodes networks by initializing the generator with different parameter configurations. We consider the default values for the exponents of the power laws describing the degree sequence (t1) and the community size distribution (t2), i.e. $t1 = -2$ and $t2 = -1$, and use the following values for the three other parameters: (I) average degree: 10, 20, 50; (II) maximum degree: 50, 100, 150; (III) mixing parameter: 0.01, 0.05, 0.25.

We perturb each network by introducing some new (outlier) nodes so that they represent the 5% of the total number of nodes of the network. We connect these nodes to the rest of the network in order to ensure their degree is equal to the average node degree and they are almost uniformly connected to all the communities.

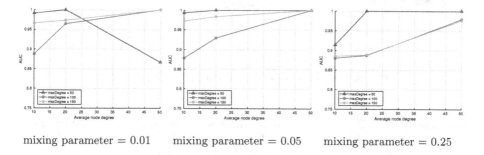

mixing parameter = 0.01   mixing parameter = 0.05   mixing parameter = 0.25

Fig. 2. Accuracy of the outlier node classification.

Figure 2 refers to networks generated with different mixing parameter values and reports the accuracy of the results we obtain on networks generated for growing values for the average node degree. The mixing parameter represents the average interconnection among communities. This analysis shows that the technique is sensitive to the presence of community outliers and identifies them with a good approximation.

Depending on the combination of the various parameters some boundary nodes assume exceptional values and this may worsen the AUC, though these nodes represent interesting knowledge in practice. E.g., for maximum and average degree 50 and mixing parameter 0.01 almost all community nodes have only internal links, so the few ones connected with the extern are detected as anomalous. We notice also that the set of top-ranked nodes includes the ones whose degree is close to the maximum, thus hubs are identified as well.

In order to discuss the knowledge we are able to highlight we consider a brief comparative analysis with ODDBall [1] which is designed to retreive nodes whose neighbors are very well connected (near-cliques) or not connected (stars).

We cover these ideas by our notion of hubs, so we are as good as ODD-Ball in detecting such structures, however the identification of bridge nodes is harder for ODDBall as the egonet of a bridge includes blocks of connected nodes from different communities and it cannot distinguish between bridge and inner community nodes when their egonet has the same size.

Consider the network example in Fig. 3. Such a network has been generated according to the procedure described above and nodes have been colored according to the scores calculated by each technique; inliers are associated with higher scores and correspond to blue nodes. In the picture referred to SBMOut (Fig. 3a) all nodes belonging to communities have a score higher than 0.5, while red nodes correctly represent bridges which are difficult to be identified by ODD-Ball (Fig. 3b). The score distribution of the two methods is reported in Fig. 3c.

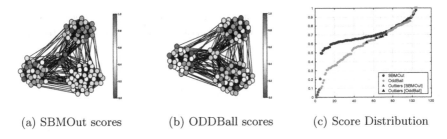

(a) SBMOut scores   (b) ODDBall scores   (c) Score Distribution

Fig. 3. Comparison between SBMOut and ODDBall.

Interestingly, the SBMOut scores of outlier nodes are clearly separated from the rest of the scores, while the ODDBall curve is smoother and the scores of bridge nodes are located towards the center of the distribution.

References

1. Akoglu, L., McGlohon, M., Faloutsos, C.: Oddball: spotting anomalies in weighted graphs. In: Zaki, M.J., Yu, J.X., Ravindran, B., Pudi, V. (eds.) PAKDD 2010. LNCS (LNAI), vol. 6119, pp. 410–421. Springer, Heidelberg (2010). https://doi.org/10.1007/978-3-642-13672-6_40
2. Akoglu, L., Tong, H., Koutra, D.: Graph based anomaly detection and description: a survey. Data Mining Knowl. Discov. **29**(3), 626–688 (2014). https://doi.org/10.1007/s10618-014-0365-y
3. Funke, T., Becker, T.: Stochastic block models: a comparison of variants and inference methods. PLoS ONE **14**(4), 0215296 (2019)
4. Guimerà, R., Sales-Pardo, M.: Missing and spurious interactions and the reconstruction of complex networks. Proc. Natl. Acad. Sci. **106**(52), 22073–22078 (2009)
5. Holland, P.W., Laskey, K.B., Leinhardt, S.: Stochastic blockmodels: first steps. Soc. Netw. **5**(2), 109–137 (1983)
6. Lancichinetti, A., Fortunato, S., Radicchi, F.: Benchmark graphs for testing community detection algorithms. Phys. Rev. E **78**(4), 046110 (2008)
7. Peixoto, T.P.: Efficient monte carlo and greedy heuristic for the inference of stochastic block models. Phys. Rev. E **89**(1), 012804 (2014)
8. Snijders, T.A., Nowicki, K.: Estimation and prediction for stochastic blockmodels for graphs with latent block structure. J. Classif. **14**(1), 75–100 (1997)
9. Sun, H., Huang, J., Han, J., Deng, H., Zhao, P., Feng, B.: gskeletonclu: density-based network clustering via structure-connected tree division or agglomeration. In: Proceedings of ICDM 2010 (2010)
10. Sun, J., Qu, H., Chakrabarti, D., Faloutsos, C.: Neighborhood formation and anomaly detection in bipartite graphs. In: Proceedings of ICDM 2005 (2005)
11. Tong, H., Lin, C.Y.: Non-negative residual matrix factorization with application to graph anomaly detection. In: Proceeding of ICDM 2011. SIAM (2011)
12. Xu, X., Yuruk, N., Feng, Z., Schweiger, T.A.: Scan: a structural clustering algorithm for networks. In: Proceedings of the 13th ACM SIGKDD KDD (2007)

Medical-Based Text Classification Using FastText Features and CNN-LSTM Model

Mohamed Walid Zeghdaoui[1,2]([✉]), Omar Boussaid[1] [iD], Fadila Bentayeb[1] [iD], and Frederik Joly[2]

[1] Université de Lyon, Université Lyon 2, ERIC Lab, Lyon, France
{walid.zeghdaoui,omar.boussaid,fadila.bentayeb}@univ-lyon2.fr
[2] Sword Group - 9, Av. Charles de Gaulle, 69370 St-Didier-au-Mont-d'Or, France
{walid.zeghdaoui,frederik.joly}@sword-group.com

Abstract. Text classification is a fundamental task that is often carried out upstream of natural language processing (NLP) techniques. Therefore, this task plays an essential role in information retrieval and extraction, and has a wide range of applications in many areas. Many text classification techniques have been proposed, with promising results. However, to propose an efficient model, the particularity of the application domain also needs to be addressed to better grasp the syntactic and semantic complexity of the texts. In this paper, we proposed a classification model for medical text classification that is based on a convolutional neural network (CNN) combined with a long short term memory (LSTM) neural networks. The proposed CNN-LSTM is using word vectors computed with FastText to achieve the highest accuracy. We compared our proposed model results with other state-of-the-art models such as CNN, support vector machines, decision trees, naive Bayes, and K-nearest neighbor. The performance of all used models were evaluated in terms of accuracy, precision, recall, and F1 score. The CNN-LSTM outperforms all other models in terms of all evaluation parameters and achieved 86.34%, 90.68%, 91.72%, 90.67% accuracy, precision, recall, and F1 score, respectively.

Keywords: Medical text classification · Deep learning · Text mining · Convolutional neural networks · LSTM · Word embedding

1 Introduction

The quantity of medical records has increased exponentially over the last two decades. This data represents a wealth of information and a rich knowledge base that must be used to improve healthcare quality, particularly in medical decision-making [1]. Text classification is an automated, natural language processing (TALN) function that can help us to take advantage of this data. The effectiveness of this task depends heavily on the representation and extraction of features, which is a particularly important step in classification because of the high dimensionality of the features that a text may have [2]. All the more

© Springer Nature Switzerland AG 2021
C. Strauss et al. (Eds.): DEXA 2021, LNCS 12923, pp. 155–167, 2021.
https://doi.org/10.1007/978-3-030-86472-9_15

so when dealing with medical report texts presented in the form of grammatically mediocre sentences. These texts contain medical terminology and numerous abbreviations and irregularities, which leads to problems with large-scale data classification as well as difficulties caused by the scarcity of such data [3].

In this paper, we propose a text classification model that combines a CNN trained from word vectors obtained with FastText [5] with a recurrent neural network (i.e., LSTM). This model benefits from both the advantages of CNNs for the extraction of local features and also from the long-term dependencies captured using the high memory capacity of LSTM networks to properly connect the extracted features, thereby ensuring better accuracy when classifying texts.

Our experimental results clearly show that our approach is successful as compared to other text classification methods. The key points of this study are as follows:

1. This study proposed an approach for medical domain text classification that combines CNN+LSTM+FastText.
2. This study performs a deep comparison with other state-of-the-art well-tuned models.
3. Evaluation of all models in terms of accuracy, precision, recall, and F1 score.

The remainder of the paper is organized as follows: Sect. 2 contains related work, and Sect. 3 contains the material and methods used in this study. Section 4 describes the experiment flow, and discusses the results. The study's conclusion is presented in Sect. 5.

2   Related Works

In recent years, deep neural networks have shown great success in many NLP tasks, most of which are based on RNN or CNN networks. Kowsari et al. [4] reviewed classification methods and compared their advantages and disadvantages. Similarly, Otter et al. [8] gave a brief presentation of deep learning methods and their applications to NLP tasks and discussed in detail the current state of these techniques. They concluded with some suggestions for future research in this area. Several approaches were based on recurrent neural networks. Tang et al. proposed a hierarchical RNN model to construct the relationship between sentences [9]. Yang et al. proposed a hierarchical attention model that incorporates an attention mechanism into the hierarchical GRU model so that the model can better capture the most discriminating parts of text [10]. Wang et al. proposed to use disconnected recurrent neural networks for Yahoo responses and Yelp reviews, which captured key phrases and long-term dependencies [11]. Collobert and Weston used CNNs for the first time in natural language processing tasks [12]. Later, they were applied to text classification [13–15]. Conneau et al. have recently proposed a VDCNN model using the deep convolution network in sentiment classification [16]. Similarly, Johnson and Zhang have proposed a deep pyramidal CNN that not only provides excellent performance but also reduces training time [17].

The challenge of automatic text classification is not new. Indeed, the most traditional method is that of the Naïve Bayes (NB) classifiers, known as generative classifiers, which model the distribution of texts in each class using a probabilistic model with strong assumptions of independence on the distributions of the different terms [6]. Support Vector Machines (SVM) are also widely used in text classification. The main idea of a SVM is to partition the data space using linear or non-linear delimitations between the different classes. The goal is to maximize the distance between the boundaries of the different classes [7].

Recently, several in-depth learning methods were successfully applied to the automatic processing of clinical texts. Beaulieu-Jones et al. proposed a neural network to construct phenotypes for classifying a patient's disease states [18]. The performance of their model outperformed decision trees (DT), random forests (RF) [19], and support vector machines (SVM). Hughes et al. presented a method for the automatic classification of sentences from clinical texts [20]. Their method relies on CNNs to represent complex features. Baker et al. used a CNN network approach for the classification of biomedical texts [21].

In summary, deep learning methods have some advantages in countless tasks and can be applied to text classification, eliminating the tedious engineering operations of traditional methods.

3   Materials and Methods

3.1   Dataset

The data that we process in this research comes from five member institutions of the French Comprehensive Cancer Centers (FCCC). Each center has several million medical reports to its credit. These reports contain a variety of medical information, such as clinical decisions made by physicians, treatments administered to patients, history (which includes, among other things, previous surgery, hospitalizations, and chronic illnesses in family members). Table 1 shows how many reports each institution gets.

Table 1. Distribution of medical reports by institution.

Center	Number of reports
Institut Paoli-Calmettes	313
Institut Curie	339
Institut du Cancer de Montpellier	247
Centre Georges-François Leclerc	214
Centre Léon-Bérard	317
Total	1430

Data Preprocessing: In our experiments after data acquiring first we have done preprocessing on the dataset to make its clean. Taking into account the heterogeneity of the data, the specific lexicon used in the medical field, and the very different text formatting, we first proceeded to clean the data to eliminate any noise that can reduce the performance of the classification, as follows: (1) removing HTML tags, (2) transforming empty character strings in a line by a single space (idem for empty line strings), (3) decapitalization, and (4) we decided to keep the stopwords because they can be semantically relevant for the analysis of clinical reports. This step made it easier for us to detect sentence boundaries, which plays a crucial role in their classification. To do so, we set up a sentence boundary detection mechanism (beginning and end) to transform each report into a sequence of sentences. With the help of experts (physicians), we identified seven classes to which a sentence in a report could belong. Table 2 lists the name, description, and the total number of sentences in each class.

Table 2. Name, description and number of the identified sentences.

Class	Description	#Sentences	Percentage
Personal history	Personal medical history	3660	8,32%
Family history	Family medical history	3453	7,86%
Headings	Information of a low medical nature	5196	11,83%
Hypothesis	Sentences expressing uncertainty	6487	14,77%
Metastasis	Metastatic information	7574	17,24%
Negation	Negative sentences	5180	11,79%
Other	Does not fit into the other categories	12368	28,16%
Total	–	43918	100%

Data Annotation: After cleaning and segmenting the reports into sentences, 51604 sentences were identified, with 43918 distinct sentences. This makes an average of 36.8 sentences per report, and a total of 1,182,949 distinct words. Each sentence consists of an average of 25 words, with an interval ranging from 1 to 293 words. The annotation consists of associating a label or tag to each sentence based on the information contained therein. The quality of this step is essential to produce a fully generalizable classification model. This must ensure completeness, so that the annotated sentences are representative of the overall corpus while also being free of annotation errors that could corrupt model learning.

3.2   Feature Extraction

The quality of text classification is mainly based on the representation and extraction of characteristics. During word embedding, to each word in the vocabulary is assigned a vector of real values to capture both syntactic and semantic information of the words. In this way, if the model is well-formed, then the close vectors in terms of distance in space correspond to synonymous words or words used in a similar context. In our experiments, we used FastText features for training of models.

FastText Features: In the first step, we generated the word vectors from a reduced volume of data (i.e., about 250,000 medical reports) and compared it with a model trained on the entire French Wikipedia corpus (more than 2,000,000 articles, often more significant than the medical reports). The first results confirmed our initial need to build word vectors adapted to the medical context. Therefore, we used all of the reports of the Institut Curie, which listed nearly 12,000,000 reports, as well as the Wikipedia texts to develop a third model. For our experiments, we used 320 terms in 11 categories (Table 3).

Table 3. Results of the distance scores of the three models over the 320 words.

Category	Wikipedia	Local data	Curie + Wikipedia
Metastasis	0,4821	0,4327	**0,2964**
Biomarker	0,9715	0,9010	**0,7587**
City	0,6619	0,8508	**0,5877**
Localisation	**0,5093**	0,7301	0,655
Physician name	0,9281	**0,7085**	0,805
Date	0,5128	0,4676	**0,1179**
Unit	0,7792	**0,6381**	0,6442
Tumor name	0,8600	0,7442	**0,7106**
Protocol	0,9513	0,7311	**0,5381**
Act	0,7849	0,4121	**0,3186**
Chemotherapy	0,9972	0,7969	**0,5386**
Total	0,8246	0,7254	**0,5977**

The results obtained after combining Wikipedia and Institute Curie data are much better than the results of the first two models. However, some categories still score better in the previous models.

3.3   Machine Learning Algorithms

In our study, we used different machine learning models with their best hyper parameters setting as shown in Table 4.

3.4   CNN-LSTM

Our text classification model can be divided into four layers: an input layer for word embedding, a one-dimensional convolutional network layer for local feature

Table 4. Description of traditional machine learning models and parameter settings.

Model	Description and parameters
Decision Trees	Decision trees are based on a hierarchical representation of data in the form of decision sequences. Feature extraction = TF-IDF, Alpha = 1 (Laplace smoothing)
Naive Bayes	A probabilistic classifier based on Bayes' theorem. Feature extraction = TF-IDF, Alpha = 1, Maximum depth = 30
Support Vector Machines	SVM establishes hyperplanes so that the distance between the dataset is maximum, Feature extraction = TF-IDF. Number of iterations = 300, Penalty C = 1, Kernel = Poly. Tolerance for stopping criterion = 10^{-2}
K-Nearest Neighbors	KNN finds the k nearest neighbors of an element and assigns a category to it based on the class of k neighbors. Feature extraction = TF-IDF, Number of neighbors = 7

extraction, an LSTM network layer for capturing long-term dependencies, and a classification layer for label prediction. The structure of our model is shown in Fig. 1.

One-Dimension Convolutional Layer: We use one-dimensional convolution layer (Conv1D) to capture the sequence information and reduce the dimensions of the input data. A convolution operation involves a convolutional kernel applied to a fixed window $W \epsilon Rnk$ of words to compute a new feature. This kernel, also called a filter, completes the feature extraction. Each filter is applied to a window of m words to obtain a single feature. To ensure the integrity of the word as the smallest granularity, the width of the filter is equal to the width of the original matrix. In the convolution layer, a matrix-vector operation is applied to each window using the W-weight matrix to obtain a characteristic map $C \in R_{n-m+1}$.

The ith element of the feature map is:

$$C_i = \sigma(\sum W \cdot [C_{i:i+m-1}]) + b \tag{1}$$

where b is the bias term used to adjust the output as well as the weighted sum of the inputs of the neuron, which allows shifting the nonlinear activation function ReLU noted here σ. We used ReLU because it reduces the number of iterations needed for convergence in deep networks. Then, the result of the convolution is pooled using the maximum pooling operation to capture essential features in the text. At the end of this step and to improve the quality of our text classification task, the different calculated features are concatenated to constitute the input of our LSTM layer.

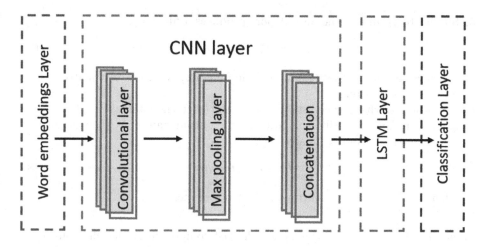

Fig. 1. Overall architecture of the proposed model.

Long Short-Term Memory Layer: LSTM is introduced to solve the vanishing gradient problem because it learns to regulate the flow of information. LSTMs can efficiently capture contextual information from the input text thanks to their high memory power.

The information is passed through two channels from one cell to the next, h, and c.

1. At time t, the recurrence relation is calculated as follows:

$$h_t, c_t = f(x_t, h_{t-1}, c_{t-1}) \qquad (2)$$

where x_t represents the current word of the sequence, h_{t-1} the hidden state of the previous cell and c_{t-1} the cell state of the previous cell. The latter vector avoids the vanishing gradient problem because it is updated additive at each step, without going through an activation function.

2. The forget gate is a dense layer with sigmoid activation that acts as a filter to «forget» certain cell state information. From h_{t-1} and x_t, this oblivion gate produces a vector:

$$f_t = \sigma(W_f \cdot [h_{t-1}, x_t] + b_f) \qquad (3)$$

3. Similarly, the front door produces a filter it from h_{t-1} and x_t.

$$i_t = \sigma(W_i \cdot [h_{t-1}, x_t] + b_i) \qquad (4)$$

4. At the same time, a g_t vector is created by a *tanh* function, which allows updating the cell state.

$$g_t = tanh(W_g \cdot [h_{t1}, x_t] + b_g) \qquad (5)$$

$$c_t = f_t \otimes c_{t-1} + i_t \otimes g_t \qquad (6)$$

5. Similar to f_t and i_t, the output port produces a filter o_t.

$$o_t = \sigma(W_o \cdot [h_{t-1}, x_t] + b_o) \tag{7}$$

where σ designates the sigmoid logistic function, W are weight matrices and b represents bias term.

6. The values of the new cell state c_t are reduced to the interval $]-1, 1[$ by $tanh$ activation function. Filtering through the output gate o_t is then performed to obtain the output h_t finally.

$$h_t = o_t \otimes tanh(c_t) \tag{8}$$

where $\otimes$ designates the vector product per element.

Classification Layer: In the model, the last component is the fully connected layer, which takes as input the characteristics generated from a sentence by the LSTM layer and then predicts the most appropriate label according to semantic and syntactic content. The probability that a sentence belongs to a category is calculated by the softmax activation function, as follows:

$$P_i = \frac{exp^{o_i}}{\sum_{j=1} exp^{o_j}} \tag{9}$$

where P_i indicates the probability of the ith category, exp^{o_i} means the corresponding value of the output of the ith category, and j indicates the total number of categories.

4   Experimentations and Results

4.1   Experimental Flow

In this section, we discuss the flow of our proposed approach experiments as illustrate in Fig. 2. All the experiments as performed on the Core i7 7th generation computer with windows operating systems. Jupyter Notebook is used as IDE to implement experiments with Python language.

We used medical reports from various cancer control centers in France in our approach. These reports can be of various types and contain a wide range of medical information, such as clinical decisions, treatments administered to patients, or administrative information such as names, addresses, and so on. To be able to use these data in an optimized way, we first performed preprocessing on the set of data as indicated in the section ref dataset. Following that, we split the data into training and testing sets in an 80:20 ratio, with 80% of the data used to train the models and 20% used to test the models. The feature extraction technique was used after data division. These characteristics are used to train the proposed and other states of the art models. In the end, we evaluated the performance of all trained models in terms of accuracy, precision, recall and F1 score.

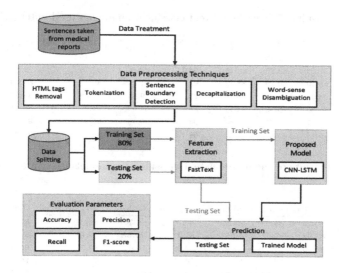

Fig. 2. Flow of experimental approach

Table 5. Final results of the different text classification models used in terms of F1 score.

Class	DT (%)	NB (%)	SVM (%)	KNN (%)	CNN (%)	FT+CNN (%)	Our model (%)
Personal hist	88.54	87.01	91.13	90.33	91.02	91.48	**91.61**
Family hist	90.51	86.37	90.37	90.54	**94.70**	91.86	93.17
Headings	87.65	85.46	87.91	86.19	88.22	88.41	**89.21**
Hypothesis	83.72	80.67	86.41	86.44	86.29	86.60	**87.26**
Metastasis	80.98	75.61	82.42	83.08	83.55	87.47	**88.13**
Negation	94.84	92.55	96.57	95.03	96.82	97.18	**97.79**
Avg. F1 score	86.83	83.55	88.34	87.90	89.14	89.98	**90.67**

Table 6. Final results of the different text classification models used in terms of precision.

Class	DT (%)	NB (%)	SVM (%)	KNN (%)	CNN (%)	FT+CNN (%)	Our model (%)
Personal hist	88.34	86.27	86.14	88.24	88.96	88.73	**91.02**
Family hist	84.32	83.2	90.47	86.69	**95.51**	90.00	93.48
Hypothesis	89.15	**90.75**	87.68	8.209	84.84	87.64	89.91
Headings	75.51	**86.19**	84.77	82.2	83.24	84.60	85.44
Metastasis	86.85	67.39	80.79	78.02	82.21	83.98	**86.34**
Negation	95.93	89.9	95.51	95.16	97.16	98.75	**97.89**
Avg. Precision	86.68	83.95	87.56	85.4	88.65	88.95	**90.68**

For our experiments, to validate the usefulness of our feature extraction model with FastText and the LSTM layer, we tested three possible model combinations. All of the results obtained with the machine learning and deep learning models are presented in the Tables 5, 6, 7 and 8:

Table 7. Final results of the different text classification models used in terms of Recall.

Class	DT (%)	NB (%)	SVM (%)	KNN (%)	CNN (%)	FT+CNN (%)	Our model (%)
Personal hist	88.73	87.76	96.72	92.51	93.17	94.40	**92.20**
Family hist	97.68	89.78	90.26	94.74	**93.89**	93.79	92.85
Hypothesis	86.19	80.75	88.13	90.71	91.87	89.18	**88.51**
Headings	93.92	75.81	88.10	91.13	89.56	88.68	**89.15**
Metastasis	75.84	86.10	84.11	88.83	84.93	91.25	**89.98**
Negation	93.77	95.35	97.64	94.89	96.48	95.65	**97.68**
Avg. Recall	89.35	85.92	90.82	92.13	91.65	92.15	**91.72**

Table 8. Final results of the different text classification models used in terms of accuracy

Class	DT (%)	NB (%)	SVM (%)	KNN (%)	CNN (%)	FT+CNN (%)	Our model (%)
Personal hist	80.16	81.03	86.43	86.16	86.30	86.97	**87.31**
Family hist	83.99	80.07	86.51	85.41	**90.68**	87.75	89.34
Hypothesis	78.60	79.43	82.47	80.63	79.85	81.38	**81.93**
Headings	73.74	70.70	79.34	80.67	77.73	79.41	**79.92**
Metastasis	70.69	67.24	75.24	73.83	73.92	82.24	**83.01**
Negation	90.83	86.43	94.01	92.05	95.22	95.55	**96.54**
Avg. Accuracy	79.66	77.48	84.00	83.12	83.95	85.55	**86.34**

Table 9. Number of correct and wrong predictions

Model	No. of CP	No. of WP
DT	6998	1786
NB	6806	1978
SVM	7379	1405
KNN	7302	1482
CNN	7375	1409
FastText+CNN	7515	1269
Our model	7585	1199

4.2 Analysis and Discussion of the Results

Based on the values of accuracy, precision, recall and F1 score, we can see that our model obtains the best classification performances on the majority of categories, even if there is no specific model adapted to all categories. These performances vary between 87.26% and 97.79% for hypothesis and negation, respectively in terms of F1 score. Our proposed models achieved the highest accuracy score (86.34%) and outperform all other models in terms of all evaluation parameters. First, these results confirm in general that deep learning models perform better than traditional machine learning models in the field of text classification. Second, we observe an improvement in the F1 score of x points with the use of FastText as a feature extraction technique. Finally, we see that our combination of a CNN and LSTM model further improves our classification scores by x points. A possible explanation for this is that the CNN network extracts local features from the input, and then the LSTM network as proposed better characterizes semantic information by providing feature representations at the sentence level. In this way, our model takes advantage of both the CNN model and the LSTM model, which allows us to obtain a higher classification accuracy than other models. The evaluation of the used models is also done in terms of correct and wrong predictions. The proposed CNN-LSTM model outperforms all other models in terms of predictions, with 7,585 correct predictions and 1199 wrong predictions, as shown in Table 9. These experiments confirm the validity of our idea.

5   Conclusion

Classification of texts is an essential step for many natural language processing tasks. This step has been the subject of several studies, and for different types of applications. In the medical field, its application presents a special case due on the one hand to the specificity and the great variety of the vocabulary used, on the other hand to the unstructured format often used in the drafting of medical reports. To address these issues, we presented a method of classifying medical texts based on a hybrid CNN-LSTM model through an experimental comparative study. Thus, our model benefits from the extraction of local characteristics thanks to the CNN model but also of long-term dependencies captured thanks to the high memory power of LSTMs. For feature extraction, we trained a FastText model to generate vectors of words after the plain text has been cleaned up and segmented into sentences. Finally, we carried out several experiments which showed that our model is very efficient for the classification of texts. Our proposed CNN-LSTM model achieved the highest study precision 86.34% and also gives the highest correct prediction ratio which shows the importance of the proposed model.

Although the use of a corpus of medical texts for the extraction of characteristics makes it possible to obtain a better classification of sentences, it constitutes a major limitation to our approach, namely a classification system closely linked to our datasets and therefore a classification model that is difficult to generalize to other fields of application. A second limitation identified is that our model is difficult to compare with already existing classification systems often tested on data in English, yet the word processing is based on both computer modeling and linguistic studies.

In future work, we will continue to study the application of deep learning methods and other methods in the field of text classification to improve our results. This also involves the integration of new sources of annotated data in order to ensure a minimum of representativeness of the various corpora in the centers. This model will then be used to facilitate the extraction of medical concepts into a system that allows the creation of homogeneous patient cohorts from a large volume of data. Additionally, future research could test our model on other datasets, both medical and non-medical.

References

1. Nguyen, P., Tran, T., Wickramasinghe, N., Venkatesh, S.: Deepr: a convolutional net for medical records. IEEE J. Biomed. Health Inform. **21**(1), 22–30 (2017). https://doi.org/10.1109/JBHI.2016.2633963
2. Biricik, G., Diri, B., Sönmez, A.C.: Abstract feature extraction for text classification. Turk. J. Electr. Eng. Comput. Sci. **20**, 1137–1159 (2012). https://doi.org/10.3906/elk-1102-1015
3. Qing, L., Linhong, W., Xuehai, D.: A novel neural network-based method for medical text classification. Future Internet **11**, 255 (2019). https://doi.org/10.3390/fi11120255
4. Kowsari, K., Jafari, M.K., Heidarysafa, M., Mendu, S., Barnes, L., Brown, D.: Text classification algorithms: a survey. Information **10**(4), 150 (2019)
5. Joulin, A., Grave, E., Bojanowski, P., Douze, M., Jégou, H., Mikolov, T.: FastText.zip: compressing text classification models (2016)
6. Frank, E., Bouckaert, R.R.: Naive Bayes for text classification with unbalanced classes. In: Fürnkranz, J., Scheffer, T., Spiliopoulou, M. (eds.) PKDD 2006. LNCS (LNAI), vol. 4213, pp. 503–510. Springer, Heidelberg (2006). https://doi.org/10.1007/11871637_49
7. Zhang, W., Yoshida, T., Tang, X.: Text classification based on multi-word with support vector machine. Knowl.-Based Syst. **21**, 879–886 (2008). https://doi.org/10.1016/j.knosys.2008.03.044
8. Otter, D., Medina, J., Kalita, J.: A survey of the usages of deep learning for natural language processing. IEEE Trans. Neural Netw. Learn. Syst. **32**(2), 604–624 (2020). https://doi.org/10.1109/TNNLS.2020.2979670
9. Lin, R., Liu, S., Yang, M., Li, M., Zhou, M., Li, S.: Hierarchical recurrent neural network for document modeling, pp. 899–907 (2015). https://doi.org/10.18653/v1/D15-1106
10. Yang, Z., Yang, D., Dyer, C., He, X., Smola, A., Hovy, E.: Hierarchical attention networks for document classification, pp. 1480–1489 (2016). https://doi.org/10.18653/v1/N16-1174
11. Wang, B.: Disconnected recurrent neural networks for text categorization, pp. 2311–2320 (2018). https://doi.org/10.18653/v1/P18-1215
12. Collobert, R., Weston, J.: A unified architecture for natural language processing: deep neural networks with multitask learning. In: Proceedings of the 25th International Conference on Machine Learning, pp. 160–167. ACM (2008). https://doi.org/10.1145/1390156.1390177
13. Kim, Y.: Convolutional neural networks for sentence classification. In: Proceedings of the 2014 Conference on Empirical Methods in Natural Language Processing (2014). https://doi.org/10.3115/v1/D14-1181
14. Kalchbrenner, N., Grefenstette, E., Blunsom, P.: A convolutional neural network for modelling sentences. In: Proceedings of the Conference on 52nd Annual Meeting of the Association for Computational Linguistics, ACL 2014 (2014). https://doi.org/10.3115/v1/P14-1062
15. Zhang, X., Zhao, J., Lecun, Y.: Character-level convolutional networks for text classification (2015)
16. Conneau, A., Schwenk, H., Barrault, L., Lecun, Y.: Very deep convolutional networks for natural language processing. In: KI - Künstliche Intelligenz, vol. 26 (2016)
17. Johnson, R., Zhang, T.: Deep pyramid convolutional neural networks for text categorization, pp. 562–570 (2017). https://doi.org/10.18653/v1/P17-1052

18. Beaulieu-Jones, B., Greene, C.: Semi-supervised learning of the electronic health record for phenotype stratification. J. Biomed. Inform. **64**, 168–178 (2016)
19. Liaw, A., Wiener, M.: Classification and regression by random forest. R News **2**, 18–22 (2002)
20. Hughes, M., Li, I., Kotoulas, S., Suzumura, T.: Medical text classification using convolutional neural networks. Stud. Health Technol. Inform. **235** (2017). https://doi.org/10.3233/978-1-61499-753-5-246
21. Baker, S., Krohonen, A., Pyysalo, S.: Cancer hallmark text classification using convolutional neural networks (2016)

Data Mining

Diversified Pattern Mining on Large Graphs

Xin Wang[1(✉)], Liang Tang[1], Yong Liu[1], Huayi Zhan[2], and Xuanzhe Feng[3]

[1] Southwest Petroleum University, Chengdu, China
{xinwang,tl}@swpu.edu.cn
[2] Sichuan ChangHong Electric Co. Ltd., Mianyang, China
huayi.zhan@changhong.com
[3] Sichuan University, Chengdu, China
2018141461166@stu.scu.edu.cn

Abstract. Frequent pattern mining (FPM) on large graph has been receiving increasing attention due to its wide applications. The FPM problem is defined as mining all the subgraphs (*a.k.a.* patterns), with frequency above a user-defined threshold in a large graph. Though a host of techniques have been developed, most of them suffers from high computational cost and inconvenient result inspection. To tackle the issues, we propose an approach to discover diversified top-k patterns from a large graph G. We formalize the distributed top-k pattern mining problem based on a diversification function. We develop an algorithm with *early termination* property, to efficiently identify diversified top-k patterns. Using real-life and synthetic graphs, we show advantages of our algorithm via intensive experimental studies.

1 Introduction

Frequent pattern mining has been at the core of data mining research for a period. Existing work considers the problem under two different settings: transactional-based and single-graph-based. In recent years, more attention has been paid to the latter setting, as it plays a crucial role in a variety of applications such as bioinformatics, cheminformatics, web analysis, social network analysis, etc. Most of prior methods follow the combinatorial pattern enumeration paradigm. In real world applications such as social network analysis, the complete enumeration of patterns is practically infeasible, as the mining results are explosive in size [19,28].

Indeed, it is often unnecessary to enumerate all the patterns. Consider a frequent pattern Q, all its subgraphs must be frequent as well. If Q is returned, why do we need to identify its sub-patterns? Moreover, users are often only interested in a set of diversified patterns, instead of the overwhelmed pattern set [29].

This highlight the need for *diversified top-k pattern mining*: given a graph G, a support threshold θ and an integer k, it is to find k patterns, that not

© Springer Nature Switzerland AG 2021
C. Strauss et al. (Eds.): DEXA 2021, LNCS 12923, pp. 171–184, 2021.
https://doi.org/10.1007/978-3-030-86472-9_16

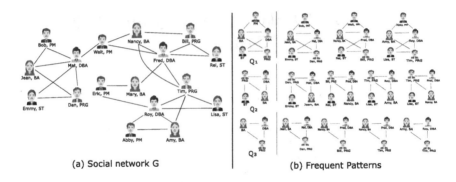

(a) Social network G   (b) Frequent Patterns

Fig. 1. A social graph G & a set of patterns along with their matches

only satisfy support constraint but also are as diverse as possible. Furthermore, if an algorithm for the problem preserves the *early termination property, i.e.,* it discovers diversified top-k patterns without identifying the entire pattern set, then we do not have to pay the price of costly pattern mining.

Example 1. A fraction of a social graph G is shown in Fig. 1(a), where each node denotes a person with name and job title (*e.g.,* project manager (PM), database administrator (DBA), programmer (PRG), business analyst (BA) and software tester (ST)); and each edge indicates friendship, *e.g.,* (Bob, Mat) indicates that Bob and Mat are friends. Observe that, number of matches of Q_1 is close to that of Q_2, Q_3 and moreover, when Q_1 is selected as one of the top-k patterns, neither Q_2 nor Q_3 is a favored pattern to be chosen, as they can be *represented by* Q_1 due to large overlap.   □

Contributions. The paper investigates the *diversified top-k pattern mining* superior performance and provides an effective approach for it.

(1) We introduce viable *support* and *distance* metrics to measure patterns. Based on the metrics, we introduce a diversification function and formalize the *diversified top-k pattern mining* (DTopkPM) problem (Sect. 3).
(2) We investigate the DTopkPM problem and develop an efficient algorithm for it (Sect. 4). The algorithm has desirable performances: its computational cost is influenced by a factor n (number of pattern extensions), that is often small in practice and it preserves *early termination property.*
(3) Using real-life and synthetic graphs, we experimentally verify the performance of our algorithms (Sect. 5). We find the following. (a) Our mining algorithm scales well with the increase of both θ and k. (b) Our algorithm works reasonably well on large graphs. For example, on a graph with 4 million nodes and 53.5 million edges, our algorithm spends less than $1000\,$s to discover high-quality diversified top-k patterns when $\theta = 1K$. (c) Our algorithm also scales well with the increase of the size of underlying graph G.

2   Related Work

There exist considerable works on the FPM and result diversification. We next review them as follows.

FPM *on Large Graphs*. Typical centralized FPM approaches can be classified into two categorizes, *i.e.*, [6,14,21] for static graphs and [4,23] for evolving graphs. A novel approach GRAMI is first introduced by [14]. GRAMI applies a minimum-image-based support metric, that preserves anti-monotonic property and models FPM problem as a constraint satisfaction problem. To address the issue on weighted graphs, [6,21] proposed approaches to mining weighted frequent subgraphs, on edge-weighted single large graphs. In recent years, distributed FPM techniques over single large graphs are intensively studied. Typical methods are listed as follows. [24] proposed a parallel subgraph listing framework PSgL, which deals with subgraph listing in a divide-and-conquer fashion. Another distributed platform Arabesque [26] employs a high-level filter-process model to facilitate mining computation. DISTGRAPH [25] uses a set of optimizations and efficient collective communication operations to minimize total amount of messages shipped among different sites. [3] proposed a scalable system ScaleMine. The system leverages the approximate and exact phases to achieve better load balance and more efficient evaluation for candidate patterns. Gemini [30] is another distributed and synchronous graph processing framework. It uses a low-overhead edge-cut partitioning strategy to distribute graph data, and applies a co-scheduling mechanism to alleviate the computing bottleneck. G-Miner [11] provides an expressive API as well as a novel task-pipeline that removes the synchronization barrier and hides the overheads of network and disk I/O and achieves high performance. [22] introduced an approach to mining the top-k uncertain frequent patterns from uncertain databases. This approach combines the mining and ranking phases as a whole and proposes effective threshold raising strategies to enhance the mining time and reduce the memory usage. [7] studied approximate k-vertex frequent pattern mining on a dynamic graph with high probability in a given time. PrefixFPM framework [27] fully utilizes the CPU cores in a multicore machine and adopts the prefix projection approach pioneered by PrefixSpan to achieve high performance. There also exist a host of techniques [8,13,20] developed on MapReduce. [8] introduced FSM-H, a novel iterative MapReduce-based frequent subgraph mining algorithm. Along the similar line, [13] introduces MR-SimLab, a scalable approach for representative subgraph selection based on MapReduce. In particular, MR-SimLab takes advantage of the similarity between node labels to support approximate isomorphism checking. [20] describes Pegasus, a graph mining system on top of MapReduce with the key component GIM-V.

Result Diversification. Result diversification is a bi-criteria optimization problem for balancing result relevance and diversity [9,16], with applications in *e.g.*, social searching [5]. It remains an open issue to efficiently mine diversified top-k patterns.

3 Graphs, Patterns and Pattern Mining

In this section, we first review graphs, patterns, graph pattern matching; we then formalize the pattern mining problem.

3.1 Graph Pattern Matching

Graph. A *data graph* (or simply *graph*) is defined as $G = (V, E, L)$, where (1) V is a set of nodes; (2) $E \subseteq V \times V$ is a set of *undirected* edges; and (3) each node v in V carries a tuple $L(v) = (A_1 = a_1, \cdots, A_n = a_n)$, where $A_i = a_i (i \in [1, n])$ represents that the node v has a value a_i for the attribute A_i, and is denoted as $v.A_i = a_i$, e.g., v.name = "Bill", v.job_title = "PM".

A graph $G' = (V', E', L')$ is a *subgraph* of $G = (V, E, L)$, denoted by $G' \subseteq G$, if $V' \subseteq V$, $E' \subseteq E$, and moreover, for each $v \in V'$, $L'(v) = L(v)$.

Pattern. A *pattern* Q is defined as a graph (V_p, E_p, f_v), where V_p and E_p are the set of nodes and edges, respectively; for each u in V_p, it is associated with a predicate $f_v(u)$ defined as a conjunction of atomic formulas of the form of '$A = a$' such that A denotes an attribute of the node u and a is a value of A. Intuitively, $f_v(u)$ specifies search conditions imposed by u.

A pattern $Q' = (V'_p, E'_p, f'_v)$ is *subsumed by* another pattern $Q = (V_p, E_p, f_v)$, denoted by $Q' \sqsubseteq Q$, if (V'_p, E'_p) is a subgraph of (V_p, E_p), and function f'_v is a restriction of f_v. Then, Q' is referred to a sub-pattern of Q if $Q' \sqsubseteq Q$.

Graph Pattern Matching. Consider graph G and pattern Q, a node v in G satisfies the search conditions of a pattern node u in Q, denoted as $v \sim u$, if for each atomic formula '$A = a$' in $f_v(u)$, there is an attribute A in $L(v)$ such that $v.A = a$.

We adopt subgraph isomorphism [12] as the matching semantic. A *match* of pattern Q in graph G is a *bijective function* ρ from the nodes of Q to the nodes of a subgraph G, such that (1) for each node $u \in V_p$, $\rho(u) \sim u$, and (2) (u, u') is an edge in Q if and only if $(\rho(u), \rho(u'))$ is an edge in G. When an isomorphism ρ from pattern Q to a subgraph G_s of G exists, we say G *matches* Q, and denote G_s as a *match* of Q in G. Abusing notations, we say v in G_s as a *match* of u in Q, when $\rho(u) = v$.

We denote by $M(Q, G)$ the set of matches G_s of Q in G. Then, for each node u in V_p, we derive a set $\{v | v \in \rho(Q), \rho(Q) \in M(Q, G), v = \rho(u)\}$ from match set $M(Q, G)$, and denote it by img(u). Intuitively, img(u) contains a set of *distinct* nodes v in G as matches of u in Q.

Example 2. Given graph G in Fig. 1 and patterns Q_1, Q_2 and Q_3 in Fig. 1(b), one may verify that Q_2, Q_3 are *subsumed by* Q_1, and moreover, $M(Q_1, G) = \{G_{11}, G_{12}, G_{13}, G_{14}\}$, $M(Q_2, G) = \{G_{21}, G_{22}, G_{23}, G_{24}\}$ and $M(Q_3, G) = \{G_{31}, G_{32}, G_{33}\}$.

DFS Tree. Given a pattern Q, its DFS tree T_Q can be built via a depth-first search in Q from a node u. Then, edges that are in T_Q are referred to as *forward edges* and the remaining edges in Q are denoted as *backward edges*.

Thus, the *forward extension* on a pattern Q essentially introduces a new edge from one node in Q; while the *backward extension* includes a new edge from two existing nodes. For example, a pattern Q_c with edge set $\{(\mathsf{BA}, \mathsf{DBA}), (\mathsf{DBA}, \mathsf{PRG})\}$ can be generated via *forward extension* from a pattern with edge $(\mathsf{BA}, \mathsf{DBA})$; with Q_c, another pattern Q_1 (shown in Fig. 3(b)) is generated via *backward extension*.

We will use the following notations. (1) The *size* $|G|$ of G (resp. $|Q|$ of Q) is $|V| + |E|$ (resp. $|V_p| + |E_p|$), the total number of nodes and edges in G (resp. Q). (2) A graph G (resp. pattern Q) is a *complete* graph (resp. pattern), if there exists an edge for each pair of nodes in it. (3) In a directed tree T, the height of a node v is the length of the longest downward path to a leaf node from v. Then the height h of T is the largest height among all tree nodes. (4) $Q_i \cap Q_j$ (resp. $Q_i \cup Q_j$) indicates the intersection (resp. union) operation on edge sets of patterns Q_i and Q_j.

3.2 Diversified Top-k Pattern Mining

We first introduce the *support* metric. We then propose a diversification function, followed by the diversified top-k pattern mining problem.

Support. The support of a pattern Q in a graph G, denoted by $\mathsf{sup}(Q, G)$, indicates the appearance frequency of Q in G. Analogous to the association rules for itemsets, the support metric for patterns should be *anti-monotonic*, i.e., for patterns Q and Q', if $Q' \sqsubseteq Q$, then $\mathsf{sup}(Q', G) \geq \mathsf{sup}(Q, G)$ for any G, to facilitate search space pruning. Several *anti-monotonic* support metrics for pattern mining exist, e.g., minimum image (mni) [14], harmful overlap [15] and maximum independent sets [18]. In this paper, mni is adopted due to its efficiency.

$$\mathsf{sup}(Q, G) = \min\{|\mathsf{img}(u)| \mid u \in V_p\}, \tag{1}$$

where $\mathsf{img}(u)$ is the image of pattern node u in G. It can be easily verified that this support measure is *anti-monotonic*.

Pattern Diversification. In practice, people are more favored with those patterns, which not only have high support but also are as diverse as possible. This calls for a function to measure the diversification of a pattern set. To this end, we first introduce a distance function to measure the difference of a pair of patterns, followed by the diversification function.

Distance Function. Given a pair of patterns Q_i and Q_j, we define a distance function to measure how "dissimilar" they are.

$$d(Q_i, Q_j) = 1 - \frac{|Q_i \cap Q_j|}{|Q_i \cup Q_j|}.$$

Intuitively, the distance between a pair of patterns Q_i and Q_j indicates their structural difference, and the larger $d(Q_i, Q_j)$ is, the more dissimilar Q_i and Q_j are.

Diversification Function. On a set of frequent patterns $\mathbb{S} = \{Q_1, \cdots, Q_k\}$, the diversification function $F(\cdot)$ is defined as follows.

$$F(\mathbb{S}) = (1 - \lambda) \sum_{Q_i \in \mathbb{S}} \mathsf{sup}(Q_i) + \frac{2 \cdot \lambda}{k - 1} \sum_{Q_i \in \mathbb{S}, Q_j \in \mathbb{S}, i \leq j} d(Q_i, Q_j),$$

where $\lambda \in [0, 1]$ is a parameter set by users.

The diversity metric is scaled down with $\frac{2 \cdot \lambda}{k-1}$, since there exist $\frac{k \cdot (k-1)}{2}$ pairs for the difference sum, while only k numbers for the support sum. The function $F(\cdot)$ is a minor revision of the max-sum diversification problem [16], which is a bicriteria objective function to capture both relevance and diversity, and strikes a balance between the two with a user-defined parameter λ.

Problem Statement. With the diversification function, the problem of *diversified top-k pattern mining* is stated as follows. Given a graph G, support threshold θ, integer k and parameter $\lambda \in [0, 1]$, it is to find a set $\mathbb{S}_k$ of k patterns such that

$$\mathbb{S}_k = \arg \max_{\mathbb{S}' \subseteq \mathbb{S}} F(\mathbb{S}'),$$

where $\mathbb{S}$ is the set of patterns with support above θ and $\mathbb{S}'$ is the k-element subset of $\mathbb{S}$. Intuitively, the problem is to find k (specified by users) patterns that are not only with high support but also as diverse as possible.

4 Diversified Top-k Pattern Mining

Intuitively, a naive algorithm (denoted as naive) for DTopkPM problem works as follows: it (1) discovers a complete set $\mathbb{S}$ of frequent patterns; (2) exhaustively enumerates k-element subset $\mathbb{S}'$ of $\mathbb{S}$ and computes $F(\mathbb{S}')$; and (3) picks one subset $\mathbb{S}_k$ with the largest $F(\cdot)$. Though straightforward, naive has to discover the complete pattern set $\mathbb{S}$ and performs $\binom{|\mathbb{S}|}{k}$ rounds comparison to pick $\mathbb{S}_k$. As $|\mathbb{S}|$ is theoretically of exponential size of $|G|$, naive is extremely costly in practice.

While one can rectify naive by using "early termination" algorithms. Compared with naive, these algorithms stop as soon as k patterns with *high quality* are identified, without identifying the complete pattern set $\mathbb{S}$.

Theorem 1. *Given a graph G, a support threshold θ, a tuning parameter λ and an integer k, there exists an algorithm for the DTopkPM problem, that finds a set $\mathbb{S}_k$ of patterns such that (a) $\mathsf{sup}(Q, G) \geq \theta$ for each Q in $\mathbb{S}_k$, (b) the support computation is in $O(|V|^n \cdot n^{n+1})$ time (n refers to the expansion times), and (c) can terminate as soon as k patterns are discovered.*

Algorithm DMiner

Input: A graph G, support threshold θ, tuning parameter λ and integer k.
Output: A set of k patterns.
1. initialize $S^{[1]}:=\emptyset$; $\mathbb{S}_k := \emptyset$; flag := false; $\mathcal{T}$ as an empty tree;
2. update $S^{[1]}$; update $\mathcal{T}$;
3. **while** (flag $\neq$ true) **do**
4. $L := \mathsf{TreeGen}(S^{[1]}, \mathcal{T})$;
5. $\mathcal{T} := \mathsf{EvaTPs}(L, \theta, \mathcal{T})$;
6.   **if** $\mathcal{T}$ is not updated **then** flag := true;
7. $\mathbb{S}_k := \mathsf{TopkSch}(\mathcal{T}, S^{[1]}, \theta, \lambda, k)$;
8. **return** $\mathbb{S}_k$.

Procedure TopkSch

Input: A tree $\mathcal{T}$, a set $S^{[1]}$, support threshold θ, tuning parameter λ, and k.
Output: A set of patterns.
1. initialize Terminate:=false; $\mathbb{S}_k := \emptyset$; h as the height of $\mathcal{T}$;
2. **while** (Terminate $\neq$ true) **do**
3.   **for each** v at level h of $\mathcal{T}$ **do**
4. L_p:=$\mathsf{NonTreeGen}(Q_{[v]}, S^{[1]}, \mathcal{T})$;
5.   **for each** Q_c in L_p **do**
6.   **if** Q_c was not generated **and** $\sup(Q_c, G) \geq \theta$ **then**
7.   **if** $|\mathbb{S}_k| < k$ **then** $\mathbb{S}_k := \mathbb{S}_k \cup \{Q_c\}$;
8.   **else** update $\mathbb{S}_k$ with Q_c s.t. $F(\mathbb{S}_k)$ is maximized;
9.   **if** termination condition is satisfied **then**
10.   Terminate:=true; **break** ;
11.   update h;
12. **return** $\mathbb{S}_k$;

Fig. 2. Algorithm DMiner

Proof. We show Theorem 1 by presenting an algorithm as a constructive proof. The algorithm is denoted as DMiner and shown in Fig. 2. In a nutshell, it incrementally identifies diversified top-k patterns and terminates as soon as the termination condition is satisfied. We next illustrate the details of DMiner.

Algorithm. DMiner takes a graph G, support threshold θ, tuning parameter λ and integer k as input and outputs a set $\mathbb{S}_k$ of patterns.

Initialization. DMiner first initializes a set of parameters: set $S^{[1]}$ for maintaining single-edge (*a.k.a.* "seed") patterns, set $\mathbb{S}_k$ for keeping track of diversified top-k patterns, a Boolean variable flag to control **while** loop and a tree $\mathcal{T}$ (line 1). It then computes support of each "seed" pattern and only maintains those patterns with support above θ in $S^{[1]}$ and expands the tree $\mathcal{T}$ with nodes, that correspond to "seed" patterns in $S^{[1]}$ (line 2).

Tree Pattern Generation. Based on $S^{[1]}$, DMiner repeatedly produces "tree" patterns and verifies their supports, following a *level-wise* strategy (lines 3–6).

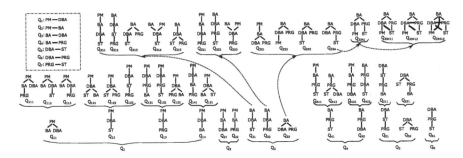

Fig. 3. A tree $\mathcal{T}$ showing the hierarchical structure of candidate patterns (Color figure online)

Starting from bottom level, DMiner repeatedly performs the following. It first generates a set L of "tree" patterns as candidates with procedure TreeGen (line 4). Note that TreeGen (not shown) generates candidate patterns by expanding frequent "tree" patterns that locate at the top level of $\mathcal{T}$ with "seed" patterns in $S^{[1]}$, following *forward expansion*. Then, DMiner invokes EvaTPs to evaluate support of each candidate pattern in L and updates $\mathcal{T}$ with qualified patterns (line 5). After above process finished, if $\mathcal{T}$ remains unchanged, the Boolean variable flag is changed to true, indicating that the **while** loop no longer needs to continue (line 6).

Example 3. On graph G of Fig. 1(a), DMiner first initializes the set $S^{[1]}$ with patterns Q_1–Q_7 (shown in Fig. 3), as their supports all equal to 3. Then, DMiner applies TreeGen to generate candidate patterns following *forward extension*, in a *level-by-level* manner. For example, using pattern Q_1, S_c generates candidate patterns by enlarging Q_1 with other frequent single-edge patterns and produces L = $\{Q_{11}, Q_{12}, Q_{13}, Q_{14}\}$. Four levels of "nontrivial" candidate patterns (patterns without duplicate node labels) are shown in Fig. 3, where patterns marked with blue are considered infrequent (with support less than 3).

Diversified Pattern Mining. Using frequent "tree" patterns, DMiner employs procedure TopkSch to discover diversified top-k patterns. More specifically, TopkSch first initializes a Boolean variable Terminate, an empty set $\mathbb{S}_k$ and an integer h as the height of $\mathcal{T}$ (line 1). It next repeatedly identifies qualified non-tree patterns as follows (lines 2–11). In each round, TopkSch selects a node v, that corresponds to a "tree" pattern $Q_{[v]}$, at level h of $\mathcal{T}$, and generates a set L_p of candidate patterns with procedure NonTreeGen (line 4). Note that NonTreeGen works in the similar way as TreeGen, but only enlarges pattern $Q_{[v]}$ with "seed" patterns via *backward expansion*. For each candidate pattern Q_c in L_p, TopkSch verifies whether Q_c has been generated before and computes its support along the same line as before (line 6). To facility existence verification of a candidate pattern, we apply a heuristic strategy, which is verified effective for fast pruning. For each pattern, that is generated before, we maintain its key features, *e.g.*, label distribution, degree distribution, diameter and use the features to prune unqualified patterns. If Q_c meets the support constraint and is unseen before, TopkSch either includes it in $\mathbb{S}_k$ if $|\mathbb{S}_k| < k$ (line 7) or identifies a pattern Q_r from $\mathbb{S}_k$ with maximum $F(\mathbb{S}_k \cup \{Q_c\} \backslash \{Q_r\}) - F(\mathbb{S}_k)$ and replaces it with Q_c (line 8). TopkSch next verifies whether the termination condition, specified by following proposition is satisfied.

Proposition 2: *Given graph G, parameters θ, λ, k and a set $\mathbb{S}^t$ of frequent "tree" patterns organized in a tree $\mathcal{T}$ structure, a k-element set $\mathbb{S}_k$ is a diversified pattern set, if (1) $\mathsf{sup}(Q, G) \geq \theta$ for each Q in $\mathbb{S}_k$, and (2) $\min\{\mathsf{sup}(Q, G)|Q \in \mathbb{S}_k\} \geq \max\{\mathsf{sup}(Q_t, G)|Q_t \in \mathbb{S}^t \backslash \mathbb{S}_b^k\}$ and $\widehat{Q_t} \cap \bigcup_{Q \in \mathbb{S}_k} Q = \emptyset$ for any $Q_t \in \mathbb{S}^t \backslash \mathbb{S}_b^k$.*
$\square$

Here $\mathbb{S}_b^k$ includes all the tree patterns that were used to generate patterns in $\mathbb{S}^k$, and $\widehat{Q_t}$ indicates a *complete* pattern that is expanded from a tree pattern Q_t in $\mathbb{S}_b^t$. Intuitively, Proposition 2 states that when the minimum support of a pattern in $\mathbb{S}_k$ is already no less than that of any pattern in $\mathbb{S}^t \backslash \mathbb{S}_b^k$ and $\widehat{Q_t}$ can not bring any new edge, then $F()$ value can not grow, hence no further investigation is needed.

If the termination condition is satisfied, TopkSch sets Terminate as true and breaks the **while** loop (line 10) and returns $\mathbb{S}_k$ as final result (line 12). Otherwise, after all the nodes on level h are processed, TopkSch decreases h by 1 for next round iteration (line 11).

Example 4. To identify the top-1 pattern, TopkSch first generates new candidate patterns, e.g., Q_{33411}, Q_{33412}, Q_{33413} by expanding Q_{3341}, at the top level ($h = 3$) of $\mathcal{T}$ via *backward extension*. Assume that these candidates have not been generated before, TopkSch then evaluates their supports following the descending order of itrs and obtains $\mathbb{S}_k = \{Q_{33413}\}$. The above process can terminate until candidates generated from "tree" patterns at level 2 are all processed, as the remaining candidates can not have higher itrs values.

When a set L of candidate patterns are generated, procedure EvaTPs (not shown) is invoked for support evaluation. Specifically, EvaTPs identifies one parent node Q'_c of Q_c in $\mathcal{T}$, whose matches will be used for support evaluation and initializes an empty set I_b, which is used for recording "virtual matches". For each match G_s of Q'_c, EvaTPs checks whether G_s can be extended as a match G'_s of Q_c and includes G'_s in match set $M(Q_c, G)$ if G'_s exists and updates $\mathcal{T}$.

Correctness. The correctness of DMiner is warranted by the following observations. (1) The pattern generation scheme will never miss any qualified pattern. (2) The support computation with *partial evaluation* is correct. (3) The strategy used by pattern selection guarantees to choose a "best" pattern in each round iteration.

Complexity. For a pattern Q_c with vertex set V_{p_c}, there may exist at most $|V|^{|V_{p_c}|}$ matches in a graph G. Thus, it takes $O(|V|^{|V_{p_c}|+1})$ time to identify matches of a candidate pattern Q'_c ($Q_c \sqsubseteq Q'_c$). If Q_c is a "tree" pattern, there may generate at most $|V_{p_c}||\mathcal{L}|$ candidate patterns after *forward extension*; otherwise, at most $O(|V_{p_c}|^2)$ candidate patterns will be generated through *backward exten-sion*. In the meanwhile, it still needs $O(|V|^{|V_{p_c}|+1} \cdot (|V_{p_c}| + 1)^{|V_{p_c}|+1})$ time to verify whether Q_c has been generated before, since at most $|V|^{|V_{p_c}|+1}$ candidate patterns may be generated before and each round isomorphism checking needs $O((|V_{p_c}| + 1)^{|V_{p_c}|+1})$ time. By induction, it takes DMiner $\sum_{i \in [1,n]} (|V_{p_c}| + i - 1)|V| \cdot |V|^{|V_{p_c}|+i}(1 + (|V_{p_c}| + i)^{|V_{p_c}|+i})$ times to verify all the subsequent patterns for a candidate Q_c, where n refers to the extension times and is bounded by $|V|$, $|\mathcal{L}|$ and $|V_{p_c}|$ are bounded by $|V|$. As the iteration starts from single-edge patterns Q_c and at most $|V|^2$ different Q_c exists, hence DMiner is bounded by $O(|V|^n \cdot n^{n+1})$ time.

The analysis above completes the proof of Theorem 1.   □

5   Experimental Study

Using real-life and synthetic data, we conducted comprehensive experimental studies to evaluate: efficiency, data shipment and scalability of algorithm DMiner.

Experimental Setting. We used three real-life graphs: (a) *Amazon* [1], a prod-uct co-purchasing network with 0.55 million nodes and 1.79 million edges. The total size of *Amazon* is 0.95 GB. (b) *Pokec* [2], a social network with 1.63 million nodes, and 30.6 million edges. Its size is 2.2 GB. (c) *Google+* [17], a social graph whose size is 2.6 GB, has 4 million entities and 53.5 million links.

We designed a generator to produce synthetic graphs $G = (V, E, L)$, con-trolled by the # of nodes $|V|$ and edges $|E|$, where L is taken from an alphabet of $1K$ labels.

Algorithms. We implemented the algorithm DMiner, compared with algo-rithm GRAMI$_D$, which implements GRAMI [14] to find all frequent patterns and selects diversified k patterns with an approximation strategy proposed by [10], in Java.

Experimental Results. We used the *logarithmic scale* for the y-axis in the figures for RT (response time). To evaluate result quality, we also define a ratio as $R_d = \frac{F(\mathbb{S}_k^1)}{F(\mathbb{S}_k^2)}$, where $\mathbb{S}_k^a$ and $\mathbb{S}_k^b$ are the top-k pattern set discovered by DMiner and GRAMI$_D$, respectively. The performance comparisons are carried out via changing of θ, k and $|G|$; while due to space constraint, we do not show the influence of parameter λ. We next report our findings.

Varying θ. Fixing $k = 50$ and $\lambda = 0.5$, we varied the support threshold θ from $0.1K$ to $0.5K$ in $0.1K$ increments, $2K$ to $4K$ in $0.5K$ increments and $0.6K$ to $1.0K$ in $0.1K$ increments on *Amazon*, *Pokec* and *Google+*, respectively.

Figures 4(a)–4(c) show results on RT and tell us the following. (1) Both the algorithms take longer with smaller θ. This is because more candidate patterns and their matches have to be verified. (2) DMiner outperforms GRAMI$_D$ in all cases and is less sensitive to the increase of θ owing to its *early termination* property.

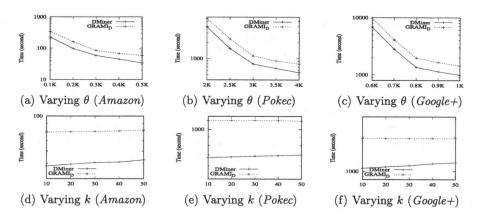

Fig. 4. Performance of DMiner on real-life graphs

Table 1 shows ratios R_d as well as influence of θ on R_d. We find the following. (1) Pattern sets identified by $GRAMI_D$ have higher diversification values, since it applies a more costly method with better performance. (2) With the increase of θ, R_d grows with minor changes. This is because for a larger θ, the corresponding pattern set $\mathbb{S}$ becomes smaller, leading to the weaker advantage of $GRAMI_D$.

Table 1. Evaluation of R_d by varying θ

	Amazon					Pokec					Google+				
	0.1K	0.2K	0.3K	0.4K	0.5K	2K	2.5K	3K	3.5K	4K	0.6K	0.7K	0.8K	0.9K	1.0K
R_d	0.85	0.87	0.87	0.88	0.88	0.75	0.75	0.77	0.78	0.78	0.81	0.83	0.84	0.85	0.85

Varying k. Fixing $\lambda = 0.5$ and $\theta = 0.3K$, $3K$ and $0.8K$ for _Amazon_, _Pokec_ and _Google+_, respectively, we varied k from 10 to 50 in 10 increments, and compared DMiner with $GRAMI_D$, _w.r.t._ RT.

Results shown in Figs. 4(d)–4(f) tell us following. (1) DMiner runs much more efficiently than $GRAMI_D$, owing to its _early termination_ property. For example, at _google+_, DMiner only takes less than 70% time of $GRAMI_D$. (2) DMiner is sensitive to the increase of k, since it has to verify more candidate patterns before termination condition can be satisfied. (3) DMiner is less sensitive to the change of k, still owing to its special strategy to early terminate.

Table 2. Evaluation of R_d by varying k

	Amazon					Pokec					Google+				
	10	20	30	40	50	10	20	30	40	50	10	20	30	40	50
R_d	0.90	0.89	0.89	0.87	0.87	0.81	0.79	0.78	0.78	0.77	0.89	0.88	0.86	0.85	0.84

Table 2 tells us that with the increase of k, GRAMI$_D$ can find even better top-k pattern set (with smaller R_d). This is because it implements an approximation algorithm, which is more costly, to identify k patterns from a complete pattern set.

Varying $|G|$ (Synthetic). Fixing $n = 4$, $k = 50$ and $\theta = 1K$, we varied $|G|$ from $(10M, 20M)$ to $(50M, 100M)$ with $10M$ and $20M$ increments on $|V|$ and $|E|$. We find that (1) both algorithms take longer time, which is as expected; and (2) DMiner is less sensitive to $|G|$ than others, *w.r.t.* RT, showing better scalability. Due to space constraint, we do not show the figure.

6   Conclusion

We have investigated the diversified top-k pattern mining problem. We first introduce the support and distance metrics and propose a diversification function based on the metrics. We then develop an algorithm with *early termination* property to efficiently discover diversified top-k patterns. Our experimental study has verified the efficiency, effectiveness and scalability of the algorithm. We hence contend that our approach yields a promising tool for big graph analysis.

The study of DTopkPM is still in its infancy. One topic for future work is to develop distributed method such that costly pattern mining can be performed in parallel. Another topic concerns early pruning for generation and supports verification of candidate patterns. The third topic is to develop techniques to efficiently maintain top-k patterns from evolving graphs.

References

1. Amazon. http://snap.stanford.edu/data/amazon-meta.html
2. Pokec social network. http://snap.stanford.edu/data/soc-pokec.html
3. Abdelhamid, E., Abdelaziz, I., Kalnis, P., Khayyat, Z., Jamour, F.T.: ScaleMine: scalable parallel frequent subgraph mining in a single large graph. In: Proceedings of the International Conference for High Performance Computing, Networking, Storage and Analysis, SC, pp. 716–727. IEEE Computer Society (2016)
4. Abdelhamid, E., Canim, M., Sadoghi, M., Bhattacharjee, B., Chang, Y., Kalnis, P.: Incremental frequent subgraph mining on large evolving graphs. IEEE Trans. Knowl. Data Eng. **29**(12), 2710–2723 (2017)
5. Alonso, O., Gamon, M., Haas, K., Pantel, P.: Diversity and relevance in social search. In: DDR (2012)
6. Ashraf, N., et al.: WeFreS: weighted frequent subgraph mining in a single large graph. In: Perner, P. (ed.) 19th Industrial Conference on Advances in Data Mining - Applications and Theoretical Aspects, ICDM, pp. 201–215. ibai Publishing (2019)
7. Aslay, Ç., Nasir, M.A.U., De Francisci Morales, G., Gionis, A.: Mining frequent patterns in evolving graphs. In: Proceedings of the 27th ACM International Conference on Information and Knowledge Management, CIKM, pp. 923–932. ACM (2018)

8. Bhuiyan, M., Hasan, M.A.: An iterative MapReduce based frequent subgraph mining algorithm. IEEE Trans. Knowl. Data Eng. **27**(3), 608–620 (2015)
9. Borodin, A., Lee, H.C., Ye, Y.: Max-sum diversification, monotone submodular functions and dynamic updates. In: PODS, pp. 155–166. ACM (2012)
10. Borodin, A., Lee, H.C., Ye, Y.: Max-sum diversification, monotone submodular functions and dynamic updates. In: Benedikt, M., Krötzsch, M., Lenzerini, M. (eds.) Proceedings of the 31st ACM SIGMOD-SIGACT-SIGART Symposium on Principles of Database Systems, PODS, pp. 155–166. ACM (2012)
11. Chen, H., Liu, M., Zhao, Y., Yan, X., Yan, D., Cheng, J.: G-Miner: an efficient task-oriented graph mining system. In: Oliveira, R., Felber, P., Hu, Y.C. (eds.) Proceedings of the Thirteenth EuroSys Conference, EuroSys, pp. 32:1–32:12. ACM (2018)
12. Cordella, L.P., Foggia, P., Sansone, C., Vento, M.: A (sub)graph isomorphism algorithm for matching large graphs. TPAMI **26**(10), 1367–1372 (2004)
13. Dhifli, W., Aridhi, S., Nguifo, E.M.: MR-SimLab: scalable subgraph selection with label similarity for big data. Inf. Syst. **69**, 155–163 (2017)
14. Elseidy, M., Abdelhamid, E., Skiadopoulos, S., Kalnis, P.: GRAMI: frequent subgraph and pattern mining in a single large graph. PVLDB **7**(7), 517–528 (2014)
15. Fiedler, M., Borgelt, C.: Subgraph support in a single large graph. In: Workshops Proceedings of the 7th IEEE International Conference on Data Mining, pp. 399–404. IEEE Computer Society (2007)
16. Gollapudi, S., Sharma, A.: An axiomatic approach for result diversification. In: Quemada, J., León, G., Maarek, Y.S., Nejdl, W. (eds.) Proceedings of the 18th International Conference on World Wide Web, pp. 381–390. ACM (2009)
17. Gong, N.Z., et al.: Evolution of social-attribute networks: measurements, modeling, and implications using Google+. In IMC (2012)
18. Gudes, E., Shimony, S.E., Vanetik, N.: Discovering frequent graph patterns using disjoint paths. IEEE Trans. Knowl. Data Eng. **18**(11), 1441–1456 (2006)
19. Huan, J., Wang, W., Prins, J., Yang, J.: SPIN: mining maximal frequent subgraphs from graph databases. In: SIGKDD (2004)
20. Kang, U., Faloutsos, C.: Big graph mining: algorithms and discoveries. SIGKDD Explor. **14**(2), 29–36 (2012)
21. Le, N., Vo, B., Nguyen, L.B.Q., Fujita, H., Le, B.: Mining weighted subgraphs in a single large graph. Inf. Sci. **514**, 149–165 (2020)
22. Le, T., Vo, B., Huynh, V., Nguyen, N.T., Baik, S.W.: Mining top-k frequent patterns from uncertain databases. Appl. Intell. **50**(5), 1487–1497 (2020). https://doi.org/10.1007/s10489-019-01622-1
23. Ray, A., Holder, L., Choudhury, S.: Frequent subgraph discovery in large attributed streaming graphs. In: Proceedings of the 3rd International Workshop on Big Data, Streams and Heterogeneous Source Mining: Algorithms, Systems, Programming Models and Applications, vol. 36, pp. 166–181. JMLR.org (2014)
24. Shao, Y., Cui, B., Chen, L., Ma, L., Yao, J., Xu, N.: Parallel subgraph listing in a large-scale graph. In: SIGMOD (2014)
25. Talukder, N., Zaki, M.J.: A distributed approach for graph mining in massive networks. Data Min. Knowl. Discov. **30**(5), 1024–1052 (2016). https://doi.org/10.1007/s10618-016-0466-x
26. Teixeira, C.H.C., Fonseca, A.J., Serafini, M., Siganos, G., Zaki, M.J., Aboulnaga, A.: Arabesque: a system for distributed graph mining. In: Miller, E.L., Hand, S. (eds.) Proceedings of the 25th Symposium on Operating Systems Principles, SOSP 2015, Monterey, CA, USA, 4–7 October 2015, pp. 425–440. ACM (2015)

27. Yan, D., Qu, W., Guo, G., Wang, X.: PrefixFPM: a parallel framework for general-purpose frequent pattern mining. In: 36th IEEE International Conference on Data Engineering, ICDE, pp. 1938–1941. IEEE (2020)
28. Yan, X., Han, J.: CloseGraph: mining closed frequent graph patterns. In: Getoor, L., Senator, T.E., Domingos, P.M., Faloutsos, C. (eds.) Proceedings of the Ninth ACM SIGKDD International Conference on Knowledge Discovery and Data Mining, pp. 286–295. ACM (2003)
29. Zhu, F., Qu, Q., Lo, D., Yan, X., Han, J., Yu, P.: Mining top-k large structural patterns in a massive network. VLDB **4**(11), 807–818 (2011)
30. Zhu, X., Chen, W., Zheng, W., Ma, X.: Gemini: a computation-centric distributed graph processing system. In: Keeton, K., Roscoe, T. (eds.) 12th USENIX Symposium on Operating Systems Design and Implementation, OSDI 2016, Savannah, GA, USA, 2–4 November 2016, pp. 301–316. USENIX Association (2016)

EHUCM: An Efficient Algorithm for Mining High Utility Co-location Patterns from Spatial Datasets with Feature-specific Utilities

Yinqiao Li, Lizhen Wang[✉], Peizhong Yang, and Junyi Li

School of Information Sciences and Engineering, Yunnan University, Kunming, China
lzhwang@ynu.edu.cn

Abstract. High utility co-location pattern mining is still computationally expensive in terms of both runtime and memory consumption. In this paper, an efficient high utility co-location pattern mining algorithm, named EHUCM, is proposed to address this problem, which introduces the ideas of neighborhood materialization, participating objects of features and filtering unpromising candidate patterns to discover high utility co-location patterns more efficiently. To reduce the cost of dataset scanning, EHUCM pre-storing spatial relationships in a data structure to facilitate the search for potential candidate patterns. In addition, two effective pruning strategies are proposed in the EHUCM algorithm to improve the running overhead due to the utility measure not satisfying the downward closure property. Extensive experiments show that the EHUCM algorithm is 10 times or even 100 times faster than the traditional high utility co-location pattern mining algorithm.

Keywords: Spatial Data Mining · High utility co-location pattern · Pruning

1 Introduction

Co-location pattern mining in spatial datasets is a knowledge discovery problem. It aims at finding a set of spatial features whose objects are frequently located in the close geographic proximity [1]. This problem is useful in business [2], environmental science [3], biology [4] and many other fields. However, an important limitation of co-location pattern mining is that all features are considered equally important. This causes some important but non-prevalent patterns are missed [5]. To address this issue, Yang et al. first proposed the problem of high utility co-location pattern mining (HUCM) in spatial datasets with feature-specific utilities [6].

In contrast to co-location pattern mining, HUCM considers the case where each feature has a utility. Therefore, it can be used to discover sets of features with high utility, i.e. high utility co-location patterns (HUCPs). Whereas the utility of a co-location pattern may be lower, equal or higher than the utility of its subsets. Hence the pruning strategies based on the anti-monotonicity of prevalence in co-location pattern mining is not applicable to HUCM.

There have been several algorithms proposed for HUCM. Yang et al. [6] proposed the EPA algorithm, Wang et al. [7] proposed a base algorithm and three pruning strategies,

© Springer Nature Switzerland AG 2021
C. Strauss et al. (Eds.): DEXA 2021, LNCS 12923, pp. 185–191, 2021.
https://doi.org/10.1007/978-3-030-86472-9_17

and the min/max feature utility ratio algorithm were designed in [8]. Despite these research efforts, HUCM is still very expensive in terms of computation and memory. As they all mine HUCPs by generating and storing row instances of candidate patterns.

In this paper, we propose an efficient algorithm EHUCM (Efficient High Utility Co-location pattern Mining) which differs from past HUCM algorithms that generate all row instances of a candidate pattern c to compute pattern utility. It simply depends on the participating objects of each feature in c. Moreover, to reduce the cost of dataset scanning, we organize the spatial relationships in a data structure of feature-object neighbor tree that can be used to find potential candidate patterns.

2 Problem Definition

In a spatial dataset S, consider a set of spatial features F and a set of spatial objects O, each object o is represented with a tuple $<$ feature type, object id, location, utility $>$. If the distance between two objects $o, o' \in O$ is not greater than a given distance threshold d, the two objects satisfy **neighbor relationship** R. A **co-location pattern** c is a subset of spatial feature set F. The size of c is the number of features in c. A **row instance** RI of c represents a subset of objects, which includes an object of each feature in c and any two objects in RI satisfy the neighbor relationship, i.e., objects in RI form a clique. For a feature f in c, we say that an object o of f participates in c if at least one row instance of c involves o. The set of **participating objects** of f in c is noted as $Obj(f, c)$. Each feature $f_i \in F$ is associated with a positive number $v(f_i)$ called **external utility** of the feature that represents its importance. Correspondingly, the **internal utility** of f_i in c is the number of participating objects of f_i in c, denoted as $q(f_i, c) = |Obj(f_i, c)|$. Given a size k co-location pattern $c = \{f_1, f_2, \ldots, f_k\}$. The **feature utility** of a feature f_i in c is defined as $v(f_i, c) = v(f_i) \times q(f_i, c)$. The **pattern utility** of c is the sum of utility of each feature in c, defined as $u(c) = \sum_{f_i \in c} v(f_i, c)$. The pattern utility ratio of c is defined as $\lambda(c) = u(c)/U(S)$, where $U(S)$ is the total utility of S. c is a HUCP if and only if $\lambda(c) \geq$ *minutil*, where *minutil* is a user-specified minimum pattern utility ratio threshold.

Problem Definition. Given a spatial dataset S with feature-specific utilities, a distance threshold d and a utility ratio threshold *minutil*, the high utility co-location pattern mining is to find all high utility co-location patterns in S.

3 The EHUCM Algorithm

As stated above, the aim of this paper is to improve the efficiency of HUCM. In this section, we present the EHUCM algorithm.

3.1 The Search Space

Since the utility measure does not satisfy the downward closure property, all patterns in the spatial dataset need to be searched. Based on the idea that the combination of features with the greater total utility of all objects in a spatial dataset is more likely to be a HUCP, we search the space in descending order of the total utility of all objects of each feature in a spatial dataset. Let $\succ$ be the descending order of the total utility of the feature in F.

Definition 1 (Extensible Feature Set of a Pattern). Given a co-location pattern c. Let $E(c)$ denote the set of features that can be used to extend c according to the depth-first search, so

$$E(c) = \{y|y \in F \wedge y \succ x, \forall x \in c\} \tag{1}$$

Definition 2 (Extension of a Pattern). Given a co-location pattern c. A pattern c' is a single-feature extension of c if $c' = c \cup \{z\}$ for $z \in E(c)$. Also, if $c' = c \cup Z$ where Z is a set of features $Z \in 2^{|E(c)|}$ with $Z \neq \emptyset$, c' is an extension of c.

3.2 Feature-Object Neighbor Tree (FONT)

The participating objects of each feature in a candidate pattern c are generated by scanning the dataset. To reduce the cost of dataset scanning, we adopt the idea of neighborhood materialization to organize spatial relationships in a feature-object neighbor tree (FONT) data structure. With the FONT we can easily find objects that have neighbor relationships with a given object.

Definition 3 (Object Neighbor Set). Given an object $o_i \in O$ with feature type $o_i.\text{feature} = f_i$, the object neighbor set of o_i is defined as

$$ONS(o_i) = \{o_j|o_j \in O \wedge R(o_i, o_j) \wedge o_j.\text{feature} \in F \setminus \{f_i\}\} \tag{2}$$

The object neighbor set of o_i includes objects that have neighbor relationships with o_i and the feature type of o_j is different from f_i.

Definition 4 (Feature-Object Neighbor Set). Given an object o_i and its object neighbor set $ONS(o_i)$, the feature-object neighbor set of o_i on feature f_i is defined as

$$FONS(o_i, f_j) = \{o_j|o_j \in ONS(o_i) \wedge (o_j.\text{feature} = f_j)\} \tag{3}$$

The feature-object neighbor set $FONS(o_i, f_j)$ is a subset of object neighbor set $ONS(o_i)$, and it includes all objects of f_j in $ONS(o_i)$.

Definition 5 (Feature-Object Neighbor Tree). Given the set of spatial features $F = \{f_1, f_2, \ldots, f_m\}$ and all neighbor relationships among spatial objects, the feature-object neighbor tree (FONT for short) is designed as follows. (1) The root of the FONT is marked as "null" and each feature is a child of the root. (2) The root of the feature f_i sub-tree is f_i, and the object neighbor set of all objects of f_i constitute the branch of f_i. Each branch records an object and its feature-object neighbor set.

3.3 Two Pruning Strategies

The search space of HUCPs has $2^{|F|}-|F|-1$ candidates, the number of candidates grows exponentially with the number of features. Therefore, in order to efficiently mine the HUCPs, two pruning strategies are proposed in this subsection, named Pattern Utility Loss Ratio and Extended Pattern Utility Ratio.

Definition 6 (Utility Loss Ratio). Given a co-location pattern c. Let $lu(c)$ represent the utility loss ratio of c, denoted as

$$lu(c) = \frac{\sum\limits_{f_i \in c} tu(f_i) - u(c)}{U(S)} \tag{4}$$

where $tu(f_i)$ is the total utility of all objects of feature f_i in a spatial dataset S.

Lemma 1. If $lu(c) > 1\text{-}minutil$, all extended patterns of c cannot be high utility patterns.

Definition 7 (Extensible Objects of Extensible Feature of a Pattern). Given a co-location pattern c. The set of objects of feature f' in $E(c)$ is defined as

$$nei(f') = \left\{ o' | o'.\text{feature} = f', \forall FONS(o', f) \neq \emptyset, f \in c \right\} \tag{5}$$

Definition 8 (Extensible Utility Ratio Upper-bound). Given a co-location pattern c. The extensible utility ratio upper-bound of c in a spatial dataset S is defined as

$$ub(c) = \frac{\sum\limits_{f_i \in E(c)} v(f_i)|nei(f_i)|}{U(S)} \tag{6}$$

Definition 9 (Extended Pattern Utility Ratio). Given a co-location pattern c. The extended pattern utility ratio of c is defined as

$$eu(c) = \lambda(c) + ub(c) \tag{7}$$

Lemma 2. If $eu(c) < minutil$, all extended patterns of c cannot be high utility patterns.

3.4 The EHUCM Algorithm

Algorithm 1 scans the dataset once to generate the feature-object neighbor tree, and reorder the set of features F according to the descending order of the total utility of each feature in F. Then, initialize the candidate pattern c to the empty set.

Algorithm 1: The EHUCM algorithm

---

method:
1. $FONT$ = gen_feature_object_neighbor_tree(O, R)
2. The features in F be listed in descending order of $tu(f_i)$
3. Initialize $c = \varnothing$, $k = 1$
4. **Search(k)**

Algorithm 2: The Search procedure

---

method:
1. **while** $k \leq F$.size **do**
2. $c = c \cup \{f_k\}$ //generates candidates
3.   **if** c.size ≤ 1 **then** $k = k+1$, **Search(k)**
4.   **else**
5. $CS\text{-}HBS(c)$ = generate all participating objects per feature in c
6.   **if** $CS\text{-}HBS(c) \neq \varnothing$ **then** calculate $\lambda(c)$
7.   **if** $\lambda(c) \geq minutil$ **then** output c, $k = k+1$, **Search(k)**
8.   **else**
9.   calculate $lu(c)$
10.   **if** $lu(c) \leq 1 - minutil$ **then** calculate $ub(c)$, $eu(c)$
11.   **if** $eu(c) \geq minutil$ **then** $k = k+1$, **Search(k)**
12.   remove f_k from c, **Search(k)**   //backtrack
13. **end**

The *Search* procedure (Algorithm 2) takes as a parameter the ordinal number of the k-th element of the set of features F. The procedure executes a loop that considers each single-feature extension of c of the form $c = c \cup \{f_k\}$, where f_k is the k-th element of F.

4   Experimental Evaluation

The experiments were conducted on a Windows 10 platform with an Intel Core i7-8700K CPU @3.70 GHz and 32 GB of RAM. We used two real spatial datasets, namely plants dataset of Three Parallel Rivers of Yunnan Protected Area and Beijing POI dataset. We compared the performance of the EHUCM algorithm with the EPA [6].

Influence of the Distance Threshold d. This experiment compares the running times of the two algorithms for mining HUCPs when the distance threshold is varied and the minimum utility ratio threshold parameter is fixed. Figure 1(a) and (b) shows the results. It can be observed that the EHUCM is always much faster than the EPA algorithm. For example, on Three Parallel Rivers with $minutil = 0.18$ and $d = 10900$, EHUCM is about 165 times faster than EPA.

Influence of the pattern utility ratio threshold $minutil$. The performance of algorithms was evaluated by fixing the value of the distance threshold in each dataset and varying the value of the minimum utility ratio threshold. The results are shown in Fig. 1(c) and (d). The running time of both algorithms decreases as $minutil$ increases, and EHUCM always runs in less time than EPA. For example, on Three Parallel Rivers with $d = 10500$ and *minutil* = 0.16, EHUCM is about 144 times faster than EPA.

The results of the EPA algorithm and the EHUCM algorithm in the above experiments are consistent.

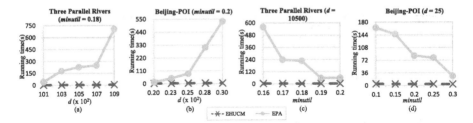

Fig. 1. Influence of the d and *minutil* on different datasets.

5   Conclusions

Since existing high utility co-location pattern mining algorithms are still very expensive in terms of both runtime and memory consumption. This paper proposes an efficient algorithm EHUCM for high utility co-location pattern mining. This algorithm differs from past algorithms that generate all row instances of candidate patterns to compute pattern utility. EHUCM simply depends on the participating objects of each feature in the candidate pattern. Because the utility measure fails to satisfy downward closure property, we propose two effective pruning strategies to prune the search space more efficiently. Extensive experimental results show that the EHUCM algorithm is efficient.

Acknowledgement. This work is supported by the National Natural Science Foundation of China (61966036, 62062066), the Project of Innovative Research Team of Yunnan Province (2018HC019).

References

1. Yang, P., Wang, L., Wang, X., Zhou, L.: SCPM-CR: a novel method for spatial co-location pattern mining with coupling relation consideration. IEEE Trans. Knowl. Data Eng. **4347**, 1–14 (2021). https://doi.org/10.1109/TKDE.2021.3060119
2. Yu, W.: Spatial co-location pattern mining for location-based services in road networks. Exp. Syst. Appl. **46**, 324–335 (2016). https://doi.org/10.1016/j.eswa.2015.10.010
3. Akbari, M., Samadzadegan, F., Weibel, R.: A generic regional spatio-temporal co-occurrence pattern mining model: a case study for air pollution. J. Geogr. Syst. **17**(3), 249–274 (2015). https://doi.org/10.1007/s10109-015-0216-4
4. Yoo, J.S., Bow, M.: A framework for generating condensed co-location sets from spatial databases. Intell. Data Anal. **23**, 333–355 (2019). https://doi.org/10.3233/IDA-173752
5. Truong, T., Duong, H., Le, B., Fournier-Viger, P.: Efficient algorithms for mining frequent high utility sequences with constraints. Inf. Sci. (Ny). **568**, 239–264 (2021)

6. Yang, S., Wang, L.: A framework for mining spatial high utility co-location patterns. In: The 12th International Conference on Fuzzy Systems and Knowledge Discovery (FSKD 2015), pp. 595–601. IEEE Press, New York (2015)
7. Wang, L., Jiang, W., Chen, H., Fang, Y.: Efficiently mining high utility co-location patterns from spatial data sets with instance-specific utilities. In: Candan, S., Chen, L., Pedersen, T.B., Chang, L., Hua, W. (eds.) DASFAA 2017. LNCS, vol. 10178, pp. 458–474. Springer, Cham (2017). https://doi.org/10.1007/978-3-319-55699-4_28
8. Wang, X., Wang, L., et al.: Mining spatial high utility co-location patterns based on feature utility ratio. Chin. J. Comput. **42**, 1721–1738 (2019)

BERT-Based Multi-Task Learning for Aspect-Based Opinion Mining

Manil Patel[✉] and C. I. Ezeife[✉]

School of Computer Science, University of Windsor, 401 Sunset Avenue, Windsor, ON N9B3P4, Canada
{patel3h,cezeife}@uwindsor.ca

Abstract. Aspect-Based Opinion Mining (ABOM) mainly focuses on mining the aspect terms (product's features) and related opinion polarities (e.g., Positive, Negative, and Neutral) from user's reviews. The most prominent neural network-based methods to perform ABOM tasks include BERT-based approaches, such as BERT-PT and BAT. These approaches build separate models to complete each ABOM subtasks, such as aspect term extraction (e.g., pizza, staff member) and aspect sentiment classification. Both approaches use different training algorithms, such as Post-Training and Adversarial Training. Also, the BERT-LSTM/Attention approach uses different pooling strategies on the intermediate layers of the BERT model to achieve better results. Moreover, they do not consider the subtasks of aspect categories (e.g., a category of aspect pizza in a review is food) and related opinion polarity. This paper proposes a new system for ABOM, called BERT-MTL, which uses Multi-Task Learning (MTL) approach and differentiates from these previous approaches by solving two tasks such as aspect terms and categories extraction simultaneously by taking advantage of similarities between tasks and enhancing the model's accuracy as well as reduce the training time. Our proposed system also builds models to identify user's opinions for aspect terms and aspect categories by applying different pooling strategies on the last layer of the BERT model. To evaluate our model's performance, we have used the SemEval-14 task 4 restaurant dataset. Our model outperforms previous models in several ABOM tasks, and the experimental results support its validity.

Keywords: Aspect-based opinion mining · BERT · Multi-task learning · Sentiment analysis · Pooling strategies

1 Introduction

Opinion mining aims to extract people's opinions or sentiments (e.g., positive or negative) and subjectivity (subjective statements are those statements which

This research was supported by the Natural Science and Engineering Research Council (NSERC) of Canada under an Operating grant (OGP-0194134) and a University of Windsor grant.

C. Strauss et al. (Eds.): DEXA 2021, LNCS 12923, pp. 192–204, 2021.
https://doi.org/10.1007/978-3-030-86472-9_18

contain opinion terms) from the texts [4]. There are large numbers of user's opinions available about a particular place (e.g., Hotel, Restaurant) and electronic products (e.g., Laptop, Phone) on different applications (e.g., TripAdvisor, Amazon). It is exceedingly difficult for a user to review/read all these available opinions and decide whether to visit a place or not, buy a product or not.

The Aspect-Based Opinion Mining (ABOM) focuses on the aspect term and term related opinion polarity (positive, negative, and neutral) [14]. In other words, instead of classifying the general opinion of the text as positive or negative, aspect-based analysis allows one to associate strong opinions with specific features of the product or service [4]. For example, "The pizza was delicious, but the staff members were horrible to us." The ABOM system detects opinion polarity for each aspect term (pizza = positive, staff members = negative). The ABOM task consists of four subtasks such as Aspect Term Extraction (ATE), Aspect Term Related Polarity (Fine-grained ABOM), Aspect Category Detection (ACD), and Aspect Category Related Polarity (Coarse-grained ABOM).

From the given sentence as an example, "The pizza here is rather good, but only if you like to wait for it." we can get aspect term such as pizza from the example, but the sentence does not have any aspect terms related to service as it shows the only context related to service. However, the aspect category model can detect food and service both as it depends on context rather than the aspect term, due to which finding of aspect terms and categories is essential while performing ABOM.

BERT [3] stands for Bidirectional Encoder Representations from Transformers. We are using the pre-trained BERT-BASE model, which contains 12 encoder layers, and each encoder layer contains Attention, Layer Norm, and Feed Forward Neural Network (FFNN). Unlike the other neural network-based approaches, BERT is pre-trained, which can be fine-tuned with just one additional output layer. The basic idea behind pre-training is that it trained the model by different datasets and used their weights as an initial weight in the model [16]. Therefore, pre-training gives the network a head start. Also, BERT uses the tokenize function to generate the tokens and each token-related ids from the input sentences. BERT tokenize function adds two unique tokens in each sentence, such as [CLS] and [SEP]. The token [CLS] is used to indicate the starting of sequences and classification, while [SEP] is used to separate the sequence from the subsequent. To generate Tokens, Input_Id, Attention_Mask, token_type_ids from a sentence, the BERT Tokenizer function converts each sentence into a list of tokens. Input_Id are token indices (numerical representations of tokens), and Attention_Mask is used to identify the tokens and padding, where tokens are represented as one and padding denotes Zero. token_type_ids are required two different sequences to be joined in a single "input_ids" entry, where the "context" corresponding to the question in the first sequence is represented by a 0, whereas the "question" in the second sequence is represented by a 1 [3].

1.1 Problem Definition

The user's reviews contain multiple aspect terms and categories as well as their related opinion polarity, which is essential to identify the user's actual concerns on specific features. Previous research (BERT-PT [17], BAT [8], BERT-LSTM/Attention [15]) require individual models to perform each subtask of ABOM (e.g., ATE and Fine-grained ABOM). The problem identified in this research is to extract the aspect terms and aspect categories for each user's reviews with minimum computation and at the same time.

1.2 Contributions

1. Our approach (BERT-MTL) can extract aspect categories and aspect terms simultaneously using Multi-Task Learning, which will identify the commonalities and differences between the tasks to enhance model's performance.
2. Also, our approach trained on a half number of the epochs as compared to previous BERT based approaches (BERT-PT [17], BAT [8], BERT-LSTM/Attention [15]) due to which we can say that our proposed approach required less amount of time to train on data.
3. In our experiments, we find that the pooling strategies work better on the final layer of BERT than the intermediate layers of the BERT model, and the results of the model validate the claim.
4. The proposed BERT-based approach (BERT-MTL) achieves better results on every subtask of ABOM compared to the previous State-of-the-art model.

The rest of the paper is summarized as follows; Sect. 2 presents Literature Review. Section 3 is the Proposed Approach called BERT-MTL. In Sect. 4, the Steps of the Proposed Algorithm are discussed, and in Sect. 5, Experimental Evaluation is discussed. In the end, in Sect. 6, the Conclusions and Future Work are presented.

2 Literature Review

Some of the past state-of-the-art approaches in the ABOM are given below.

MGAN: The Multi Granularity Alignment Network (MGAN) [10] can perform the Coarse-grained ABOM and Fine-grained ABOM. To achieve both the task, the author's designed the Coarse2Fine Attention module, which can transfer the aspect knowledge to the coarse-grained and fine-grained networks.

IMN: The authors [5] proposed Interactive Multi-Task Learning Network (IMN), which can jointly perform aspect and opinion term extraction as well as Fine-grained ABOM using Convolutional Neural Network (CNN) and Attention model. It can also learn from multiple training data to exploit the correlation between Fine-Grained ABOM and Document-level Opinion Mining.

BERT-Post Training: The authors [17] developed three different BERT-based models to perform the ABOM task. In the first model (ATE), they perform Named Entity Recognition (NER) to extract the aspect terms from the user reviews using BERT. The second model (ATSC) performs a classification task to predict opinion polarity for the aspect terms. The last model (RRC) is unique, and it is based on question answering where the user asked a question, and the BERT [3] model will predict answers from the user reviews. Also, they proposed Post Training algorithm to train all three BERT-based approaches.

AAL: The Aspect Aware Learning (AAL) model [21] uses three components to perform the Coarse-grained ABOM task. The first component (Network input) converts the words in a sentence to their respective vector form using Word2vec, and Bi-LSTM [6]. The second component (AAL) captures the correlation between aspect term and category. The third component (sentiment classification) performs the Coarse-grained ABOM using the attention layer.

BERT-LSTM/Attention: The authors [15] used intermediate layers of BERT and design two knowledge pooling strategies such as LSTM and Attention-based to perform ABOM task on different datasets. The authors [15] applied the post-training concept on the proposed approached and during evaluation, the post-training based BERT-LSTM/Attention achieved higher results.

BAT: BERT Adversarial Training (BAT) approach [8], first create the adversarial example by applying small perturbations to the original inputs. Although these examples are not actual sentences, they have been shown to serve as a regularization mechanism that can enhance the robustness of neural networks. The adversarial examples were then fed into the BERT encoder model for training, and the model was able to achieve high accuracy with the aid of the adversarial training model.

3 Proposed Approach

The previous BERT-based approaches require different models to perform all the four subtasks of ABOM from user reviews. To resolve that problem, we suggested a novel BERT-based model (BERT-MTL) for ABOM. In which, we have build three BERT-based models. The First model performs BERT-based Multi-Task Learning (MTL) approach to extract the aspect terms and aspect categories from the user reviews. The second model performs the Fine-grained ABOM, and the third model performs the Coarse-grained ABOM using different pooling straties on last layer of the BERT model.

3.1 Multi-Task Learning Model

To extract all the aspect terms and aspect categories present in the user reviews, we will use Multi-Task Learning (MTL) approach using BERT. In MTL, various tasks are learned together in a single network, each task having its own output [1]. MTL captures the similarities between tasks and improves model generalization capacity in certain situations by learning semantically similar tasks in parallel using a shared representation [5]. The BERT-MTL network has a common input, and layers of the BERT model are shared between the two tasks, such as Aspect Term Extraction (ATE) and Aspect Category Detection (ACD).

We performed two different methods in the BERT-MTL model, such as Sequence Labeling to extract the aspect terms and Multi-Label Classification to identify aspect categories. We have build a model which can take user reviews as an input, and it can predict aspect term and aspect category at the same time from those reviews by sharing the model knowledge, which means if the model learns about the aspect term, then it will help the aspect category task to learn, and vice-versa [1]. For example, If the user's review contains the name of a dish (e.g., pizza), it is easy to infer from the context that it is about the aspect category Food.

According to the previous research, the use of MTL reduces the chances of overfitting the model, which is a bigger problem in most neural network approaches[13]. Our proposed approach showed in Algorithm 1; by giving user review as an input to the model, it can predict the aspect term and aspect category simultaneously by applying MTL. In the BERT-MTL approach, we calculate the final loss function as sum of both the task-related loss function such as $Loss = Loss_{ACD} + Loss_{ATE}$. where $Loss_{ACD}$ calculated by BCEWithLogitsLoss() function and $Loss_{ATE}$ calculated by CrossEntropyLoss() [20] function.

Algorithm 1: BERT-MTL for ATE and ACD

Input: Training Sentences (s) from Dataset
Output: Aspect Terms and Categories
Initialize: Initialize the hyper-parameter learning rate, dropout probability
Loop until the terminal condition is met. Maximum Training Epochs:

> $Sentences_{batch} \leftarrow sample(Sentences; b)$;   // sample a minibatch of size b
> $Input\_Id, Attention\_Mask, token\_type\_ids = BERT\,Tokenizer(Sentences)$;   //
> $x_i = BERT\,Embedding\,Function(Input\_Id, Attention\_Mask, token\_type\_ids)$;
> //
> $Final, \_ = BERT\,Encoder\,Function(x_i)$;   // Output from last layer of BERT
> $h_t = Pooling\,Strategy(Final)$;   // LSTM and GRU NN
> $prediction1(ATE) = Feed - Forward\,NN(h_t)$;   // Classification layer 1
> $prediction2(ACD) = Feed - Forward\,NN(h_t)$;   // Classification layer 2
> $Loss1 = CrossEntropyLoss(prediction1, target1)$;   // Loss for ATE
> $Loss2 = BCEWithLogitsLoss(prediction2, target2)$;   // Loss for ACD
> $Loss = Loss1 + Loss2$;   // sum the Loss
> $Back - propagation\,algorithm\,is\,used\,to\,changed\,the\,weights\,of\,the\,approach$

end

Sequence Labeling (ATE (Task 1)): In Sequence Labeling, each word in the sentence has a label in the BIO format. Where B stands for the Beginning of aspect terms, I stands for Inside (continue) of aspect terms, and O stands for Outside of aspect terms. Some of the aspect terms are phrase level (two or more words, e.g., Cheese Garlic Bread), due to which we need the Inside label to identify all the words that can be considered as aspect terms. For the given sentence "Garlic Bread is good.", the output of sequence labeling task for each word of input will be Garlic = B, Bread = I, is = O, and good = O.

Multi-Label Classification (ACD (Task 2)): In multi-label classification, each sentence has its list of labels based on how many categories are displayed in one sentence. In our case, we consider that the sentence can be classified into five aspect categories as Service, Food, Price, Ambiance, and Miscellaneous. For the given sentence "pizza is good, but staff members were horrible to us.", the output of the multi-Label classification task for input will be Food and Service, as it indicates context related to these categories.

3.2 Aspect Term and Category Related Opinion Polarity

Fine-Grained ABOM: After extracting the aspect term and aspect category from the user's reviews, we aim to determine the opinion polarity (Positive, Negative, and Neutral) related to each of the aspect terms and categories. We will use the sentence pair classification approach using BERT [3] model to perform that task. In this task, we will give user review and aspect term as an input to the model (e.g., (user review, aspect term) a pair given as input) due to which model can predict the opinion related to that aspect from a review.

In sentence pair classification, the input for the BERT model will be sentenced aspect pair as represented below:

$$Input_2 = ([CLS], w_1, w_2, ..., w_n, [SEP], a_1, a_2, ..., a_n, [SEP]);$$
$$Final, \_ = BERT(Input_2);$$
$$h_t = Pooling\ Strategy(Final);$$
$$Y = softmax(W_3 \cdot h_t + b_3);$$

where w_1, w_2, w_3, ..., w_n are words in a sentence which contains the aspect terms while a_1, a_2, ..., a_n are the aspect term present in a sentence. When a sentence contains multiple aspect terms at that time sentence is repeated with each unique aspect term present in a sentence as a pair of input to the BERT model. The weight $W_3 \in \mathbb{R}^{3*d_I}$ and bias $b_3 \in \mathbb{R}^3$ (3 is a total number of classes (Positive, Negative, and Neutral) and d_I is a hidden dimension of Pooling Strategy).

Coarse-Grained ABOM: To perform the Coarse-grained ABOM, we perform the same task as we performed for Fine-grained ABOM. We are passing sentence and aspect category to detect opinion polarity related to that aspect category in input.

$$Input_3 = ([CLS], w_1, w_2, ..., w_n, [SEP], c_1, c_2, ..., c_n, [SEP])$$
$$Final, \_ = BERT(Input_3);$$
$$h_t = Pooling\ Strategy(Final);$$
$$Y = softmax(W_4 \cdot h_t + b_4);$$

where w_1, w_2, w_3, ..., w_n are words in a sentence which contains the aspect categories or related words while c_1, c_2, ..., c_n are the aspect category present in a sentence. When a sentence contains multiple aspect categories, the sentence is repeated with each unique aspect category present in a sentence as a pair of inputs to the BERT model. The computation will be same as Fine-grained ABOM approach.

Pooling Strategy: In this paper, we applied three different network-based pooling strategies, such as Dot Product Attention [11], LSTM [6], and GRU [2], on the last layer of the BERT model. As LSTM and GRU networks are suitable for processing sequence information, we used both the network to connect the last layer of the BERT model [15]. The last cell of the LSTM and GRU will be the output of the network, which will be given to the FFNN layer to predict the output.

LSTM Pooling Strategy: The LSTM pooling strategy applied on the last layer of BERT are discussed below:

$$Final, \_ = BERT(Input_i)$$
$$h_t = \overrightarrow{LSTM}(Final)$$

where $Input_i$ in BERT denote either input is $Input_1$ (ATE and ACD), $Input_2$ (Fine-grained ABOM) or $Input_3$ (Coarse-grained ABOM). Also, h_t represents the LSTM network output.

GRU Pooling Strategy: The GRU pooling strategy applied on the last layer of BERT are discussed below:

$$Final, \_ = BERT(Input_i)$$
$$h_t = \overrightarrow{GRU}(Final)$$

where $Input_i$ in BERT denote either input is $Input_1$ (ATE and ACD), $Input_2$ (Fine-grained ABOM) or $Input_3$ (Coarse-grained ABOM). Also, h_t represents the GRU network output.

Attention Pooling Strategy: We have also used the dot product attention on the last layer of the BERT model as a pooling strategy. The computational formula for the attention layer will be given below:

$$Final, \_ = BERT(Input_i)$$
$$h_t = W_h softmax(a \cdot Final^T)Final$$

where $Input_i$ in BERT denote either input is $Input_2$ (Fine-grained ABOM) or $Input_3$ (Coarse-grained ABOM). w_h and a are the learnable parameters of the Attention network, and softmax is the activation function. The output (h_t) generated by all the three pooling strategies will be given to the FFNN layer as input to generate the final prediction probability.

4   Steps of BERT-MTL

4.1   Steps of BERT-MTL for ATE and ACD Algorithm

1. Dataset contains sentences or reviews, and each sentence has a target. For example, "The pizza is good." and the target value for a sentence is "pizza" and "food" to determine the aspect term and category, respectively.
2. First, the sentences are given to the BERT Tokenize Function to generate the tokens, Input_Id, Attention_Mask, Token_Type_Ids. For example, from above sentence, (Tokens = ['[CLS]', 'the', 'pizza', 'is', 'good', '.', '[SEP]']), (Input_Id = [101, 1996, 10,733, 2003, 2204, 1012, 102]), (Attention_Mask = [1, 1, 1, 1, 1, 1, 1]), (token_type_ids = [0, 0, 0, 0, 0, 0, 0]).
3. The above generated Input_Id, Attention_Mask, and token_type_ids will be given in BERT Embedding Function to convert each token into respective vectors called as embedding vector. After that the embedding vector related to each token fed to the BERT Encoder Function which will generate a 768 dimension vector for each token of the sentence.
4. Each token related to a 768 dimension vector will be given to the pooling strategy (discussed in above Sect. 3.2). The pooling strategy is another neural network attached to the BERT Encoder function to identify the context and generate the final result.
5. The output of the pooling strategy given to the two different FFNN to perform the different tasks, such as sequence labeling to extract aspect terms and Multi-label classification to detect aspect categories, where it converts the vectors into each class probabilities. For example, class probability generated by FFNN for word pizza in task ATE will be [0.6, 0.1, 0.2] and for sentence in task ACD will be [0.15, 0.6, 0.2, 0.05, 0.17].
6. The class probabilities generated by each FFNN and their respective actual target from the dataset will be given to the loss function. For example, actual target in task ATE for word pizza will be [1,0,0] (1 at first position indicate Beginning of the aspect term) and for ACD [0,1,0,0,0] (1 at second position indicate Food as aspect category). The loss will be calculated for both the task as displayed below:

$$Loss_{ATE} = -\sum t_i \cdot log(p_i)$$
$$Loss_{ACD} = -w_n[t_i \cdot log(\sigma(p_i)) + (1 - t_i) \cdot log(\sigma(1 - p_i))]$$

where w_n denote the weights by default its set to none, t_i denotes target value and p_i denotes the predicted value.
7. After the calculation of loss for two different tasks such as ATE and ACD, both loss will be sum to generate the final loss function. $Loss = Loss_{ATE} + Loss_{ACD}$. After that, to reduce the final loss from both the tasks, the backpropagation algorithm was used to change the weights of the approach.

4.2   Steps of BERT-MTL for Fine and Coarse-Grained Algorithm

1. Dataset contains sentences or reviews with aspect, and each (sentence, aspect) pair have targets. For example, ("The pizza is good.", pizza/food) pair will be

input, and the target value for a pair is "positive" to determine the opinion polarity from the sentence and aspect.

2. First, the (sentence,aspect) pair is given to the BERT Tokenize Function to generate the tokens, Input_Id, Attention_Mask, and token_type_ids. And Later, it will be processed through the BERT Embedding and Encoder function and generate a 768 dimension vector for each token of the sentence.

3. Each token related a 768 dimension vector generated by BERT Encoder would be given to the pooling strategy. The pooling strategy is another neural network attached to the BERT Encoder function to identify the context and generate the final result.

4. The output of the pooling strategy given to the FFNN to perform the classification task, such as Multi-class classification to detect aspect term/category related opinion polarity, where it convert the vectors into class probabilities.

5. The class probabilities probability generated by each FFNN and their respective actual target from the dataset will be given to the BCEWithLogitsLoss function, where a loss will be calculated. Then, to reduce the loss of the task, the backpropagation algorithm was used.

5   Experimental Evaluation

In this section, we discuss the experiment and its results in detail. We tested our model in Google Colab - GPU[1]. We used the BERT-based Aspect-Based Sentiment Analysis (ABSA) code created by Avinash Sai[2] and performed changes in the code according to our approach. We implemented our code in NumPy 1.19.5, PyTorch 1.7.1, and Hugging Face transformers 4.3.2 environment.

5.1   Dataset

We conduct experiments on the SemEval-2014 task 4 [12] dataset, which contains customer reviews on restaurants. The statistics related to the dataset are represented in Table 1 for each task. In Table 1, each number in the Train and Test row represents the number of user reviews present in the dataset for every task. We have removed the sentences from the dataset, which leads to the conflict opinion polarity because the number of user reviews is small with conflict opinion polarity. The Fine-grained ABOM (training and testing combined) dataset contains 2892 positive, 1001 negative, and 829 neutral sentences, while Coarse-grained ABOM (training and testing combined) data contains 2836 positive, 1061 Negative, and 594 Neutral sentences. In the dataset, not every review contains the aspect terms due to which it may affect the statistics of the data related to Fine-grained ABOM and Coarse-grained ABOM tasks. To evaluate our model's performance, we will consider two evaluation strategies: Accuracy and F1-score. In evaluation, we are using the Macro F1-score because it is used to deal with the problem of unbalanced class, and Macro F1-score is calculated as the average F1-score of each class [8].

---

[1] https://colab.research.google.com.

[2] https://github.com/avinashsai/BERT-Aspect.

Table 1. Statistics of the dataset

Dataset	ATE and ACD	Fine-grained ABOM	Coarse-grained ABOM
Train	3,044	3,602	3518
Test	800	1,120	973

5.2 Hyper-Parameters

The selection of hyper-parameters is essential during the evaluation of model performance. We have used BERT-BASE uncased model to conduct experiments on the dataset. We have executed our models several times to figure out what number of epochs and how much dropout probability yields the highest results for our approach. Epochs indicate how many iterations are required by model to train on dataset. After the execution of the models on different parameter, we found that the dropout probability and epochs for the proposed approach should be 0.3 and 4, respectively in all three approaches. Also, we have used the Adam optimizer for better learning, and the learning rate is set to be $2e - 5$ in all three approaches. During training, batch size is set to 16 and during testing it is 4 in ATE and ACD tasks and 8 in Fine and Coarse-grained ABOM tasks.

5.3 Result Analysis

In this section, we compare our results with several state-of-the-art models on the SemEval-2014 task 4 restaurant dataset. Also, not all the models include the aspect category and aspect term related opinion polarity, due to which comparison is based on every task. The result for each subtask, such as Aspect Term Extraction (ATE), Aspect Category Detection (ACD), Fine-grained ABOM, and Coarse-grained ABOM, are displayed in Tables 2, 3a, and 3b, respectively. In all the comparisons of the tables ACC represents the accuracy. The higher value of accuracy (ACC) and Macro - F1 denotes the better model. As we can see in Table 2, the BERT-MTL achieve good results on ACD and ATE tasks in terms of Macro-F1. Also, BERT-MTL with GRU pooling strategy achieves higher results on ATE task compared to other methods. Our approach outperforms most of the previous BERT-based models such as BERT-PT, BAT, DomBERT, and BERT-LSTM/Attention in Fine-grained ABOM with three different pooling strategies. Also, All these BERT-based approaches trained their models up to 10 epochs while our models train up to only 4 epochs, due to which we achieve a better result with less amount of computation. In Coarse-grained ABOM, our approach BERT-MTL achieve state-of-the-art results with all the three pooling strategies and outperforms all the previous approaches in terms of accuracy and Macro-F1.

Table 2. Result of ACD and ATE

Model	ACD (Macro F1)	ATE (Macro F1)	Epochs
MTNA [20]	88.91	84.01	-
NRC-Canada [9]	88.58	80.19	-
BERT-PT [17]	-	77.97	10
IMN [5]	-	84.01	-
DomBERT [18]	-	77.21	10
BERT-MTL	**90.18 (± 1.00)**	**84.86 (± 1.00)**	4
BERT-MTL-LSTM	**88.136 (± 1.00)**	**84.64 (± 1.00)**	4
BERT-MTL-GRU	**89.47 (± 1.00)**	**86.19 (± 1.00)**	4

Table 3. Opinion Polarity Detection

(a) Fine-grained ABOM

Model	ACC	Macro F1	Epochs
MGAN [10]	81.49	71.48	-
BERT-PT [17]	84.95	76.96	10
BERT-LSTM [15]	82.21	72.52	10
BERT-Attention [15]	82.22	73.38	10
BAT [8]	82.27	73.7	10
DomBERT [18]	83.14	75.00	10
BERT-MTL-LSTM	**85.0**	**78.15**	4
BERT-MTL-GRU	**84.11**	**77.61**	4
BERT-MTL-Attention	**82.52**	**75.83**	4

(b) Coarse-Grained ABOM

Model	ACC	Macro F1	Epochs
GCN [19]	79.67	-	-
GCAE [19]	79.35	-	30
CapsNet-BERT [7]	86.55	-	-
AAL [21]	85.61	75.54	10
BERT-MTL-LSTM	**89.31**	**82.54**	4
BERT-MTL-GRU	**87.72**	**80.61**	4
BERT-MTL-Attention	**87.72**	**80.40**	4

6   Conclusions and Future Work

This paper proposes a novel BERT-based approach (BERT-MTL) for ABOM, which includes a Multi-Task learning model to extract the aspect terms and aspect categories simultaneously from the user reviews. Furthermore, we also perform Fine-grained and coarse-grained ABOM using the BERT model and applied three different pooling strategies on the last layer of the BERT to achieve high accuracy and take less amount of time to train. We achieved better results during the evaluation of the model than the previous approaches on the SemEval 2014 restaurant dataset. Also, some possible future works are: (a) performs soft parameter sharing between Fine-grained and coarse-grained ABOM models to build a more optimized model. (b) The adversarial or post-training in the MTL model can also increase the accuracy and F1-score.

References

1. Caruana, R.: Multitask learning. Mach. Learn. **28**(1), 41–75 (1997)
2. Cho, K., et al.: Learning phrase representations using RNN encoder-decoder for statistical machine translation. arXiv preprint arXiv:1406.1078 (2014)
3. Devlin, J., Chang, M.W., Lee, K., Toutanova, K.: Bert: Pre-training of deep bidirectional transformers for language understanding. arXiv preprint arXiv:1810.04805 (2018)
4. Ejieh, C., Ezeife, C., Chaturvedi, R.: Mining product opinions with most frequent clusters of aspect terms. In: Proceedings of the 34th ACM/SIGAPP Symposium on Applied Computing (2019)
5. He, R., Lee, W.S., Ng, H.T., Dahlmeier, D.: An interactive multi-task learning network for end-to-end aspect-based sentiment analysis. In: Proceedings of the 57th Annual Meeting of the Association for Computational Linguistics. Association for Computational Linguistics (2019)
6. Hochreiter, S., Schmidhuber, J.: LSTM can solve hard long time lag problems. In: Advances in Neural Information Processing Systems, pp. 473–479 (1997)
7. Jiang, Q., Chen, L., Xu, R., Ao, X., Yang, M.: A challenge dataset and effective models for aspect-based sentiment analysis. In: Proceedings of the 2019 Conference on EMNLP-IJCNLP, pp. 6280–6285. Association for Computational Linguistics, Hong Kong, China, Novembers 2019
8. Karimi, A., Rossi, L., Prati, A., Full, K.: Adversarial training for aspect-based sentiment analysis with BERT. CoRR abs/2001.11316 (2020)
9. Kiritchenko, S., Zhu, X., Cherry, C., Mohammad, S.M.: Nrc-canada-2014: detecting aspects and sentiment in customer reviews. SemEval. **2014**, 437 (2014)
10. Li, Z., Wei, Y., Zhang, Y., Zhang, X., Li, X.: Exploiting coarse-to-fine task transfer for aspect-level sentiment classification. In: Proceedings of the AAAI Conference on Artificial Intelligence, vol. 33, pp. 4253–4260 (2019)
11. Luong, M.T., Pham, H., Manning, C.D.: Effective approaches to attention-based neural machine translation. arXiv preprint arXiv:1508.04025 (2015)
12. Pontiki, M., Galanis, D., Pavlopoulos, J., Papageorgiou, H., Androutsopoulos, I., Manandhar, S.: Semeval-2014 task 4: aspect based sentiment analysis. Proceedings of the 8th International Workshop on Semantic Evaluation (SemEval 2014), pp. 27–35 (2014). https://doi.org/10.3115/v1/S14-2004
13. Ruder, S.: An overview of multi-task learning in deep neural networks. arXiv preprint arXiv:1706.05098 (2017)
14. Schouten, K., Frasincar, F.: Survey on aspect-level sentiment analysis. IEEE Trans. Knowl. Data Eng. **28**(3), 813–830 (2015)
15. Sun, C., Huang, L., Qiu, X.: Utilizing BERT for aspect-based sentiment analysis via constructing auxiliary sentence. arXiv preprint arXiv:1903.09588 (2019)
16. Trușcă, M.M., Wassenberg, D., Frasincar, F., Dekker, R.: A hybrid approach for aspect-based sentiment analysis using deep contextual word embeddings and hierarchical attention. In: International Conference on Web Engineering. pp. 365–380. Springer (2020). https://doi.org/10.1007/978-3-030-50578-3
17. Xu, H., Liu, B., Shu, L., Yu, P.S.: BERT post-training for review reading comprehension and aspect-based sentiment analysis. arXiv preprint arXiv:1904.02232 (2019)
18. Xu, H., Liu, B., Shu, L., Yu, P.S.: Dombert: Domain-oriented language model for aspect-based sentiment analysis (2020)

19. Xue, W., Li, T.: Aspect based sentiment analysis with gated convolutional networks. arXiv preprint arXiv:1805.07043 (2018)
20. Xue, W., Zhou, W., Li, T., Wang, Q.: MTNA: a neural multi-task model for aspect category classification and aspect term extraction on restaurant reviews. IJCNLP **2017**, 151 (2017)
21. Zhu, P., Chen, Z., Zheng, H., Qian, T.: Aspect aware learning for aspect category sentiment analysis. ACM Trans. Knowl. Discov. Data **13**(6), 1–21 (2019)

GPU-Accelerated Vertex Orbit Counting for 5-Vertex Subgraphs

Shuya Suganami[1(✉)] and Toshiyuki Amagasa[2]

[1] Graduate School of Science and Technology, University of Tsukuba,
Tsukuba, Japan
suganami@kde.cs.tsukuba.ac.jp
[2] Center for Computational Sciences, University of Tsukuba, Tsukuba, Japan
amagasa@cs.tsukuba.ac.jp

Abstract. In this paper, we propose a parallel 5-vertex orbit counting method using GPUs. Given a graph and a set of subgraph patterns, the vertex orbit counting problem is to output, for each vertex in the graph, the number of subgraph patterns that involve the vertex. It is useful to characterize the graph and has been used in many graph applications. However, existing methods for 5-vertex orbit counting require long time to complete in particular for large graphs. To this problem, we employ GPUs to accelerate it. Our method is based on one of the state-of-the-art algorithms, and we extend it to run in parallel using GPU's massive threads. To evaluate, we conduct experiments using real-world datasets, and the results show that the proposed method outperforms the existing methods.

Keywords: Orbit counting · Subgraph counting · GPU computing

1 Introduction

The graph structure is useful in representing the relationship among different entities in the real world, and have successfully been used in various fields including computer science and biology. The subgraph counting (or motif counting) is a fundamental tool to analyze a graph, where the number of occurrences of the subgraph patterns consisting of several vertices is counted, and it has been used in various applications like other features, such as the diameter.

Recently, *local count* (or *orbit count*) is gaining attentions as a feature of a graph, where, for each vertex (or edge) in the target graph, we count the (small) pattern subgraphs that involve the vertex (or edge) [16]. (We will give its concrete definition in Sect. 2.1.) By the orbit counting, we can capture more local feature of the target graph than global features, such as subgraph counting, and is therefore used for protein residues classification [18] and keyword identification [9].

One of the problems of orbit counting is its high computational complexity due to the combinatorial explosion. According to [11], even for graphs with

© Springer Nature Switzerland AG 2021
C. Strauss et al. (Eds.): DEXA 2021, LNCS 12923, pp. 205–217, 2021.
https://doi.org/10.1007/978-3-030-86472-9_19

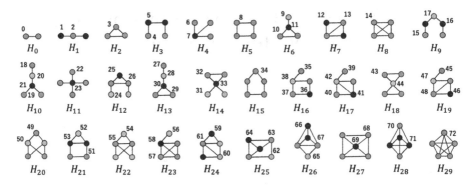

Fig. 1. All connected patterns up to five vertices. Within all patterns, vertices with the same orbit have the same color. (Color figure online)

hundreds of thousands of vertices, the frequency of 5-vertex subgraphs could reach billions to trillions. Furthermore, we need to maintain larger number of candidate answers on the way to the final result, which makes it extremely challenging to find the exact answer to the problem.

To this problem, Pashanasangi et al. have proposed a vertex orbit counting method called EVOKE [10], whereby it computes all vertex orbit counts for the subgraphs up to size five in the target graph. Basically, EVOKE extends the idea of ESCAPE [11], which is known to be the state-of-the-art methods for exact 5-vertex subgraph counting method. The idea is *pattern cutting* where the pattern is decomposed into smaller patterns, and the frequency is calculated by the combination of the frequencies of smaller patterns. EVOKE significantly outperforms other methods, but it still requires long running time for large graphs (e.g., 4,800 s for web-baidu-baike dataset with 2.14M vertices and 17.1M edges).

To further speed up EVOKE, we propose a GPU-accelerated method based on EVOKE where we can process the counting in parallel using massive threads offered by a GPU. More concretely, EVOKE's process comprises precompute and counting phases, and we execute both of them in parallel on GPU. To evaluate the proposed method, we have conducted experiments using real datasets. The results show that the proposed method is up to 16× faster than EVOKE running on CPU.

2   Preliminaries

In this section, we formally define *orbit counting* and briefly introduce the EVOKE [10]. Let $G = (V(G), E(G))$ be a connected, unweighted, and undirected graph, where $V(G)$ and $E(G)$ are the set of vertices and edges, respectively. In a similar way, we denote by $H = (V(H), E(H))$ a pattern H where $V(H)$ and $E(H)$ are the set of vertices and edges, respectively.

2.1   Orbit Counting

Next, we define *orbit counting*. Although the target of this study is *vertex orbit counting*, we also define *edge orbit counting* for the later discussion. Before orbit counting, we define some terminologies used in the sequel discussion.

Definition 1 (Automorphism). *Given a graph $G = (V(G), E(G))$, an automorphism is a bijection $\phi\colon V(G) \to V(G)$ such that $(u, v) \in E(G)$ iff $(\phi(u), \phi(v)) \in E(G)$. The set of automorphisms of G is denoted by $\mathrm{AUT}(G)$.*

We then define a relation among vertices, called *equivalence*.

Definition 2 (Equivalence). *Let $G = (V(G), E(G))$ be a graph. Two vertices $u, v \in V(G)$ are said to be equivalent if there exist one or more automorphisms that map u to v. Similarly, we call two edges $e = (u, v), e' = (u', v') \in E(G)$ equivalent if there exist one of more automorphisms that map e to e'.*

Next, we define orbits.

Definition 3 (Orbits). *Given a graph $G = (V(G), E(G))$, $V(G)$ $(E(G))$ can be classified into specific classes according to the vertex (edge) equivalence relation. We call such classes vertex (edge) orbits.*

In the following, we call a vertex orbit an orbit if there is no ambiguity. Figure 2 shows an example of orbits. In Fig. 2(a), there exist two automorphisms as shown in Fig. 2(b). Since vertex 2 is mapped to vertex 3, they have the same orbit (Fig. 2(c)). For two edges $(0, 2)$ and $(0, 3)$, there is an automorphism such that $\phi_2(0) = 0$, $\phi_2(2) = 3$. Due to the edge equivalence, $(0, 2)$ and $(0, 3)$ are of the same edge orbit.

In this paper, we represent by a pair of H and S an orbit where H is the respective pattern and S is the vertex set in H that form the orbit. For example, if the graph in Fig. 2(a) is H, each orbit in Fig. 2(c) is represented as $(H, \{0\})$, $(H, \{1\}), (H, \{2, 3\})$, respectively.

Next, we define the match.

Definition 4 (Match). *A match of H in G is defined as a bijection $\pi\colon T \to V(H)$ where $T \subseteq V(G)$ and $\forall u, v \in T$; i.e., if $(\pi(u), \pi(v))$ is an edge of H, (u, v) is an edge of G.*

Definition 5 (Vertex orbit match). *Given a graph $G = (V(G), E(G))$, a vertex $v \in V(G)$, and an orbit $\theta = (H, S)$, the set of all distinct matches is defined as $\pi\colon T \to V(H)$ where $T \subseteq V(G)$, such that $v \in T$. Let us denoted by $\mathcal{M}(v, \theta)$ $\pi(v) \in S$.*

We use $\mathrm{DM}(v, \theta)$ to denote the $|\mathcal{DM}(v, \theta))|$.

We now introduce the vertex orbit counting.

Definition 6 (Vertex orbit counting). *The vertex orbit counting of θ over $v \in V(G)$ is to find the size of $\mathcal{DM}(v, \theta)$.*

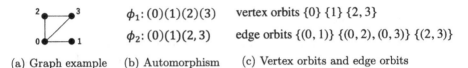

(a) Graph example   (b) Automorphism   (c) Vertex orbits and edge orbits

Fig. 2. Orbits example

Problem Statement. Given a graph G, the problem addressed in this paper is to compute the vertex orbit counting of all orbits up to five vertices (as shown in Fig. 1) w.r.t. all vertices $v \in V(G)$. As can be seen from the definition, a pattern generally has multiple orbits. Within every pattern, vertices with the same orbit are colored by the same color in Fig. 1. For example, H_9 in Fig. 1 has three orbits (blue, red, and green). In the vertex orbit counting, we separately count different orbits of the same pattern. Since there are 73 different orbits with at most five vertices, the number of outputs of vertex obit counting is $73 \cdot |V(G)|$. We denote by θ_i the ith orbit in Fig. 1; e.g., we refer to the orbit 34 in H_{15} by θ_{34}.

Finally, we define the edge orbit counting.

Definition 7 (Edge orbit counting). *Given a graph $G = (V(G), E(G))$, an edge $(u, v) \in E(G)$, and edge orbit i of pattern $H = (V(H), E(H))$, let us denote by $\pi : T \to V(H)$ the set of all distinct matches where $T \subseteq V(G)$ such that $u, v \in T$. Besides, let $(\pi(u), \pi(v))$ be an edge in edge orbit i, and let us denote it by $\mathcal{E}_i(u, v)$. The edge orbit counting of i over $(u, v) \in E(G)$ is to find the size of $\mathcal{E}_i(u, v)$, i.e., $|\mathcal{E}_i(u, v)|$.*

2.2 EVOKE

EVOKE, proposed by Pashanasangi et al. [10], is currently the state-of-the-art algorithm for 5-vertex orbit counting. The main idea is *graph orientation* and *pattern cutting*. In fact, they are pioneered by Pinar et al. [11] as a part of a subgraph counting method. Pashanasangi et al. extended it for vertex orbit counting problem.

The graph orientation is a technique used in many triangle counting methods for reducing the search space. Specifically, we convert an undirected graph into a directed graph according to a specific rule, thereby removing redundant counting.

The key insight underlies pattern cutting is that a pattern can be divided into several smaller patterns by removing some vertices (called cutset). Then, we can calculate the target vertex orbit counts (VOCs) from the enumerations of subpatterns. Based on this idea, in EVOKE, they constructed a counting framework and proved that it allows us to count exact VOCs for all orbits up to five vertices. EVOKE provides the formulas for counting all 73 orbits using this framework. For instance, we can calculate the VOCs of θ_{33} as follows. For a vertex v, $DM(v, \theta_{33}) = T(v)\binom{d(v)-2}{2}$. $T(v)$ and $d(v)$ represent the number of triangles of v and degree of v, respectively. Due to the space limitation, we omit the concrete description, and interested readers may refer to the original paper.

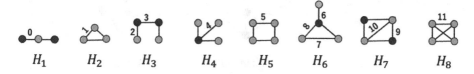

Fig. 3. All edge orbits up to four vertices (Color figure online)

3 Proposed Method

In this section, we introduce our proposed method. The proposed method is based on the counting framework of EVOKE. The difference between EVOKE and our method is that the original EVOKE assumes sequential execution on a CPU while the proposed method assumes parallel execution on a GPU.

3.1 {3, 4}-Vertex Orbit Counting

We can easily calculate all 4-VOCs in parallel except for 4-cycle (θ_8). More precisely, we utilize the equations for 4-VOCs given by EVOKE, which can be easily parallelized. Meanwhile, for the 4-cycle, we can do the counting by using the technique described below. We provide some examples of 4-vertex orbit counting due to the page limitation.

Counting Orbit 6. The equation of θ_6 is $\mathrm{DM}(u, \theta_6) = \sum_{v \in N(u)} \binom{d(v)-1}{2}$ for a vertex u. In θ_6, we can run in parallel for each node.

Counting Orbit 7. We can compute the θ_7 as follows: $\mathrm{DM}(u, \theta_7) = \binom{d(u)}{3}$, for a vertex u. In θ_7, we can execute in parallel for each node.

3.2 5-Vertex Orbit Counting

Unlike most vertex orbit counting up to four vertices, there is a challenge to execute 5-vertex orbit counting in parallel. In EVOKE, the process of 5-VOCs is roughly divided into two steps: the precompute step and the counting step.

Precompute Step. The precompute step contains the following three things to do: (1) {3, 4}-VOCs, (2) edge orbit counts up to four vertices, and (3) the additional counts of the three patterns shown in Fig. 4, which is one of the most challenging parts with a GPU. (We will describe details later.) We can get 3, 4-VOCs as described in Sect. 3.1. The edge orbits up to four vertices are shown in Fig. 3. There are eleven edges orbits, and EVOKE also provides equations to calculate them, which can be easily parallelized. In (3) which is the hardest part for parallelization, we extract from the graph three patterns: wedge, diamond, and 4-clique. Regarding wedges and diamonds, for each vertex, we maintain all vertices that form a wedge and a diamond, respectively. Regarding 4-cliques,

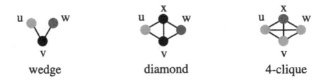

Fig. 4. Fundamental patterns (Color figure online)

for each triangle, we store all vertices that form a 4-clique. For instance, in Fig. 4, vertex u (green) in wedge forms a wedge with the vertex corresponding to vertex w (blue). Similarly, in a diamond, vertex u (green) forms a diamond with the vertex corresponding to vertex w. Note that, in both cases, we do not care about the connection between u and w. Note also that, as mentioned in Sect. 2.2, EVOKE utilizes graph orientation to reduce the search space. Likewise, we apply graph orientation in our work, but we do not present the edge direction in the discussion for understandability. It is trivial to modify the method to take into account the edge order.

In the following, we will first explain why (3) is difficult to parallelize and then describe our solutions.

The Problem of Parallelism. The main reason for the difficulty in parallelization is that it requires large memory space for storing the information about all vertices. EVOKE addresses this problem by doing it piecemeal, i.e., they only have this information for the one vertex they wish to count the orbit. Let us explain this using orbit 63 (in H_{25}) as an example. If we focus on the red nodes in H_{25}, we can divide them into a wedge and a diamond. In EVOKE, we first store the wedges and then search for the diamond to obtain orbit 63. Algorithm 1 shows the algorithm for counting orbit 63. W and $N(v)$ represent an array to maintain the wedges and the set of neighbors of v, respectively. In Algorithm 1, for a vertex u, we reserve w forming a wedge with u (Lines 5–7). After that, we search for a diamond for u and update the $DM(u, \theta_{63})$ (Lines 8–11). Note that the counts obtained by this algorithm contain overcount, which we can eliminate by the graph orientation.

However, we cannot simply parallelize this algorithm. If the search is sequential, we can reuse W for all vertices as shown in Algorithm 1. In parallel execution, multiple vertices are executed simultaneously. So, each thread needs to have its own W, requiring huge memory.

Data Structure. To address the above problem, the proposed method adopts a (large) array based on CSR (compressed sparse row) format, which is shared by multiple threads. CSR format is a well-known data structure to store graph data and is represented using the *ptr* and *to* arrays for unweighted graphs. In the wedge example, *ptr* array stores the offsets where the indices of the vertex that forms the wedge for a specific vertex while *to* maintains vertices that form a wedge. For example, the wedges in the graph in Fig. 5(a) can be represented as

Algorithm 1. Counting orbit 63 by EVOKE

Input: $G = (V(G), E(G)), W$
Output: $\mathrm{DM}(u, \theta_{63})$ for all vertex $u \in V(G)$
1: **for each** $i \in V(G)$ **do**
2:     $W[i] \leftarrow 0$
3:     $\mathrm{DM}(i, \theta_{63}) \leftarrow 0$
4: **for each** $u \in V(G)$ **do**
5:     **for each** $v \in N(u)$ **do**
6:         **for each** $w \in N(v)$ **do**
7:             $W[w] \leftarrow W[w] + 1$
8:     **for each** $v \in N(u)$ **do**
9:         **for each** $w \in T(u, v)$ **do** $\triangleright$ $T(u, v)$ represents the set of triangles incident to (u, v)
10:             **for each** $x \in T(v, w)$ **do** $\triangleright$ $\{u, v, w, x\}$ form a diamond
11:                 $\mathrm{DM}(u, \theta_{63}) \leftarrow \mathrm{DM}(u, \theta_{63}) + W[x] - 2$
12:         **for each** $i \in V(G)$ **do**
13:             $W[i] \leftarrow 0$ $\triangleright$ Clear

---

shown in Fig. 5(b). For vertex 2, $ptr[2] = 3$ and $ptr[3] = 6$ (where $ptr[i]$ represents the value of ith ptr), so the wedge for a vertex 2 is in the range $[3, 6)$ of the to array. Besides, vertex 2 forms three wedges, (2, 1, 0), (2, 3, 5), and (2, 4, 5). So, $to[3] = 0$, $to[4] = 5$, and $to[5] = 5$. (We assume that to array is sorted.)

In the proposed method, we create the ptr and to arrays in parallel using a GPU. To do this, we need to deal with two issues. The first is how to determine the appropriate partition. In the aforementioned data structure, we store the to array for every vertex in advance. However, as mentioned earlier, it is infeasible to maintain to array for all vertices in practice with the limited memory of a GPU. So, we divide the vertices into several groups so that to array fits in the GPU memory and run each group in turn.

The second is how to avoid write conflicts. Since each thread shares ptr array and to array, each thread writes to the shared arrays in parallel, causing write conflicts. We use a technique called *inspection-execution* [3] to address this issue. Inspection-execution consists of three steps: (1) We calculate the write amount of each thread, e.g., in the wedge, the write amount for vertex (thread) v is given by $\sum_{u \in N(v)} (d(u) - 1)$ as each vertex corresponds to a thread. (2) We take the prefix sum (a.k.a. scan) to compute the write start position of each thread in to array. (The result of prefix sum is ptr array.) We can get the required size of to array from the ptr array. (The last element of ptr array is the required size of the to array.) If the size does not fit into the GPU memory, we can easily divide the vertices into appropriate groups using ptr array. (3) Each thread writes to to array based on the calculated start indices, enabling each thread to write in parallel without conflicts.

Figure 6 shows the inspection-execution of the wedge for the graph in Fig. 5(a). In Step (1), for instance in vertex 3, since the neighbors are vertex 0, 2, and 5, the write amount of this vertex is $(d(0) - 1) + (d(2) - 1) + (d(5) - 1) = 3$. In Step (2), we compute the prefix sum for the result of Step (1). The output is the same as the ptr of Fig. 5(b). Finally, we search for wedges and store them for each thread.

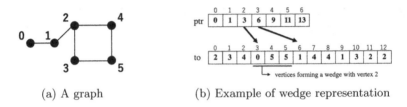

(a) A graph   (b) Example of wedge representation

Fig. 5. Example of data representation

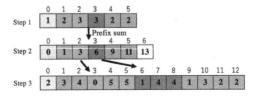

Fig. 6. Inspection-execution

Creating a ptr Array and a to Array Using GPUs. Inspection-execution incurs additional costs compared to the sequential generation of *ptr* array and *to* array. However, we can easily parallelize Step (1) and perform prefix sum quickly by taking advantage of the parallelism of a GPU [4]. On the whole, we can achieve the speedup by running this process in parallel on a GPU, because the impact of the acceleration obtained by parallelization is more extensive than these overheads. So far, we have used wedge as an example, but we can use the same procedure for diamond and 4-cycle.

Counting Step. In the counting step, we compute all 5-VOCs using the results of precompute step. Note that we assume the 4-VOCs and 4-vertex edge orbit counts have already been obtained. In fact, we have constructed the parallel counting algorithm for all 5-vertex orbits, but we only present orbit 63 as an example due to the space limit.

Counting Orbit 63. Algorithm 2 shows the algorithms of orbit 63 by the proposed method. In Algorithm 2, function Count($begin, end, x$) returns the number of elements in the range [begin, end) which are equal to x and function segmentedSort takes two arrays $S = [s_0, s_1, \ldots, s_m]$ and $A = [a_0, a_1, \ldots, a_{n-1}]$ as input and sorts each sub-array $[a_{s_i}, a_{s_i+1}, \ldots, a_{s_{i+1}-1}]$ of A. We first execute inspection-execution to generate *ptr* and W (corresponding to *to*) (Lines 3–12). We then search for diamond in parallel for each vertex and update the counts (Lines 14–19). Algorithm 2 creates W for all nodes at once (Lines 7–12), but, as mentioned above, we divide the vertices into appropriate groups that fit in memory if W is too large.

Algorithm 2. Counting orbit 63 by proposed method

Input: $G = (V(G), E(G)), W, ptr$
Output: $DM(u, \theta_{63})$ for all vertex $u \in V(G)$
1: **for each** $i \in V(G)$ **do in parallel**
2:     $ptr[i] \leftarrow 0$
3: **for each** $u \in V(G)$ **do in parallel**   ▷ Step 1
4:     **for each** $v \in N(u)$ **do**
5:         $ptr[u] \leftarrow ptr[u] + (|N(v)| - 1)$
6: $ptr \leftarrow \text{prefixSum}(ptr)$ ▷ Step 2
7: **for each** $u \in V(G)$ **do in parallel**   ▷ Step 3
8:     $index \leftarrow ptr[u]$
9:     **for each** $v \in N(u)$ **do**
10:         **for each** $w \in N(v)$ **do**
11:             $W[index] \leftarrow w$
12:             $index \leftarrow index + 1$
13: $\text{segmentedSort}(ptr, W)$
14: **for each** $u \in V(G)$ **do in parallel**
15:     **for each** $v \in N(u)$ **do**
16:         **for each** $w \in T(u, v)$ **do**   ▷ $T(u, v)$ represents the set of triangles incident to (u, v)
17:             **for each** $x \in T(v, w)$ **do**   ▷ $\{u, v, w, x\}$ form a diamond
18:                 $cnt \leftarrow \text{Count}(W + ptr[u], W + ptr[u + 1], x)$
19:                 $DM(u, \theta_{63}) \leftarrow DM(u, \theta_{63}) + cnt - 2$

---

4   Experiments

In this section we evaluate the effectiveness of the proposed method using several real-world graphs.

4.1   Experimental Setup

We use EVOKE, the state-of-the-art method of the vertex orbit counting up to 5-vertex as the comparison methods. The authors of EVOKE focus on the fact that the search for each orbit is independent of each other, and some orbits occupy most of the execution time. Based on this insight, they divided the orbits into four groups: 5-clique, 5-cycle, the orbits of H_{25} and H_{27} (that require diamond enumerations), and others, and executed each group in parallel to achieve speedup. We also compare the performance of this parallelized EVOKE (called parallel EVOKE). Note that EVOKE runs the four groups in parallel, and the counting in each orbit is executed sequentially. The code is provided by the authors of [1], and we used it.

Experimental Environments. All experiments were conducted on a Linux server with Intel(R) Xeon(R) CPU E5-2660 v4 (2.00 GHz) and 64 GB memory and NVIDIA Tesla V100 as the GPU. Our proposed method was implemented using C++ (OpenACC) and CUDA C++ and was compiled using pgc++ (PGI) 18.5 and nvcc 9.2.148, while EVOKE was implemented using C++ and compiled using g++ (GCC) 4.8.5.

Datasets. We used the Citation Network Dataset [14] and the SNAP [6] as datasets. We removed the duplicate edges and self-loops from the datasets. Besides, for a directed graph, we ignore the edge direction. Table 1 summarizes the characteristics of the datasets.

Table 1. Elapsed time for orbit counting (second).

Dataset	$\lvert V \rvert$	$\lvert E \rvert$	$\lvert T \rvert$	EVOKE	P-EVOKE	Proposal
ca-AstroPh	18.7K	396K	135K	18.1	7.99	4.68
soc-brightkite	56.7K	426K	494K	9.11	4.26	2.12
soc-lastfm	1.19M	9.04M	3.95M	298	202	22.0
soc-pokec-relationships	1.63M	22.3M	32.6M	2.57K	2.12K	159
soc-flixster	2.52M	15.8M	7.90M	298	208	37.7
web-wiki-ch-internal	1.93M	8.5M	18.2M	2.00K	1.31K	136.0
web-hudong	1.98M	14.43M	21.6M	4.04K	1.67K	819
web-baidu-baike	2.14M	17.01M	25.2M	4.80K	2.87K	303
tech-as-skitter	1.69M	28.8M	28.8M	2.34K	1.03K	253
wiki-en-cat	1.85M	7.59M	2.54K	54.9	34.7	6.79
wiki-Talk	2.39M	9.32M	9.20M	1.47K	809.4	95.1
com-amazon	335K	1.85M	667K	4.66	2.44	1.21
com-youtube	1.13M	5.97M	3.06M	192.6	102	21.1
socfb-B-anon	2.94M	41.9M	52.0M	3.27K	2.58K	205

4.2 Execution Time Comparison

We compared the execution time for orbit counting up to 5 vertices of EVOKE and our proposed method. We did not include the time for outputting the results of orbit counting, because the output of the orbit counting very large but is out of our scope.

Comparison with EVOKE. Table 1 shows the execution time of orbit counting up to 5 vertices for EVOKE and the proposed method. We can see from the results that the proposed method achieves higher performance compared to EVOKE for all the datasets. Our proposed method was about 16× faster than EVOKE for `soc-pokec-relationships` with 1.63M vertices and 22.3M edge.

Comparison with Parallel EVOKE. We show in Table 1 the run times of parallel EVOKE (denoted as P-EVOKE) and the proposed method to compute all orbits up to 5 vertices. The proposed method also outperforms parallel EVOKE on all datasets, with an average speedup of 6×. On average, our proposed method was about 6× faster than parallel-EVOKE. Our method showed the highest speedup rate with `soc-pokec-relationship`, achieving a speedup of about 13×.

Similar to EVOKE, in our proposed method, the counting of each orbit is independent each other, and the execution time of some orbit is dominant. So, we believe that we can further accelerate our method by dividing the orbit into several groups and executing these groups in parallel, i.e., multi-GPU. We consider it as future work.

5 Related Work

Subgraph counting is an important task in many fields, and various methods have been proposed for both exact and approximate methods [12, 16]. Here, we briefly describe several successful exact algorithms related to our method.

Subgraph Counting Methods. In the early works [8, 19], to perform k-vertex subgraph counting, they search for all connected k-vertex patterns in the graph and then classify each pattern using a graph isomorphism tool like nauty [7]. In this approach, counting becomes difficult as the size of the graph and k increase. In order to handle this problem, several methods have been proposed, such as a trie-based method [13], a heuristic graph isomorphism-based algorithm [20], and others [12, 16]. However, these methods still required much time to execute, and for graphs with millions of vertices, they can only perform counting for patterns with a size of four vertices.

Recently, Pinar et al. proposed ESCAPE [11], which is a 5-vertex subgraph counting method. ESCAPE uses a smart cutting framework to count patterns up to five vertices. This method can compute all 5-vertex subgraph counting in a few hours for graphs of several million vertices. As a GPU-based method, we parallelize a part of ESCAPE using GPUs to accelerate it [17].

Orbit Counting Methods. In recent years, orbit counting has been gaining attention because of its usefulness. Ahmed et al. proposed PGD [2], the first orbit counting method for large graphs. PGD counts some patterns for each edge and performs orbit counting using the results. Rossi et al. extended PGD to execute on multi-GPU and CPU [15]. However, these can only count subgraphs up to 4 vertices. For 5-vertex orbit counting, there was ORCA [5] proposed by Hočeva et al. ORCA counts some simple patterns and then computes the counts of all patterns using these results through matrix operations. However, this method took a much time to run, and 5-vertex orbit counting was considered to be infeasible for large graphs. A breakthrough method for 5-vertex orbit counting, called EVOKE [10], was proposed by Pashanasangi et al. As mentioned before, EVOKE extended the graph cutting used in ESCAPE to orbit counting. However, none of the existing work addressed GPU acceleration of EVOKE.

6 Conclusion

In this paper we have proposed a GPU-based parallel 5-vertex orbit counting method based on EVOKE, a CPU-based 5-vertex orbit counting scheme. The proposed method consists of two steps, a precompute step and a counting step, and we execute both of them in parallel using a GPU. Our experiments on real-world graphs have shown that our proposed method outperform the state-of-the-art method. In the future we will extend our method to multi-GPU to improve the performance.

Acknowledgement. This work was partly supported by the Project Commissioned by New Energy and Industrial Technology Development Organization (JPNP20006).

References

1. EVOKE. https://bitbucket.org/nojan-p/orbit-counting/src/master/
2. Ahmed, N.K., Neville, J., Rossi, R.A., Duffield, N.: Efficient graphlet counting for large networks. In: 2015 IEEE International Conference on Data Mining, pp. 1–10. IEEE (2015)
3. Chen, X., Dathathri, R., Gill, G., Pingali, K.: Pangolin: an efficient and flexible graph mining system on CPU and GPU. Proc. VLDB Endow. **13**(10), 1190–1205 (2020)
4. Harris, M., Sengupta, S., Owens, J.D.: Parallel prefix sum (scan) with CUDA. In: GPU Gems, vol. 3, no. 39, pp. 851–876 (2007)
5. Hočevar, T., Demšar, J.: A combinatorial approach to graphlet counting. Bioinformatics **30**(4), 559–565 (2014)
6. Leskovec, J., Krevl, A.: SNAP datasets: stanford large network dataset collection, June 2014. http://snap.stanford.edu/data
7. McKay, B.D., et al.: Practical graph isomorphism (1981)
8. Milo, R., Shen-Orr, S., Itzkovitz, S., Kashtan, N., Chklovskii, D., Alon, U.: Network motifs: simple building blocks of complex networks. Science **298**(5594), 824–827 (2002)
9. Nabhan, A.R., Shaalan, K.: Keyword identification using text graphlet patterns. In: Métais, E., Meziane, F., Saraee, M., Sugumaran, V., Vadera, S. (eds.) NLDB 2016. LNCS, vol. 9612, pp. 152–161. Springer, Cham (2016). https://doi.org/10.1007/978-3-319-41754-7_13
10. Pashanasangi, N., Seshadhri, C.: Efficiently counting vertex orbits of all 5-vertex subgraphs, by EVOKE. In: Proceedings of the 13th International Conference on Web Search and Data Mining, pp. 447–455 (2020)
11. Pinar, A., Seshadhri, C., Vishal, V.: ESCAPE: efficiently counting all 5-vertex subgraphs. In: Proceedings of the 26th International Conference on World Wide Web, pp. 1431–1440 (2017)
12. Ribeiro, P., Paredes, P., Silva, M.E., Aparicio, D., Silva, F.: A survey on subgraph counting: concepts, algorithms, and applications to network motifs and graphlets. ACM Comput. Surv. (CSUR) **54**(2), 1–36 (2021)
13. Ribeiro, P., Silva, F.: G-Tries: a data structure for storing and finding subgraphs. Data Min. Knowl. Discov. **28**(2), 337–377 (2014). https://doi.org/10.1007/s10618-013-0303-4
14. Rossi, R.A., Ahmed, N.K.: The network data repository with interactive graph analytics and visualization. In: AAAI (2015). http://networkrepository.com
15. Rossi, R.A., Zhou, R.: Leveraging multiple GPUs and CPUs for graphlet counting in large networks. In: Proceedings of the 25th ACM International on Conference on Information and Knowledge Management, pp. 1783–1792. ACM (2016)
16. Seshadhri, C., Tirthapura, S.: Scalable subgraph counting: the methods behind the madness. In: Companion Proceedings of The 2019 World Wide Web Conference, pp. 1317–1318. ACM (2019)
17. Suganami, S., Amagasa, T., Kitagawa, H.: Accelerating all 5-vertex subgraphs counting using GPUs. In: Hartmann, S., Küng, J., Kotsis, G., Tjoa, A.M., Khalil, I. (eds.) DEXA 2020. LNCS, vol. 12391, pp. 55–70. Springer, Cham (2020). https://doi.org/10.1007/978-3-030-59003-1_4

18. Vacic, V., Iakoucheva, L.M., Lonardi, S., Radivojac, P.: Graphlet kernels for pre-diction of functional residues in protein structures. J. Comput. Biol. **17**(1), 55–72 (2010)
19. Wernicke, S., Rasche, F.: FANMOD: a tool for fast network motif detection. Bioin-formatics **22**(9), 1152–1153 (2006)
20. Zhang, Q., Xu, Y.: Motif mining based on network space compression. BioData Min. **8**(1), 1–13 (2015)

Databases and Data Management

Efficient Discovery of Partial Periodic-Frequent Patterns in Temporal Databases

So Nakamura[1]([✉]), R. Uday Kiran[1,2,3], P. Likhitha[4], P. Ravikumar[1,4], Yutaka Watanobe[1], Minh Son Dao[2], Koji Zettsu[2], and Masashi Toyoda[3]

[1] The University of Aizu, Fukushima, Japan
{s1270204,udayrage,d8222110,yutaka}@u-aizu.ac.jp
[2] National Institute of Information and Communications Technology, Tokyo, Japan
{dao,zettsu}@nict.go.jp
[3] The University of Tokyo, Tokyo, Japan
toyoda@tkl.iis.u-tokyo.ac.jp
[4] IIIT-RK Valley, RGUKT-AP, Ongole, India

Abstract. Partial periodic-frequent pattern mining is an important knowledge discovery technique in data mining. It involves identifying all frequent patterns that have exhibited partial periodic behavior in a temporal database. The following two limitations have hindered the successful industrial application of this technique: (*i*) there exists no algorithm to find the desired patterns in columnar temporal databases, and (*ii*) existing algorithms are computationally expensive both in terms of runtime and memory consumption. This paper tackles these two challenging problems by proposing a novel algorithm known as partial periodic-frequent depth-first search (PPF-DFS). The proposed algorithm compresses a given row or columnar temporal database into a unified dictionary structure and mines this structure recursively to find all desired patterns. Experimental results demonstrate that PPF-DFS is 2 to 8.8 times faster and 5 to 31 times more memory efficient than the state-of-the-art algorithm.

Keywords: Data mining · Periodic patterns · Pattern mining

1 Introduction

1.1 Background and Related Work

Partial periodic-frequent pattern mining is an important model in data mining. It involves discovering all frequent patterns that were occurring periodically in a temporal database. The basic model of partial periodic-frequent pattern is as follows [2]: Let I be the set of items. Let $X \subseteq I$ be a **pattern** (or an itemset). A pattern containing β, $\beta \geq 1$, number of items is called a β-**pattern**. A **transaction**, $t_k = (ts, Y)$ is a tuple, where $ts \in \mathbb{R}^+$ represents the timestamp at which

---

First three authors have equally contributed to 90% of the work

© Springer Nature Switzerland AG 2021
C. Strauss et al. (Eds.): DEXA 2021, LNCS 12923, pp. 221–227, 2021.
https://doi.org/10.1007/978-3-030-86472-9_20

the pattern Y has occurred. A **temporal database** TDB over I is a set of transactions, i.e., $TDB = \{t_1, \cdots, t_m\}$, $m = |TDB|$, where $|TDB|$ can be defined as the number of transactions in TDB. For a transaction $t_k = (ts, Y)$, $k \geq 1$, such that $X \subseteq Y$, it is said that X occurs in t_k (or t_k contains X) and such a timestamp is denoted as ts^X. Let $TS^X = \{ts_j^X, \cdots, ts_k^X\}$, $j, k \in [1, m]$ and $j \leq k$, be an **ordered set of timestamps** where X has occurred in TDB. The number of transactions containing X in TDB is defined as the **support** of X and denoted as $sup(X)$. That is, $sup(X) = |TS^X|$. The pattern X is said to be a **frequent pattern** if $sup(X) \geq minSup$, where $minSup$ refers to the user-specified *minimum support* value. Let ts_q^X and ts_r^X, $j \leq q < r \leq k$, be the two consecutive timestamps in TS^X. The time difference between ts_r^X and ts_q^X is defined as a **period** of X, say p_a^X. That is, $p_a^X = ts_r^X - ts_q^X$. Let $P^X = (p_1^X, p_2^X, \cdots, p_r^X)$ be the set of all *periods* for pattern X. Let $P^X = \{p_1^X, p_2^X, \cdots, p_q^X\}$, $q = sup(X) + 1$, be the set of all periods of X in TDB. A period $p_k^X \in P^X$ is said to be an interesting period if $p_k^X \leq maxPer$, where $maxPer$ represents the user-defined maximum period. Let $IP^X \subseteq P^X$ denote the complete set of interesting periods in P^X. That is, if $\exists p_k^X \in P^X$ with $p_k^X \leq maxPer$, then $p_k^X \in IP^X$. The periodic-ratio of X, denoted as $PR(X)$, represents the proportion of periodic occurrences of a pattern in the database. That is, $PR(X) = \dfrac{|IP^X|}{|P^X|}$. Given a temporal database (TDB) and the user-specified minimum support $(minSup)$, maximum period $(maxPer)$, and minimum periodic-ratio $(minPR)$ values, the problem of partial periodic-frequent pattern mining involves discovering all patterns in TDB that have *support* no less than $minSup$ and *periodic-ratio* no less than $minPR$.

1.2 Motivation

Uday et al. [3] described a pattern-growth algorithm, called Generalized Periodic-Frequent Pattern-growth (GPF-growth), to find desired patterns in a database. Unfortunately, this algorithm suffers from the following two limitations:

- GPF-growth can discover partial periodic-frequent patterns only in row databases. In other words, this algorithm cannot find desired patterns in columnar databases. Moreover, high computational costs limit us from transforming a big columnar database into a row database.
- The process of recursive mining the constructed *tree* increases the memory and runtime requirements of the GPF-growth algorithm.

With this motivation, this paper proposes a memory and runtime efficient algorithm that can find the desired patterns in a row or columnar database.

1.3 The Contributions of This Paper

A novel algorithm, called Partial Periodic-Frequent Depth-First Search (PPF-DFS) algorithm, has been proposed in the paper to discover desired patterns in both row and columnar databases. Briefly, the proposed algorithm compresses the given row or columnar database into a common data structure, called Partial

Algorithm 1. PPF-List(temporal database (TDB), minimum support $(minSup)$, maximum periodicity $(maxPer)$, minimum periodic-ratio$(minPR)$:

1: Let $PPF\text{-}list = (X, TS\text{-}list(X))$ be a dictionary that records the temporal occurrence information of a pattern in a TDB. Let TS_l be a temporary list to record the *timestamp* of the last occurrence of an item in the database. Let IP be a temporary list to record the number of interesting periods of an item in the database.

2: **for** each transaction $t_{cur} \in TDB$ **do**

3:     Set $ts_{cur} = t_{cur}.ts$;

4:     **if** an item i occurs for the first time **then**

5:         Add i to the PPF-list and $TS_l^i = ts_{cur}$. Set $IP^i = (ts_{cur} \leq maxPer?1:0)$

6:     **else**

7:         **if** $(ts_{cur} - TS_l^i) \leq maxPer$ **then**

8:             Update IP^i++;

9:         Set $s^i = 1$ and $TS_l^i = ts_{cur}$

10: **for** each item i in PPF-list **do**

11:     **if** $(ts_{cur} - TS_l^i) \leq maxPer$ **then**

12:         Update s^i++ and IP^i++;

13: **for** each item in PPF-list **do**

14:     **if** $(\dfrac{IP^i}{minSup+1} < minPR)$ and $(s^i < minSup)$ **then**

15:         Remove i from PPF-list

16: Sort the remaining items in the PPF-list in *support* descending order of items. Call PPF-DFS(PPF-List,$\emptyset$).

Periodic-Frequent dictionary, and mines it recursively in the depth-first search order to discover the desired patterns.

The rest of the paper is organized as follows. Section 2 presents the proposed algorithm. Section 3 reports the experimental results. Section 4 concludes the paper with future research directions.

2   Proposed Algorithm

Since partial periodic-frequent patterns do not satisfy the *downward closure property*, finding candidate items (or 1-patterns) play a crucial role in discovering the complete set of partial periodic-frequent patterns. Using the set of candidate items, we later find the complete set of partial periodic k-patterns, $k \geq 1$, using the depth-first search technique. Algorithm 1 describes the procedure to construct PPF-list constituting of candidate items. Algorithm 2 describes the procedure for finding all partial periodic-frequent patterns in a database.

3   Experimental Results

In this section, we show that the proposed PPF-DFS algorithm outperforms the GPF-growth by a very large margin with respect to both runtime and memory.

Algorithm 2. PPF-DFS(PPF-List,CP)

1: **for** each item i in PPF-List **do**
2:    Set $CP = CP \cup i$;
3:    **if** $(IP(TS^Y)/(minSup+1) \geq minPR)$ and $(S(TS^Y) \geq minSup)$ **then**
4:        Add Y to CP and Y is considered as candidate.
5:        **if** $PR(TS^Y) \geq minPR$ **then**
6:            Y is considered as Partial Periodic-Ratio itemset;
7:            $PPF\text{-}DFS(PPF - Lista fteri, CP)$
8: **Note:** Since the calculation of $IP(TS^Y)$ and $PS(TS^Y)$ are simple and straight forward procedures, we are not discussing them in this paper due to page limitation.

3.1 Experimental Setup

The algorithms, GPF-growth and PPF-DFS, were developed in Python 3.9 and executed on server machine containing two AMD EPYC 7452 32-Core Processors, 64 GB RAM running CentOS 7 operating system. The experiments have been conducted using synthetic (T10I4D100K and T20I6D100K) and real-world (Congestion) databases.

The T10I4D100K and T20I6D100K are sparse synthetic databases generated using the procedure described in [1]. The T10I4D100K database contains 871 items and 100,000 transactions. The *minimum*, *average*, and *maximum* transaction lengths of T10I4D100K database are 2, 11, and 29. The T20I6D100K database contains 893 items and 99922 transactions. The *minimum*, *average*, and *maximum* transaction lengths of T20I6D100K database are 1, 20, and 47.

The **Congestion** database is a high dimensional real-world sparse database provided by an anonymous company for Kobe, Japan. The database contains 1,414 items and 8,928 transactions. The *minimum*, *average*, and *maximum* transaction lengths are 1, 57.73, and 338, respectively. The T10I4D100K and T20I6D100K databases have been provided in the GitHub[1] to verify our experiments' repeatability. Please note that we are not providing Congestion database in the Github due to confidentiality reasons.

Figure 1(a)–(c) respectively show the number of partial periodic-frequent patterns generated in T10I4D100K, T20I6D100K and Congestion data at different $minPR$ values. The $maxPer$ in T10I4D100K, T20I6100K and Congestion databases have been set at 5000, 2000, and 200, respectively. The following observations can be drawn from these figures: (i) Increase in $minPR$ decreases the number of partial periodic-frequent patterns. It is because many patterns have failed to satisfy the increased $minPR$ value. More important, it can be observed that the number of patterns getting generated at very high $minPR$ values are close to zero as it is difficult for many patterns to exhibit perfect periodic behavior in the database. **Please note that both GPF-growth and PPF-DFS generate the same partial periodic-frequent patterns.**

Figure 2(a)–(c) show the runtime requirements of GPF-growth and PPF-DFS algorithms on T10I4D100K, T20I6D100K and Congestion databases at different

---

[1] https://github.com/s1270204/PPF-DFS.

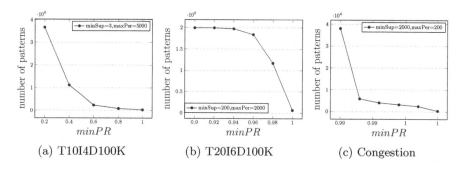

Fig. 1. Partial periodic-frequent patterns generated in various databases

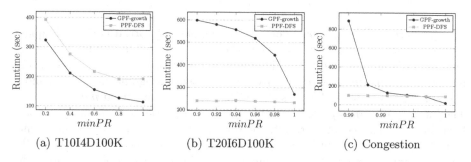

Fig. 2. Runtime comparision of both the algorithms

$minPR$ values. The following two observations can be drawn from these figures: (i) Increase in $minPR$ had reduced the runtime requirements of both the algorithms. It is because both algorithms have to generated fewer partial periodic-frequent patterns with the increase in $minPR$ value. (ii) It can be observed that PPF-DFS outperformed GPF-growth in T20I6D100K and Congestion databases. It is because these databases contained long transactions. In T10I4D100K database, GPF-growth slightly outperformed our algorithm because many transactions in this database contained fewer items. In other words, PPF-DFS is better for find long patterns, while GPF-growth is better for finding short patterns. (iii) More important, it can be observed that the proposed PPF-DFS algorithm outperformed GPF-growth algorithm by a very large margin in both synthetic and real-world databases containing long transactions.

Figure 3(a)–(c) show the memory requirements of GPF-growth and PPF-DFS algorithms on T10I4D100K, T20I6D100K and Congestion databases at different $minPR$ values. The following two observations can be drawn from these figures: (i) PPF-DFS outperformed GPF-growth on every database at any given $minPR$ value. (ii) Memory requirements of PPF-DFS is several hundreds of times less than the memory requirements of a GPF-growth. This experiment demonstrates that PPF-DFS scales better than GPF-growth on both sparse and dense databases containing either short or long transactions.

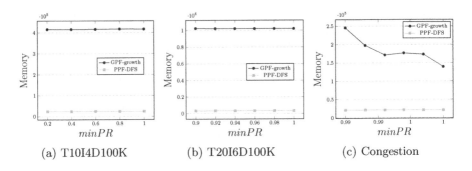

Fig. 3. Memory comparision of both the algorithms

The increase in $maxPer$), increases the number of partial periodic-frequent patterns being generated. As a result, the total runtime and memory requirements of GPF-growth and PPF-DFS algorithms will also increase. However, similar to above experimental results, PPF-DFS consumes significantly less memory and runtime as compared against the GPF-growth at any given $maxPer$ value. The increase in $minSup$, decreases the number of partial periodic-frequent patterns being generated. As a result, the total runtime and memory requirements of GPF-growth and PPF-DFS algorithms will also decrease. Moreover, PPF-DFS consumes significantly less memory and runtime as compared against the GPF-growth at any given $minSup$ value. Unfortunately, we were able to present these results in this paper due to page limitation.

4   Conclusions and Future Work

This paper has proposed an efficient algorithm named PPF-DFS to find partial periodic patterns in columnar temporal databases. An advantage of our algorithm is that it can also be used to find desired patterns in row databases. The performance of the PPF-DFS is verified by comparing it with a GPF-growth algorithm on different real-world and synthetic databases. Experimental analysis shows that PPF-DFS exhibits high performance in partial periodic-frequent pattern mining and can obtain all desired patterns faster and with less memory usage against the state-of-the-art algorithm.

As part of future work, we would like to investigate parallel and distributed algorithms to find periodic and fuzzy patterns in very large temporal databases.

References

1. Agrawal, R., Imieliński, T., Swami, A.: Mining association rules between sets of items in large databases. In: SIGMOD, pp. 207–216 (1993)
2. Kiran, R.U., Venkatesh, J.N., Fournier-Viger, P., Toyoda, M., Reddy, P.K., Kitsuregawa, M.: Discovering periodic patterns in non-uniform temporal databases. In: PAKDD, pp. 604–617 (2017)
3. Kiran, R.U., Venkatesh, J.N., Toyoda, M., Kitsuregawa, M., Reddy, P.K.: Discovering partial periodic-frequent patterns in a transactional database. J. Syst. Softw. **125**, 170–182 (2017)

Database Framework for Supporting Retention Policies

Nick Scope[1]([✉]), Alexander Rasin[1], James Wagner[2], Ben Lenard[1],
and Karen Heart[1]

[1] DePaul University, Chicago, IL 60604, USA
[2] University of New Orleans, New Orleans, LA 70148, USA

Abstract. Compliance with data retention laws and legislation is an important aspect of data management. As new laws governing personal data management are introduced (e.g., California Consumer Privacy Act enacted in 2020) and a greater emphasis is placed on enforcing data privacy law compliance, data retention support must be an inherent part of data management systems. However, relational databases do not currently offer functionality to enforce retention compliance.

In this paper, we propose a framework that integrates data retention support into any relational database. Using SQL-based mechanisms, our system supports an intuitive definition of data retention policies. We demonstrate that our approach meets the legal requirements of retention and can be implemented to transparently guarantee compliance. Our framework streamlines compliance support without requiring database schema changes, while incurring an average 6.7% overhead compared to the current state-of-the-art solution.

Keywords: Retention compliance · Databases · Privacy

1 Introduction

Laws intended to protect privacy, prevent fraud, or support financial audits require companies to implement data retention policies. Companies may also establish internal data retention policies for confidential data (e.g., for routine business operation or audits) and to minimize risks (e.g., data destruction to prevent theft). Thus, companies can be subject to multiple data retention policies requiring preservation of some data and deletion of other data. For example, the US Health Insurance Portability and Accountability Act [13] requires medical data to be retained for at least 6 years, the Children's Online Privacy Protection Act states that personal information for children is retained "for only as long as is reasonably necessary to fulfill the purpose for which the information was collected" [15]. Moreover, recent laws such as European General Data Protection Regulation and California Consumer Privacy Act [7] established the "right to be forgotten", which entitles individuals to request deletion of their personal data.

Relational database management systems (DBMS) do not support mechanisms to enforce data retention requirements. As a result, organizations build

C. Strauss et al. (Eds.): DEXA 2021, LNCS 12923, pp. 228–236, 2021.
https://doi.org/10.1007/978-3-030-86472-9_21

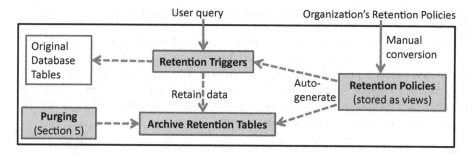

Fig. 1. Retention workflow overview. Gray boxes represent new components; dashed line represents automated framework steps; solid line represents manual steps.

ad-hoc solutions manually. As retention laws are created, databases will need to support automated retention compliance. Additionally, the solution must be intuitive for database curators to set up, and transparent from user's perspective.

In this paper, we describe a framework implementation that can guarantee data retention compliance in a relational database. Our approach builds on the work of Ataullah et al. [8] by expanding DBMS functionality to facilitate compliance with legal requirements of data retention. For example, we transparently move deleted (but retained) data to an archive (reflecting its new status) rather than block the delete operation.

Figure 1 provides an overview of our approach for enforcing retention policies. User delete (or update) transactions are allowed to proceed normally, but data that must be retained are automatically and transparently copied into the archive. As long as the retention policies are correctly defined (see Sect. 4), our database triggers can guarantee compliance by reacting to changes in data. SELECT queries are not affected, because deleted data is always removed from the original database tables. Archive tables store all deleted-but-retained data, mirroring the active table with two additional columns: *archivePolicy* and *transactionID* (see Sect. 4). Our contributions in this paper are:

- We define the requirements a database must support and enforce to comply with data retention policies (Sect. 2).
- We outline an add-on framework for complying with retention requirements within any relational database (Sect. 4).
- We detail how our framework meets the various requirements to facilitate data retention compliance (Sect. 5).

Section 6 evaluates our framework performance. We demonstrate that the retention policy implementation overhead is proportional to the number of tables and the number rows archived per-transaction. We further demonstrate that our extended functionality incurs only a 6.7% overhead over Ataullah et al. [8].

2   Retention Definitions and Requirements

Business Records: Data retention policies operate in terms of a business record. Federal law refers to a business record broadly as any "memorandum, writing, entry, print, representation or combination thereof, of any act, transaction, occurrence, or event" that is "kept or recorded" by any "business institution, member of a profession or calling, or any department or agency of government" "in the regular course of business or activity" [9].

Defining Policy: Business records may span multiple tables; therefore a comprehensive data retention framework must allow the mapping of policies across tables. Although some requirements may only preserve the records against deletion, other domains (such as the medical field) require that a complete history of record updates is retained as well. Our approach relies on SQL view syntax to define policies, making it intuitive for a database administrator (DBA) to formulate and verify policy settings. As long as the view correctly defines the protected business records, our framework will correctly identify them.

Enforcing Policy Compliance: Data in a database can be deleted or updated either by user directly (using SQL) or indirectly (e.g., by trigger effects). A comprehensive data retention framework must ensure that data is retained and purged according to the policy definition. It should not be possible for any action to bypass or interfere with retention rules; at the same time, retention should not interfere with normal database operation. We rely on triggers to ensure that all data changes (direct or indirect) are checked against the policies. Archiving the data *before* the change takes place guarantees non-interference. Retention requirements typically preserve data for a period of time (although some may be permanently such as "for life of the company"). We use a DBMS-specific built-in scheduler to perform a regular purge as requested by a policy (see Sect. 5). Finally, our approach relies completely on DBMS functionality to minimize external dependencies; consequently, transactional ACID guarantees are maintained as supplied by the DBMS.

3   Related Work

Public Research: Ataullah et al. described some of the challenges associated with record retention implementation in relational databases [8]; the authors propose an approach that uses a view-based structure to define business records for retention rules, similar to our solution. However, instead of interrupting queries, we allow them to proceed as-is after we archive retained business records (conceptually similar to a write ahead log). As discussed in Sects. 4 and 5, our approach may generate redundancy but avoids risk of retention policy conflict.

Private Sector Tools: Amazon S3 offers an object life-cycle management tool. The DoD's "Electronic Records Management Software Applications Design Criteria Standard" (DoD 5015-02-STD) defines requirements for record-keeping systems storing DoD data; a DoD compliant retention system must support retention thresholds such as time or event (C2.2.2.7). S3's object life-cycle management is limited to time criteria only. Moreover, S3 is file-based and therefore lacks sufficient granularity.

Oracle's Golden Gate (GG) [14] or IBM's Change Data Capture (CDC) [1] allow changes to be replicated from one database to another. However, these software packages are not specifically designed to support data retention requirements (although they could be expanded to support it, similarly to our approach). GG operates by inspecting REDO logs, making it difficult to incorporate the concept of business records.

Mimeo [5] provides similar functionality for Postgres; similar to CDC and GG, Mimeo would have to be revised to support retention in terms of business records since it operates on a per-table basis. IBM InfoSphere Optim Archive [11] has archiving functionality which can be used for retention. It archives data from an active database, removing data from active storage. Users define business records for archiving using Select-Project-Join (SPJ) queries, same as our and Ataullah et al. [8] approaches. A major limitation of IBM's solution is that archiving must be initiated manually or by a script.

4   Policy Setting

Policy Mapping: In practice, DBAs work with domain experts and legal counsel to implement retention policies. We assume that DBAs can expresses a business record as a view and that relevant event data is available in the database (e.g., the date of receiving a subpoena to preserve certain data). Initially mapping the business records and retention policies to database tuples is a manual process; thereafter, our SQL-based system automates enforcement of the policies.

Creating Retention Policies: A business record is mapped as a Select-Project-Join (SPJ) view. SPJ queries are sufficient to define regulations and contractual terms business records (used by both IBM [11] and Ataullah et al. [8]). We propose new SQL syntax, CREATE RETAIN, to implement views that express the business records that must be protected from deletion.

CREATE RETAIN requires the SELECT clause to contain the primary key of every table appearing in the FROM clause of the defined policy. Moreover, any columns referenced in the WHERE clause must be included in the SELECT clause. These constraints are required to verify the retained copy in the archive against the relevant policy criteria. Additionally, each retention policy is handled independently, which may incur redundancy when policies overlap.

Suppose that a company imposes a retention rule that requires retaining all applications and their interview history data (excluding interview notes) for any hired applicant. When a DELETE is issued, the target data is checked against existing policies to see if it belongs to a business record(s) that must be archived. For each active retention policy, our system automatically creates a mirror policy for the archive. Archive protection policies guarantee that business records in the archive will not be purged until the retention criteria has expired. The names of the archive policy and archive tables are prepended with "archive_". The archive tables and policies also include additional columns: *transactionID* to purge retained data in instances of aborted transactions and *archivePolicy* to purge policy-specific records in the archive. Because the same table row may be archived by two different overlapping policies, *archivePolicy* also serves to uniquely identify archive rows.

Because some retention requirements mandate a complete history of updates, we propose an optional EXACT keyword to the RETAIN syntax in order to implement views that identify business records that must retain a complete history. RETAIN EXACT policy ensures that the business record is archived before an update. Preserving a comprehensive history is a common requirement in the medical field where every update to a patient's record must be preserved.

Any column subject to an active retention policy (RETAIN or RETAIN EXACT) cannot be removed or altered in the schema. In order to drop or change a column, the DBA would first remove all retention policies that apply to that column. Our approach is designed to behave like any other database constraint (e.g., a foreign key) and therefore it must be addressed before schema changes can be applied.

5   Policy Execution

Enforcing Retention Policies: Our system uses triggers to check which of the to-be-deleted rows fall under retention policies. To ensure transactional consistency, we archive retained rows before proceeding with deletion; should the DELETE or UPDATE transaction abort, we will (eventually) delete unneeded records from the archive. Archive clean up can be executed any time after the DELETE transaction was aborted because all archive entries are uniquely identified by a transaction id and retention policy. Anytime a DELETE is run on a table with retention protections, a BEFORE DELETE trigger would fire and insert all data protected by retention policies into the archive table(s) before DELETE executes. Columns not covered under a retention policy default to NULL in the archive. Using triggers ensures that we protect *all* data, including data that is indirectly targeted by cascading DELETEs and UPDATEs. Similar to how RETAIN protects data against deletions, policies defined using RETAIN EXACT additionally archive all protected business records when an UPDATE is made to the underlying data. If a user were to update a table that is protected by RETAIN EXACT, the policy (at a minimum) would also include referenced primary keys of other tables. If additional update queries target the same business record, the data would again be copied to the archive before the UPDATE query executes.

Interacting with the Archive: The retention archive tables contain deleted data and should not be accessible without special permissions. We propose the SQL syntax SELECT ARCHIVE to retrieve data from the archive (similar to how IBM's InfoSphere Optim Archive operates). INSERT, DELETE, and UPDATE operations against the archive are prohibited to protect data integrity. To comply with the data purging requirements, we propose a PURGE command that deletes all eligible (i.e., no longer protected by a retention policy) records from the archive. The PURGE command translates the name of a provided policy into a series of

DELETE queries. Because the records in the archive incorporate the policy name, PURGE does not require checks for overlapping policies. Organizations may wish to automatically and regularly purge all data which is no longer required to be retained. For example, HIPAA requires medical information to be retained as long as it is used, after which it must immediately be removed [13]. A regular and automatic purging would remove all unprotected records. The purging process would be executed by the DBMS scheduler ([6] in Oracle [2] in Db2 [4] in PostgreSQL [3] in MariaDB). Most DBMSes include a cron-like scheduler to execute tasks on a set interval.

6 Experiments

In order to evaluate the performance of our framework, we measure the per-transaction runtime overhead. We assess linear schemas (a linked series of tables), where each table is linked to a single child table with a foreign key. All tables include columns for a primary key (char), a foreign key (char), a retention criteria (boolean), a delete criteria (boolean), and lorem (varchar), except the top-most parent table of the schema which does not contain a foreign key column.

The child-most table in the schema always contains approximately 50M rows (roughly 3.38 GB). Each primary key from a parent table is joined to an average of 2.5 rows in its child table. The dataset for each schema was built independently to fit these parameters. We chose this schema type as the most expensive (to measure the upper-bound overhead) for implementing retention; a star schema would offer additional choices for optimizing join queries. Our experiments were performed on a server with an Intel i7 7700k processor with 16 GB of RAM on spinning disk drive using PostgreSQL 12.3 with default settings on Windows 10. Both the Python 3.7 script (used to collect the runtimes) and PostgreSQL ran on the same machine.

The goal of these experiments is to determine the driving factors for our framework's overhead and quantify the performance penalties. The runtimes are evaluated on a per-query single-transaction basis. We evaluate delete transactions that affect between 1 and 100 rows (we also verified that update overhead is equivalent to delete overhead). The size of the average transaction is based on the evaluation performed by Hsu et al. [10] who quantified the number of pages written by real-world database workloads. The median number of pages written by a transaction was shown as 1.1 on average, with variations between domains (e.g., Bank, Retail, Insurance) [10]. In our analysis, we therefore assume that the number of rows written by one transaction is frequently less than 10.

Framework Overhead Analysis: In this experiment, we tested combinations of the policy size (0–100 rows), delete size (1–100 rows), and overlapping percents (0–100%). Overlap percent refers to the intersection of the retention policy and the delete query (e.g., a DELETE of 50 rows and a policy covering 50 rows with overlap of 50% corresponds to an overlap of 25 rows). Overlap rows refers to the number of rows that are ultimately archived (e.g., 25 in this example).

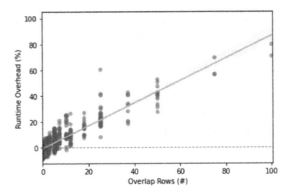

Fig. 2. Single delete Txn overhead (two tables)

To establish a performance baseline, we executed DELETEs without our retention framework. In subsequent experiments, we subtracted the runtime of baseline DELETEs from the runtime with retention enabled. We then normalized the results by computing the percentage overhead introduced by our system.

There was a 0.894 correlation between the number of rows requiring archiving and the performance overhead (illustrated in Fig. 2). This was the strongest relationship between variable combination. On the other hand, the size of the policy had a correlation of 0.192 with the runtime overhead percent, and the delete size had a correlation of 0.279. Checking if rows require archiving minimally impacts the runtime. Therefore, the overhead is driven by archiving the business records.

Furthermore, the overhead is modeled using a linear regression (illustrated as the line in Fig. 2). The model shows the overhead of our framework is a function of the number of archived rows. Our experiments found that for fewer than 10 overlap rows, the runtime overhead with archiving was statistically insignificant compared to the runtime without archiving (Fig. 2). In practice, most DELETEs and UPDATEs do not target a large quantity of rows [10]. Therefore we conclude that our framework overhead is acceptable in practice.

Comparison of Archiving to State-of-the-Art: The major difference between our proposed approach and Ataullah et al.'s work [8] is that we automatically and transparently archive retained data instead of blocking the transaction with an error. In this experiment, each query is executed as a separate transaction; therefore, whenever an exception is returned, that single query is undone (no additional rollback of previous transactions). As with Ataullah et al. [8], we use triggers to check queries against defined policies. In this experiment, we compare the runtime of these transactions when stopping the transaction using an exception versus archiving the data and letting the transaction proceed. We used the same process and data as the previous experiment.

Overall, our process averaged a runtime overhead of 6.7% compared to Ataullah et al. [8]. Although our framework introduces overhead, it eliminates potential conflicts between the retention system and existing user queries and triggers. Therefore, it ensures that organization processes are continued without the concern of violating retention policy requirements.

7   Conclusion and Future Work

In this paper, we presented and evaluated a database framework for retention policy compliance. We use views to define business records and policy conditions, thereby ensuring accurate retention (as long as the view correctly reflects the business records). When records are targeted by a delete or an update query, they are automatically and transparently retained in the archive before the data is modified. Our framework has the significant benefit of using SQL-based commands to define policies and automates archiving business records through triggers. Our experiments demonstrate that our framework can guarantee retention compliance requirements with an acceptable performance overhead.

Although our framework ensures that requested data is deleted, some data will remain in the underlying database storage as well as in previously created backups [12]. Further research must address these sources of remaining data to fully facilitate retention compliance in purging databases. Additionally, we plan to investigate migrating the archive tables to an external DBMS instance. Finally, we plan to extend a similar framework to NoSQL databases.

References

1. https://www.ibm.com/support/knowledgecenter/SSTRGZ_10.2.1/com.ibm. cdcdoc.cdcforzos.doc/concepts/infospherechangedatacaptureoverview.html
2. Admin_task_add procedure - schedule a new task. https://www.ibm.com/ support/knowledgecenter/SSEPGG_11.1.0/com.ibm.db2.luw.sql.rtn.doc/doc/ r0054371.html
3. Event scheduler. https://mariadb.com/kb/en/event-scheduler/
4. pg_cron. https://github.com/citusdata/pg_cron
5. Pgxn. https://pgxn.org/dist/mimeo/1.2.3/doc/mimeo.html
6. Scheduling jobs with oracle scheduler. https://docs.oracle.com/cd/E11882_01/ server.112/e25494/scheduse.htm#ADMIN034
7. California consumer privacy act, July 2020. https://oag.ca.gov/privacy/ccpa
8. Ataullah, A.A., Aboulnaga, A., Tompa, F.W.: Records retention in relational database systems. In: Proceedings of the 17th ACM Conference on Information and Knowledge Management, pp. 873–882 (2008)
9. Congress, U.S.: 28 u.s. code §1732 (1948)
10. Hsu, W.W., Smith, A.J., Young, H.C.: Characteristics of production database workloads and the TPC benchmarks. IBM Syst. J. **40**(3), 781–802 (2001)
11. IBM: Infosphere optim archive. https://www.ibm.com/products/infosphere-optim-archive
12. Lenard, B., Rasin, A., Scope, N., Wagner, J.: What is lurking in your backups? In: Jøsang, A., Futcher, L., Hagen, J. (eds.) SEC 2021. IAICT, vol. 625, pp. 401–415. Springer, Cham (2021). https://doi.org/10.1007/978-3-030-78120-0_26

13. Center for Medicare & Medicaid Services, et al.: The health insurance portability and accountability act of 1996, p. 158 (1996). http://www.cms.hhs.gov/hipaa
14. Oracle, September 2018. https://docs.oracle.com/goldengate/c1230/gg-winux/GGCON/introduction-oracle-goldengate.htm#GGCON-GUID-EF513E68-4237-4CB3-98B3-2E203A68CBD4
15. Wright State University: Retention guidelines for protected data (2020). https://www.libraries.wright.edu/special/recordsmanagement/retention

Internal Data Imputation in Data Warehouse Dimensions

Yuzhao Yang[1]([✉]), Fatma Abdelhédi[3], Jérôme Darmont[2], Franck Ravat[1], and Olivier Teste[1]

[1] IRIT-CNRS (UMR 5505), Université de Toulouse, Toulouse, France
{Yuzhao.Yang,Franck.Ravat,Olivier.Teste}@irit.fr
[2] Université de Lyon, Lyon 2, UR ERIC, Lyon, France
jerome.darmont@univ-lyon2.fr
[3] CBI[2] – TRIMANE, Paris, France
Fatma.Abdelhedi@trimane.fr

Abstract. Missing data occur commonly in data warehouses and may generate data usefulness problems. Thus, it is essential to address missing data to carry out a better analysis. There exists data imputation methods for missing data in fact tables, but not for dimension tables. Hence, we propose in this paper a data imputation method for data warehouse dimensions that is based on existing data and takes both intra- and inter-dimension relationships into account.

Keywords: Data warehouses · Data imputation · Dimensions

1 Introduction

Data warehouses (DWs) are widely used in companies and organizations to help building decision support systems. Data in DWs are usually modeled in a multidimensional way, which helps users consult and analyze aggregated data with On-Line Analytical Processing (OLAP). In a DW, there are non-NULL constraints on keys, but not always on the other attributes, so there may be missing data. Missing data may come from the DW's sources (operational data sources, data of other DWs) if not treated during the Extract-Transform-Load (ETL) process. In DWs, we classify missing data into factual missing data and dimensional missing data with respect to their occurrence in facts and dimensions. Factual missing data are usually quantitative, making analysis results incomplete and preventing users from getting reliable aggregates. Dimensional missing data are usually qualitative, making aggregated data incomplete and making it hard to analyse them with respect to hierarchy levels. Therefore, it is significant to complete the missing data for the sake of a better data analysis.

Data imputation is the process of filling in missing data by plausible values based on information available in the dataset [5]. Imputation of missing data focuses on factual data, with statistic-based [11], K-Nearest Neighbour (KNN)-based [3], linear programming-based [2] and hybrid (KNN and constraint programming) [1] methods. There is no research about dimensional missing data.

© Springer Nature Switzerland AG 2021
C. Strauss et al. (Eds.): DEXA 2021, LNCS 12923, pp. 237–244, 2021.
https://doi.org/10.1007/978-3-030-86472-9_22

However, dimensional data are mostly qualitative and there are methods for qualitative data imputation. Some methods replace missing values through business rules [4,9] or association rules [8,10]. Yet business rules are not always available in practice, and association rules require to define support and confidence thresholds, which is not always easy. External sources can also be employed, e.g., through crowdsourcing [6] or taking advantage of Web information [12]. Yet, suitable external sources may be difficult to find. Eventually, data imputation in DWs should consider the different structural elements in an OLAP systems, such as dimensions and hierarchies. As a result, we propose in this article an internal, i.e., based on existing data, data imputation method for dimensional missing data in DWs, by considering inter- and intra-dimension relationships.

The rest of the paper is organized as follow. In Sect. 2, we formalize the OLAP model. In Sect. 3, we detail our imputation method and provide the corresponding algorithms. In Sect. 4, we validate our proposal through a series of experimental assessments.

2   Preliminaries

We introduce here the multidimensional DW concepts and notations used in this paper [7].

Definition 1. *A data warehouse, denoted DW, is defined as $(N^{DW}, F^{DW}, D^{DW}, Star^{DW})$, where N^{DW} is the data warehouse's name, $F^{DW} = \{F_1^{DW}, ..., F_m^{DW}\}$ is a set of facts, $D^{DW} = \{D_1^{DW}, ..., D_n^{DW}\}$ is a set of dimensions and $Star^{DW} : F^{DW} \rightarrow D^{DW}$ is a mapping associating each fact to its linked dimensions.*

Definition 2. *A dimension, denoted $D \in D^{DW}$, is defined as (N^D, A^D, H^D, I^D), where N^D is the dimension's name, $A^D = \{a_1^D, ..., a_u^D\} \cup \{id^D\}$ is a set of attributes, where id^D represents the dimension's identifier. $H^D = \{H_1^D, ..., H_v^D\}$ is a set of hierarchies. $I^D = \{i_1^D, ..., i_e^D\}$ is a set of dimension instances. The value of instance i_e^D for attribute a_u^D is denoted as $i_e^D.a_u^D$.*

Definition 3. *A hierarchy of dimension D, denoted $H \in H^D$, is defined as $(N^H, Param^H)$, where N^H is the hierarchy's name. $Param^H = <id^D, p_2^H, ..., p_v^H>$ is an ordered set of dimension attributes, called parameters, which set granularity levels along the dimensions: $\forall k \in [1...v], p_k^H \in A^D$. The case where p_1^H rolls up to p_2^H in H is denoted by $p_1^H \preceq_H p_2^H$. $Weak^H = Param^H \rightarrow (A^D - Param^H)$ is a mapping possibly associating each parameter with one or several weak attributes, which are also dimension attributes providing additional information. $Weak^H[p_x^H] = \{w_1^{p_x^H} ..., w_y^{p_x^H}\}$ is the weak attribute set for parameter p_x^H.*

3   Internal Data Imputation for Dimensions

In our context, internal data imputation consists in replacing missing data in dimensions with the aid of existing data. Existing data imputation is convincible because we use accurate data and not predictions or otherwise computed values.

Imputation can be achieved through intra- and inter-dimensional relationships. Let us introduce these two types of data imputation.

Intra-dimensional Imputation. Intra-dimension imputation relies on data from the same dimension. There are indeed functional dependencies between attributes in the same hierarchy. If an attribute is a parameter, its values depend on the values of lower-granularity parameters. Our intra-dimension imputation method is presented in Algorithm 1. We first check each parameter in hierarchies of the DW. If there exists missing data for this parameter (Lines 1–2), we search for an instance with same value in a lower-granularity parameter and whose value exists (Lines 3–4). Then, we can then fill in the missing data with this value (Line 5).

Algorithm 1: Intra-dimension Imputation

1   **for** each $p_v^H \in Param^H$, where $H \in H^D, D \in D^{DW}$ **do**

2   **for** each $i_e^D \in I^D$, where $i_e^D.p_v^H$ is null **do**

3   **while** $p_{v_2}^H \in Param^H \wedge p_{v_2}^H \preceq_H p_v^H$ **do**

4   **if** $\exists i_{e_2}^D \in I^D, i_{e_2}^D.p_{v_2}^H = i_e^D.p_{v_2}^H \wedge i_{e_2}^D.p_v^H$ is not null **then**

5          $i_e^D.p_v^H \leftarrow i_{e_2}^D.p_v^H$

6   **for** each $i_{e_3}^D \in I^D$, where $i_{e_3}^D.w_y^{p_v^H}$ is null, $w_y^{p_v^H} \in Weak^H[p_v^H]$ **do**

7   **while** $p_{v_3}^H \in Param^H \wedge (p_{v_3}^H \preceq_H p_v^H \vee p_{v_3}^H = p_v^H))$ **do**

8   **if** $\exists i_{e_4}^D \in I^D, i_{e_4}^D.p_{v_3}^H = i_{e_3}^D.p_v^H \wedge i_{e_4}^D.p_v^H$ is not null **then**

9          $i_{e_3}^D.w_y^{p_v^H} \leftarrow i_{e_4}^D.w_y^{p_v^H}$

The value of a weak attribute depends on the values of its parameter. Then, for each weak attribute of the parameter we check, if there are missing data (Line 6), we search for the instance that has the same value of its parameter or a lower-granularity parameter whose value exists (Lines 7–8). The missing weak attribute data can then be supplied by this value (Line 9). It is important to note that, since the parameter sets of hierarchy are ordered sets, checking parameters is sequential (from the lowest-granularity to the highest-granularity parameter). This ensures that imputation is maximal, as the value of a higher-granularity parameter depends on its lower-granularity parameters.

Inter-dimensional Imputation. In a DW, there may be attributes that are common to different dimensions. Therefore, we can replace missing data with such inter-dimensional common attributes. The main idea of inter-dimension imputation is similar to intra-dimension imputation's, except that instead of searching for parameters in the same hierarchy, we search for common parameters of hierarchies in other dimensions (Algorithm 2, Lines 3–4 and 9–10). When performing the imputation of weak attributes, we must make sure that, in the searched dimension, the searched parameter is semantically identical with the parameter of the weak attribute to be completed; and that it bears a semantically identical weak attribute (Lines 10–11). We say "semantically identical" because in a DW, common attributes may be presented differently in different dimensions. Since in a DW, the designer would normally not use two vocabularies to describe a same entity, but may use the different prefixes or suffixes to distinguish the same entity in different dimensions, we must therefore use string similarity to match attribute names.

Algorithm 2: Inter-dimension Imputation

---

1 **while** $p_v^H \in Param^H$, *where* $H \in H^D, D \in D^{DW}$ **do**

2   **for** *each* $i_e^D \in I^D$, *where* $i_e^D.p_v^H$ *is null* **do**

3   **for** *each* $p_{v_2}^{H_2} \in Param^{H_2}$, *where* $H_2 \in H^{D_2}, D_2 \in D^{DW} \wedge D_2 \neq D$
     do

4   **if** $p_{v_2}^{H_2} \simeq p_v^H$ **then**

5   **while** $p_{v_3}^{H_2} \in Param^{H_2} \wedge p_{v_3}^{H_2} \preceq_{H_2} p_{v_2}^{H_2}$ **do**

6   **if** $\exists i_{e_2}^{D_2} \in I^{D_2} \exists p_{v_4}^H \in Param^H, p_{v_3}^{H_2} \simeq p_{v_4}^H \wedge i_{e_2}^{D_2}.p_{v_3}^{H_2} = i_e^D.p_{v_4}^H \wedge i_{e_2}^{D_2}.p_{v_2}^{H_2}$ *is not null* **then**

7              $i_e^D.p_v^H \leftarrow i_{e_2}^{D_2}.p_{v_2}^{H_2}$

8   **for** *each* $i_{e_3}^D \in I^D$, *where* $w_y^{p_v^H} \in Weak^H[p_v^H], i_{e_3}^D.w_y^{p_v^H}$ *is null* **do**

9   **for** *each* $p_{v_5}^{H_3} \in H_3$, *where* $H_3 \in H^{D_3}, D_3 \in D^{DW} \wedge D_3 \neq D$ **do**

10   **if** $p_{v_5}^{H_3} \simeq p_v^H \wedge \exists w_{y_2}^{p_{v_5}^{H_3}} \in Weak^{H_3}[p_{v_5}^{H_3}], w_{y_2}^{p_{v_5}^{H_3}} \simeq w_y^{p_v^H}$ **then**

11   **while** $p_{v_6}^{H_3} \in Param^{H_3} \wedge (p_{v_6}^{H_3} \preceq_{H_3} p_{v_5}^{H_3} \vee p_{v_6}^{H_3} \simeq p_v^H)$ **do**

12   **if** $\exists i_{e_4}^{D_3} \in I^{D_3} \exists p_{v_7}^H \in Param^H, p_{v_6}^{H_3} \simeq p_{v_7}^H \wedge i_{e_4}^{D_3}.p_{v_6}^{H_3} = i_{e_3}^D.p_{v_7}^H \wedge i_{e_4}^{D_3}.w_{y_2}^{p_{v_6}^{H_3}}$ *is not null* **then**

13              $i_{e_3}^D.w_y^{p_v^H} \leftarrow i_{e_4}^{D_3}.w_{y_2}^{p_{v_6}^{H_3}}$

---

4   Experimental Assessments

We implement our algorithms[1] and conduct experiments with different datasets. Our code is developed in Python 3.7 and is executed on a Intel(R) Core(TM) i5-10210U 1.60 GHz CPU with a 16 GB RAM. Data are integrated in R-OLAP format with Oracle 11g.

---

[1] https://github.com/BI4PEOPLE/Internal-Data-Imputationin-Data-Warehouse-Dimensions/.

4.1   Datasets and Experimental Method

Our experiments are based on one benchmark dataset and three real-world datasets. The TPC-H benchmark (**TPCH**) provides a relational schema[2] with 8 tables and a data generator we use to produce 100 MB of data. The first real-world dataset is a customer-centric dataset (**GlobalStore**) of a global super store[3]. It contains the order data of different customers and products. The second real world dataset is a regional sale dataset (**RegionalSales**) storing sales data for a company across US regions[4]. The third dataset (**GeoFrance**) contains information about French cities, departments and regions from the French government open data site[5]. We create a DW for each real-world dataset.

In our experiments, the parameter of the first granularity level in dimension hierarchies is the primary key of the dimension table. Its values are not repetitive, so weak attributes of the first granularity level and parameters of the second granularity level cannot be completed. Therefore, we generate missing data for parameters from the third granularity level of dimension hierarchies and for weak attributes from the second granularity level. Moreover, we apply different missing rates (1%, 5%, 10%, 20%, 30%, 40% and 50%). To generate a certain percentage of missing data for an attribute, we sort randomly all the tuples and remove attribute data of the first certain percentage of tuples. For each dataset, we carry out 20 tests and get the average imputation rate, accuracy and runtime. Imputation rate is the number of replaced values divided by the number of missing values. Accuracy is the number of correctly replaced values divided by the number of all replaced values.

4.2   Intra-dimensional Imputation Experiments

The datasets **TPCH**, **GlobalStore** and **RegionalSales** are employed in this experiment intra-dimensional imputation experiment. The imputation rate ranges between 61.73% and 100%; the accuracy between 97.08% and 100%.

Imputation Rate. In Fig. 1, imputation rates (X-axis) vary with respect to missing rates (Y-axis) from 1% to 50%. We observe that, for dataset **TPCH**, the imputation rate is always 100%, while the imputation rates of the other datasets decrease when the missing rate increases. The imputation rate of **Regional-Sales** is much lower than the two others. Since missing data are replaced by the tuple having the same value on a lower-granularity parameter, the imputation rate of an attribute depends on the ratio of the distinct values and the coefficient of variation of each distinct value of its lower-granularity parameters. For example, in the dimension *Part* of **TPCH**, the ratio of the distinct values and the

---

[2] http://tpc.org/tpc_documents_current_versions/pdf/tpc-h_v2.18.0.pdf.

[3] https://data.world/vikas-0731/global-super-store.

[4] https://data.world/dataman-udit/us-regional-sales-data.

[5] https://www.data.gouv.fr/fr/datasets/communes-de-france-base-des-codes-post aux/.

coefficient of variation of each distinct value are 0.125% and 0.027 for the second granularity level parameter, respectively; while in dimension *StoreLocation* of **RegionalSales**, they are 62.13% and 1.1, respectively.

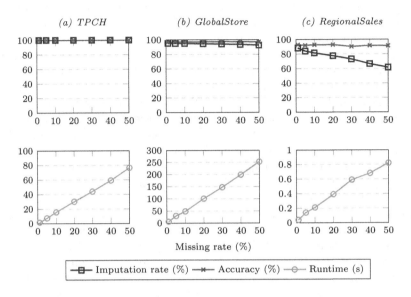

Fig. 1. Intra-dimensional imputation experiment results

Accuracy. We can see in Fig. 1 that the accuracy of **TPCH** is always 100%, while the accuracy of the two other datasets is always less than 100%. Since our imputation method is based on the hierarchy relationships, non-strict and incomplete hierarchies may impact accuracy. By analysing the data, we find that there are non-strict hierarchies in these datasets **GlobalStore** and **Regional-Sales**, i.e., in the dimension *Customer* of **GlobalStore**, there are some tuples whose *City* values are the same, but they belong to different *States*. There is a similar case in dimension *StoreLocation* of **RegionalSales**.

Runtime. The evolution of runtime with respect to missing rate is linear (Fig. 1), which is in line with the complexity of Algorithm 1, which is $O(n)$, where n is the missing rate.

4.3   Inter-dimensional Imputation Experiments

We use **TPCH** and **GeoFrance** in this inter-dimensional imputation experiment. There are two **TPCH** dimensions, *Customer* and *Supplier*, which have same geographical attributes. In **GeoFrance**, we create two dimensions and randomly divide the original data into two partitions with the same number of tuples. Then, we load the each partition into one of the dimensions. The imputation rate ranges between 48.67% and 100%. The accuracy always remains at 100% with respect to missing rate.

Imputation Rate. Yet again, the imputation rate of **TPCH** is always 100% (Fig. 2), for the same reason as intra-dimensional imputation. **GeoFrance**'s imputation rate is low when the missing rate is very low, then it increases with the missing rate. After analysing the data, we find that there is a tuple where the values of *RegionCode* and *RegioName* are originally missing. The lower-granularity parameter *DepartmentCode* being unique, *RegionCode* and *RegionName* missing data cannot be imputed. When the missing rate is low, the number of total missing data is low, too. Thus, the missing data in this tuple account for a large proportion of total missing data, which can explain why **GeoFrance**'s imputation rate is low when the missing rate is very low.

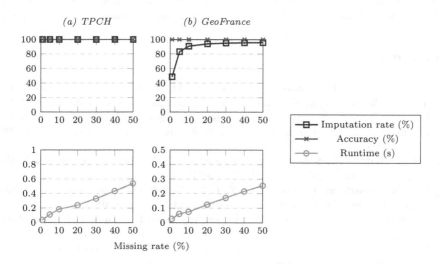

Fig. 2. Inter-dimensional imputation experiment results

Accuracy. There is no incomplete nor non-strict hierarchy in **TPCH**'s and **GeoFrance**'s DWs. Hence, the accuracy for these two datasets is always 100% (Fig. 2).

Runtime. Again, the evolution of runtime with respect to missing rate is linear (Fig. 2), which is in line with the complexity of Algorithm 2, which is $O(n)$, where n is the missing rate.

5 Conclusion and Future Work

In this article, we propose an internal data imputation method for dimensional missing data in DWs. Our method is based on the existing data found in both intra- and inter-dimensional relationships. We take in charge the imputation of both parameters and weak attributes. The solutions are formalized as algorithms and are actually implemented. Our method is validated by a series of experiments with the different percentages of missing data of the different attributes. For intra-dimensional imputations, the imputation rate ranges between 61.73% and 100% and the accuracy varies from 97.08% to 100%. For inter-dimensional imputations, the imputation rate ranges between 48.67% and 100% and the accuracy remains at 100%. However, not all missing data can be completed by the existing data, thus in the future we will also combine our method with web-based methods to achieve a better imputation

Acknowledgement. This research is funded by the French National Research Agency (ANR), project ANR-19-CE23-0005 BI4people (Business Intelligence for the people).

References

1. Amanzougarene, F., Zeitouni, K., Chachoua, M.: Predicting missing values in a data warehouse by combining constraint programming and KNN. In: EDA (2014)
2. Bimonte, S., Ren, L., Koueya, N.: A linear programming-based framework for handling missing data in multi-granular data warehouses. Data Knowl. Eng. **128**, 101832 (2020)
3. Ribeiro, L.deS., Goldschmidt, R.R., Cavalcanti, M.C.: Complementing data in the ETL process. In: DaWaK, pp. 112–123 (2011)
4. Fan, W., Jianzhong, L., Shuai, M., Nan, T., Wenyuan, Y.: Towards certain fixes with editing rules and master data. VLDB J. **21**, 173–184 (2010)
5. Li, D., Deogun, J., Spaulding, W., Shuart, B.: Towards missing data imputation: a study of fuzzy k-means clustering method. In: RSCTC, pp. 573–579 (2004)
6. Lofi, C., El Maarry, K., Balke, W.-T.: Skyline queries over incomplete data-error models for focused crowd-sourcing. In: Conceptual Modeling, pp. 298–312 (2013)
7. Ravat, F., Teste, O., Tournier, R., Zurfluh, G.: Algebraic and graphic languages for OLAP manipulations. Int. J. Data Warehousing Mining **4**, 17–46 (2008)
8. Shen, J.-J., Chang, C.-C., Li, Y.-C.: Combined association rules for dealing with missing values. J. Inf. Sci. **33**(4), 468–480 (2007)
9. Song, S., Zhang, A., Chen, L., Wang, J.: Enriching data imputation with extensive similarity neighbors. VLDB Endow. **8**(11), 1286–1297 (2015)
10. Wu, C.-H., Wun, C.-H., Chou, H.-J.: Using association rules for completing missing data. In: HIS, pp. 236–241 (2004)
11. Wu, X., Barbará, D.: Modeling and imputation of large incomplete multidimensional datasets. In: DaWak, pp. 286–295 (2002)
12. Yakout, M., Ganjam, K., Chakrabarti, K., Chaudhuri, S.: InfoGather: entity augmentation and attribute discovery by holistic matching with web tables. In: SIGMOD, pp. 97–108 (2012)

Purging Data from Backups
by Encryption

Nick Scope[1]([✉]), Alexander Rasin[1], James Wagner[2], Ben Lenard[1],
and Karen Heart[1]

[1] DePaul University, Chicago, IL 60604, USA
[2] The University of New Orleans, New Orleans, LA 70148, USA

Abstract. Data retention laws establish rules intended to protect privacy. These define both retention durations (how long data must be kept) and purging deadlines (when the data must be destroyed in storage). To comply with the laws and to minimize liability, companies should destroy data that must be purged or is no longer needed. However, database backups generally cannot be edited to purge "expired" data and erasing the entire backup is impractical. To maintain compliance, data curators need a mechanism to support targeted destruction of data in backups.

In this paper, we present a cryptographic erasure framework that can purge data from all database backups. Our approach can be transparently integrated into existing database backup processes. We demonstrate how different purge policies can be defined through views and enforced by triggers without violating database constraints.

Keywords: Purging compliance · Databases · Privacy · Encryption

1 Introduction

Efforts to protect user data privacy and give people control over their data have led to passage of laws such as the European General Data Protection Regulation (GDPR) [6] and California Consumer Privacy Act (CCPA) [11]. With the increased emphasis on proper data governance, many organizations are working to implement the data retention requirements into their databases. Laws can dictate how long data must be retained (e.g., United States Income Revenue Service tax document retention [8]), the consent required from individuals on how their data may be used (e.g., GDPR Article 6), or purging policies for when data must be destroyed (e.g., GDPR Article 17).

In this paper, we consider the problem of data purging in a database. Prior research has only considered the problems of retention and purging policies for data in an active (i.e., current instance) database [1]. Nevertheless, to fully comply with the laws mandating data purging, a system must purge data from the active database as well as from backups. Although backups are not part of the active database, they can be restored into an active database at any time.

© Springer Nature Switzerland AG 2021
C. Strauss et al. (Eds.): DEXA 2021, LNCS 12923, pp. 245–258, 2021.
https://doi.org/10.1007/978-3-030-86472-9_23

1.1   Motivation

A variety of factors make purging data from backups difficult. Backups may potentially be edited by 1) restoring the backup, 2) making changes in the restored database, and then 3) creating a new ("edited") backup. Outside of this cumbersome process, there is no other method of safely editing a backup. Only a full (i.e., non-incremental, see Sect. 2.2) backup can be altered in this manner. Furthermore, editing a full backup would invalidate all of its dependent incremental backups. Additionally, backups may be be stored remotely (e.g., off-site) and on sequential access media (e.g., on tape). Therefore, the ability to make changes to any data within backups is both limited and costly.

In order to solve this problem, we propose to implement data purging through cryptographic erasure [2]. Intuitively, a cryptographic erasure [2] approach encrypts the data and then purges that data by deleting decryption keys. The advantage of this approach is that it deletes the data "remotely" without having to access the backups. When a backup is restored, the irrecoverable data is purged while the recoverable and non-encrypted data are fully restored into the active database. Furthermore, this process does not invalidate partial backups.

Our framework creates *shadow tables* which contain an encrypted copy of all data subject to purging policies. These shadow tables are backed up instead of the original tables; we then use cryptographic erasure to simultaneously purge values across all existing backups. Our approach requires no changes to current backup practices and is compatible with both full and incremental backups. One challenge of implementing cryptographic erasure is in balancing different policy requirements across a relational database schema. A single row in a table may have columns subject to different retention and purging requirements.

Our framework only applies encryption to data which is updated or inserted after the purge policy is defined and does not retroactively apply encryption to the already-present data (e.g., if an existing policy is changed). Our approach focuses on addressing compliance rather than security. It will guarantee data destruction based on defined policies; thwarting a malicious insider who previously copied data or decryption keys is beyond the scope of this paper. Furthermore, purging data that remains recoverable via forensic tools is out of scope for this paper. Our contributions are:

– We outline the requirements for defining and enforcing data purge policies
– We describe an implementation (and present a prototype) for backup data purging that can be seamlessly integrated into existing DBMSes during backup and restore
– We design a key selection mechanism that balances multiple policies and retention period requirements

2   Background

2.1   Compliance Terminology

Business Record: Organizational rules and requirements for data management are defined in units of business records. United States federal law refers to a business record broadly as any "memorandum, writing, entry, print, representation or combination thereof, of any act, transaction, occurrence, or event [that is] kept or recorded [by any] business institution, member of a profession or calling, or any department or agency of government [...] in the regular course of business or activity" [4]. A business record may consist of a single document for an organization (e.g., an email message). In a database, a business record may span combinations of rows across multiple tables (e.g., a purchase order consisting of a buyer, a product, and the purchase transaction from three tables).

Policy: A policy is any formally established rule for organizations dictating the lifetime of data. Retention policies can dictate how long data must be saved while purge policies dictate when data must be destroyed. Policies can originate from a variety of sources such as legislation or a byproduct of a court ruling. Companies may also establish their own internal data retention policies to protect confidential data. In practice, database curators work with domain experts and sometimes with legal counsel to define business records and retention requirements based on the written policy.

Purging: In data retention, purging is the permanent and irreversible destruction of data in a business record [7]. A business record purge can be accomplished by physically destroying the device which stored the data, encrypting and erasing the decryption key (although the ciphertext still exists, destroying the decryption key makes it inaccessible and irrecoverable), or by fully erasing the data from all storage.

2.2   Database Backups and Types

Backups are an integral part of business continuity practices to support disaster recovery. There are many mechanisms for backing up a database [5] both at the file system level and internal to the DBMS. File system backups range from a full backup with an offline, or quiesced, database to a partial backup at file system level that incrementally backs up changed files. Most DBMS platforms provide backup utilities for both full and partial backups, which create backup in units of pages (rather than individual rows).

Some utilities provide block-level backups with either a *full* database backup or a partial backup capturing pages that changed since the last backup. Partial backups can be *incremental* or *delta*. For example, if we took a full backup on Sunday and daily partial backups and needed to recover on Thursday, database utilities would restore the full backup from Sunday and then either 1) apply delta backups from Monday, Tuesday, and Wednesday or 2) apply Wednesday's incremental backup. Because most organizations use multiple types of backups, any purging system must work on full, incremental, and delta backups [10].

2.3   Related Work

Kamara and Lauter's research has shown that using cryptography can increase storage protections [9]. Furthermore, their research has shown that erasing an encryption key can fulfill purging requirements. Our system expands on their

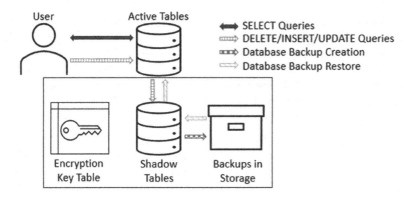

Fig. 1. Framework overview

research by using policy definitions to assign different encryption keys relative to their policy and expiration date.

Reardon et al. provided a comprehensive overview of secure deletion [12]. The authors defined three user-level approaches to secure deletion: 1) execute a secure delete feature on the physical medium 2) overwrite the data before unlinking or 3) unlink the data to the OS and fill the empty capacity of the physical device's storage. All methods require the ability to directly interact with the physical storage device, which may not be possible for database backups in storage.

Boneh et al. used cryptographic erasure, but each physical device had a single key [2]. We introduce an encryption key assignment system to facilitate targeted cryptographic erasure of business records across all backups. In order to fully destroy the data, users must also securely delete the encryption keys used for cryptographic erasure. Reardon et al. [12] provide a summary for how to destroy encryption keys to guarantee a secure delete. Physically erasing the keys depends on the storage medium and is beyond the scope of this paper. However, unlike backups, encryption keys are stored on storage medium that is easily accessible.

Ataullah et al. described some of the challenges associated with record retention implementation in relational databases [1]. The authors proposed an approach that uses view-based structure to define business records (similar to our approach); they used view definitions to prohibit deletion of data that should be retained. Ataullah et al. only consider purging data in an active database; they did not consider how their approach would interact with backups.

3   Our Process

Our proposed framework automatically applies encryption to data that is subject to purge policy requirements whenever data are inserted or updated. An overview of this process is presented in Fig. 1. We maintain and backup a shadow (encrypted) copy of the tables; other tables not subject to purging rules are not affected. SELECT queries always interact with the non-encrypted database tables (rather than shadow tables) and are not impacted by our approach. We translate (using triggers) DELETE, INSERT, and UPDATE queries into a corresponding operation on the encrypted shadow copy of the table. Our framework is designed to remain transparent to the user. For example, one can use client-side encryption without affecting conflicting with our data purging approach. A change in purge policy has to be manually triggered to encrypt existing data.

In our system, shadow tables are backed up instead of the corresponding user-facing tables; tables that are not subject to purging policies are backed up normally. When the shadow tables are restored from backup, our system decrypts all data except for data purged per policy. For encryption keys that expired due to a purge policy, the underlying data would be replaced with NULL (unfortunately, purging of data unavoidably creates ambiguity with "real" NULLs in the database). In cases where the entire row must be purged (due to a purged primary key), the tuple would not be restored. Evaluation of possible conflicts (e.g., purge policy on a column that is restricted to NOT NULL) is resolved during the policy definition step.

Our default implementation uses a table called encryptionOverview (with column definition shown in Table 1) to manage encryption keys. This table is marked to never be backed up to avoid the problem of having the encryption keys stored with the backup; otherwise the encryption keys could not be truly purged. In our proof-of-concept experiments, the encryptionOverview table is stored in the database. However, in a production system the key management tables will be stored in a separate database. Access to these tables could be established via a database link or in a federated fashion, allowing the keys to be kept completely separate from the actual data.

Our framework uses time-based policy criteria for purging, bucketed per-day by default. A bucket represents a collection of data grouped by a time range and policy that is purged together as a single unit. All data in the same bucket for the same policy uses the same encryption key. Our default bucket size is set to one day because, for most purge policies, daily purging satisfies the requirements (e.g., GDPR: Article 25 [6]). We intend to study the performance and granularity trade off (by changing bucket size) in future work. The encryption keys can be deleted by a cron-like scheduler, available in most DBMSes. However, since we intend to separate the encryptionOverview table from the database in production, we did not evaluate that functionality in our experiments.

Tables may contain data belonging to multiple business records; columns in a single row may be subject to different policies. In the shadow tables, each original column explicitly includes its [column name]EncryptionID, which serves as its encryption key identifier (chosen based on which policy takes precedence).

Table 1. encryptionOverview table column definitions

encryptionOverview	
encryptionID	Int
policy	Varchar(50)
expirationDate	Date
encryptionKey	Varchar(50)

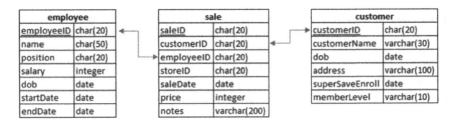

Fig. 2. Sample company schema

3.1 Defining Policies

Our method of defining purge policies uses SQL views to define the underlying business records and the purge criteria. We require defining a time-based purging period (which, at insert time, must provide at least one non-NULL value); if any one primary key attribute is included in a purge policy, all other columns must be included. The purge definition must also include all child foreign keys of the table to maintain referential integrity. For example, in the schema in Fig. 2, if the `customerID` in the `customer` table was included under a purge policy, both `customer.*` columns and `sale.customerID` must be included. During the restore process, a purged column value will be restored as a NULL. Thus, non-primary-key columns subject to a purge policy must not prohibit NULLs, including any foreign key columns. When all columns are purged from a row, the entire tuple will not be restored (i.e., ignored on restore).

Consider a policy for a company (Fig. 2) that requires purging all customer data where the Super Save enrollment date is over twenty years old:

```
CREATE PURGE customerPurge AS SELECT customer.*, sale.customerID
FROM customer LEFT JOIN sale ON customer.customerID = sale.customerID
WHERE datediff(year, customer.superSaverEnroll,
              date_part('year', CURRENT_DATE)) > 20;
```

In this example, the `superSaveEnroll` column will not contain NULL; therefore, at least one column can be used to determine the purge expiration date, satisfying our definition requirements.

3.2 Encryption Process

When a new record is inserted, we use triggers to determine if any of the columns fall under a purge policy; if so, the trigger computes the relevant policies and when the business records must be purged. For example, consider a new employee record inserted into the employee table:

```
INSERT INTO customer
    (customerID,customerName,dob,address,superSaveEnroll,memberLevel)
VALUES (1,'Johnson,Isabel','2/1/1990','Chicago','1/1/2021','Premium');
```

Under the previously defined `customerPurge` policy, Isabel Johnson's data would have a purge date of January 1, 2041. We first check if an encryption key for this date bucket and policy already exists in the `encryptionOverview` table. If an encryption key already exists, we use it to encrypt the values covered by the purge policy; if not, a new key is generated and stored in the `encryptionOverview` table. The encrypted row and the matching encryption key ID is inserted into the `customerShadow` table. If a column is not covered by a purge policy, a value of −1 is inserted into the corresponding `EncryptionID` column. The value of −1 signals that the column has not been encrypted and contains the original value. In this example, each column in the shadow table is encrypted with the same key, but our proposed framework allows policies to be applied on a per-column basis. Therefore, our framework tracks each column independently in cases where a row is either partially covered or covered by different policies.

To support multiple purge policies, we must determine which policies apply to the new data. A record in a table may fall under multiple policies (potentially with different purge periods). Furthermore, a single value may belong to different business records with different purge period lengths. In data retention the longest retention period has priority; on the other hand, in data purging, the shortest period has priority. Therefore, we encrypt each column using the encryption key corresponding to the shortest purge period policy.

It is always possible to shorten the purging period of a policy by purging the data earlier. However, our approach does not support extending the purge period since lengthening a purge period risks violating another existing policy. Thus, if a policy is dropped, data already encrypted under that policy will maintain the original expiration date.

Continuing with our example, another policy dictates a purge of all "Premium+" customer address information ten years after their enrollment date. Because this policy applies to a subset of columns on the `customer` table, some columns are encrypted using the encryption key for `customerPurge` policy while other columns are encrypted using the `premiumPlusPurge` policy. For example, if a new Premium+ member were enrolled, the `premiumPlusPurge` policy would take priority on the address field, with remaining fields encrypted using the `customerPurge` policy key.

3.3   Encryption on Update

Similarly to `INSERT`, we encrypt all data subject to purge policy during an `UPDATE`. Normally, the updated value would simply be re-encrypted and stored in the shadow table. However, if an update changes the date and alters the applicable purge policy (e.g., changing the start or the end date of the employee), the record may have to be re-encrypted with a different key or decrypted (if purge policy no longer applies) and stored unencrypted in the shadow table. Our prototype system decrypts the primary key columns in the shadow table to identify the updated row. This is a PostgreSQL-specific implementation requirement, which may not be needed in other databases (see Sect. 4). Our system automatically deletes the original row from the shadow table and inserts the new record (with encryption applied as necessary), emulating `UPDATE` by `DELETE`+`INSERT`.

Continuing with our example, let's say Isabel Johnson is promoted to the "Premium+" level, changing the purge policies for her records. We can identify her row in the shadow table using the `customerID` primary key combined with the previously used `customerIDEncryptionID`. We would then apply the corresponding updates to encrypt the fields covered by the policy, based on the new policy's encryption key.

3.4 Purging Process

Purging is automated through a cron-like DBMS job ([3] in Postgres) that removes expired encryption keys from `encryptionOverview` with a simple delete. Our framework is designed to support purge policies and not for support of retention policies (i.e., prevent deletions before the retention period expires). Retention requires a separate mechanism, similar to work in [1,13]. Moreover, key deletion will need to be supplemented by a secure deletion of the encryption keys on the underlying hardware [2,12], guaranteeing the encryption keys are permanently irrecoverable (which is outside the scope of this paper).

3.5 Restore Process

Our framework restores the backup with shadow tables that contain encrypted as well as unencrypted values. Recall that the shadow tables include additional columns with encryption ID for each value. A −1 entry in the `encryptionID` column indicates that the column is not encrypted and, therefore, does not require decryption and would be restored as-is. Our system decrypts all values with non-expired encryption keys into the corresponding active table. For any encrypted value associated with a purged `encryptionID` our system restores the value as a `NULL` in the active table. If the entire row has been purged, the tuples would not be restored into the active table.

4 Experiments

We implemented a prototype system in PostgreSQL 12.6 database to demonstrate how our method supplements backup process with purge rules and effectively purges data from backups. The database VM server consists of 8 GB of RAM, 4 vCPUs, 1 × vNIC and a 25 GB VMDK file. The VMDK file was partitioned into: 350 MB/boot, 2 GB swap, and the remaining storage was used for the/partition; this was done with standard partitioning and ext4 filesystem running CentOS 7 on VMware Workstation 16 Pro. We demonstrate the viability of our approach by showing that it can be implemented without changing the original schema or standard backup procedures, while guaranteeing data purging compliance.

We use two tables, `Alpha` and `Beta`, with `Beta` containing children rows of `Alpha`. As shown in Fig. 3, shadow tables contain the encrypted value for each attribute and the encryption key used. Shadow tables use the datatype bytea

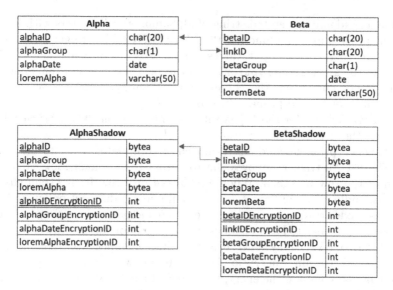

Fig. 3. Tables used in our experiments

(binary array) to store the encrypted value regardless of the underlying data type as well as an integer field that contains the encryption key ID used to encrypt the field. We tested the most common datatypes such as char, varchar and date.

This experiment used two different purge policies. The first policy requires purging data from both tables where the `alphaDate` is older than five years old and `alphaGroup`='a' (randomly generated value occurring in approximately 25% of the rows). The second policy requires purging only from the `Beta` table where `betaDate` (generated independently from `alphaDate`) is older than five years old and `betaGroup`='a' (separately generated with the same probabilities).

Our trigger on each table fires upon `INSERT`, `UPDATE`, or `DELETE` to propagate the change into the shadow table(s). When the insertion trigger fires, it first checks for an encryption key in the `encryptionOverview` table for the given policy and expiration date; if one does not exist, the key is created and stored automatically.

We pre-populated `Alpha` table with 1,000 rows and `Beta` table with 1,490 rows. We also generated a random workload of inserts (25), deletes (25), and updates (25) for the time period between 1/1/2014 to 2/1/2019. Because we used two different policies, we generated the data so that some of the business records were subject to one of the purge policies and some records were subject to both purge policies. Roughly 75% of the data generated was subject to a purge policy. Finally, not all records requiring encryption will be purged during this experiment due to the purge policy date not having passed. Records generated with dates from 2017–2019 would have not expired in running of this experiment. We then perform updates and deletes on the tables to verify that our implementation is accurately enforcing compliance.

Using a randomly generated string of alphanumeric characters with a length of 50, our process uses the function PGP_SYM_ENCRYPT to generate encryption keys to encrypt the input values. alphaID is the primary key of Alpha and (alphaID, alphaIDEncryptionID) is the primary key of AlphaShadow. If alphaID is not encrypted, the column alphaIDEncryptionID is set to −1 to maintain uniqueness and primary key constraint.

The UPDATE trigger for the Alpha table is similar to the INSERT trigger, but it first deletes the existing row in the shadow table. Next, we determine the current applicable encryption key and insert an encrypted updated row into the shadow table. The DELETE trigger removes the row from the AlphaShadow table upon deletion of the row in Alpha. When alphaDate in a row from Alpha changes, the corresponding rows in Beta table may fall under a different policy and must be re-encrypted accordingly. Furthermore, when a Alpha row is deleted, the child Beta row must be deleted as well along with the shadow table entries. Note that PGP_SYM_ENCRYPT may generate several different ciphertext values given the same value and the same encryption key. Therefore, we cannot encrypt the value from Alpha and compare the encrypted values. Instead, we must scan the table and match the decrypted value in the predicate (assuming the key is encrypted):

```
DELETE FROM alphaShadow
WHERE PGP_SYM_DECRYPT(alphaID, v_encryption_key)=old.alphaID
    AND alphaIDKey=v_key_id;
```

Changes to Beta table are a little more interesting since there is a foreign key relationship between Beta rows and Alpha rows. When a row is inserted or updated in the Beta table, in addition to the Alpha trigger processes, the Beta table triggers must compare the expiration date of the Beta row to the expiration date of the Alpha parent row and select the encryption bucket with the shorter of the two periods.

Initialization: We first import data into the Alpha and Beta tables. We then ran loadAlphaShadow() and loadBetaShadow() to populate the shadow tables using the corresponding key; the dates in the encryptionOverview table are initialized based on our expiration dates. Next, we enabled the triggers and incremented dates in encryptionOverview by five years to simulate the policy's expiration at a later time.

Validation: We wrote a procedure, RestoreTables(), to restore Alpha and Beta tables after shadow tables were restored from backup. In a production database, the backup method would depend on the Recovery Time Objective (RTO) and Recovery Point Objective (RPO) which would determine the backup methodology implemented, such as with PostgreSQL's pg_dump and excluding the tables with sensitive data. We tested the basic backup and restore process by exporting and importing the shadow tables, then truncating Alpha and Beta, and finally invoking our RestoreTables() procedure. We then modified the procedure to restore the tables to (temporarily created) Alpha' and Beta' so that we could compare restored tables to Alpha and Beta. We then verified that the values for the restored tables match the original tables' non-purged records.

Evaluation: We have verified that by deleting encryption keys to simulate the expiration of data, the restore process correctly handled the absence of a key to eliminate purged data. In total, there were 61 rows purged from `Alpha` and 182 rows purged from `Beta`, as well as the same rows purged from `AlphaShadow` and `BetaShadow`. Therefore, we have demonstrated that our framework achieves purging compliance in a relational database without altering tables in the existing schema or modifying the standard backup procedures.

Encrypting and maintaining a shadow copy of sensitive data to support purging incurs processing overheads for every operation that changes database content (read operations are not affected). Optimizing the performance of this approach is going to be considered in our future work. During an `INSERT` on the `Alpha` table, our system opens a cursor to check if an encryption key is available in the `encryptionOverview` table. If the applicable key exists we fetch it, otherwise we create a new one. Once a key is retrieved or a new key is generated, the values that are under a purge policy are encrypted with `PGP_SYM_ENCRYPT`. Next, we insert encrypted data into the shadow table as part of the transaction. For an `UPDATE`, we follow the same steps but also delete the prior version of the row from the shadow table (and may have to take additional steps if the update to the row changes the applicable purge policy). If the policy condition changes, we insert the shadow row into `AlphaShadow` and then evaluate the data in the `BetaShadow` table to see if the encryption key needs to change on the encrypted rows of the `BetaShadow` where the `linkID` refers to the `Alpha` row that changed.

The restore process is subject to decryption overheads. For example, in Postgres, in addition to the normal restore operation that restores the shadow table, we recreate the unencrypted (active) version of the table. For each encrypted column, we look up the key, then apply `PGP_SYM_DECRYPT`, and finally insert the row into the active table (unless the row already expired). Because the restore process creates an additional insert for every decrypted row, this also increases the space used for the transaction logs. The performance overhead for a restore will be correlated with doubling the size of each encrypted table (due to the shadow copy addition) plus the decryption costs. During deletion, each time we decrypt a row, the process of executing `PGP_SYM_ENCRYPT` and evaluating each row of the table incurs a CPU cost in addition to the I/O cost of deleting an additional row for each deleted row. The performance for an update statement incurs a higher overhead since an update is effectively a delete plus insert. Some of these I/O costs, such as fetching the key, can be mitigated with caching.

5 Discussion

5.1 Implementation

In our experiments we exported and imported the shadow tables to show that the system worked as expected; in practice, backup methodology would depend on the RTO and RPO on the application [10]. There are a plethora of options that can be implemented depending on the needs of the application. One could use `pg_dump` and exclude the tables containing sensitive data, so that these tables are excluded from the backup file. If the size of the database is too large for a periodic pg_dump, or if the RTO and RPO warrant a faster backup, one could replicate the database to another database, and exclude the tables with sensitive data from replication. Using the clone of the database, one could do filesystem level backups or a traditional pg_dump. While the clone is a copy, a clone is not versioned in time like backups would be. For example, if someone dropped a table, the drop would replicate to the clone and not protect data against this change, whereas a backup would allow restoring a dropped table.

5.2 ACID Guarantees

If a trigger abends at any point, the transaction is rolled back. Since we attach triggers to the base tables, we are able to provide ACID guarantees. These guarantees are also extended to the shadow tables because all retention triggers execute within the same transaction. Overall, for any table dependencies (either between the active tables or with the shadow tables), our framework executes all steps in a single transaction, fully guaranteeing ACID compliance. This guarantee requires additional steps if we replicate the changes outside of the database since the database is no longer in control of the transaction.

For example, if the remote database disconnects due to a failure (network or server), the implementation would have to choose the correct business logic for the primary database. If the primary database goes into a read-only mode, the primary can keep accepting transactions or keep a journal to replay on the remote database. If the implementation kept a journal to replay, organizations must determine if is it acceptable to break ACID guarantees. Oracle DataGuard and IBM Db2 HADR provide varying levels of replication guaranties; similar guaranties would need to be built into our framework and verbosely explained as to the implications. Similarly, supporting asynchronous propagation and encryption of data into shadow tables would require additional investigation.

5.3 Future Work

We plan to consider asynchronous propagation (instead of triggers) to shadow tables; although that would require additional synchronization mechanisms, it has the potential to reduce overheads for user queries. Because scalability is a concern, tools such as Oracle Goldengate or IBM Change Data Capture, provide a framework to replicate changes, apply business logic, and replicate the changes to the same database or other heterogeneous databases. We also intend to explore developing our framework to replicate changes outside of a single database.

Our approach can easily incorporate new policies without requiring any changes to the already defined policies. However, when a policy is removed, all data in the shadow tables will stay bucketed under the previous policy. Further research is needed to automatically re-map all data points to the newest policy after a policy has been replaced or altered, to facilitate up-to-date compliance.

6 Conclusion

Organizations are increasingly subject to new requirements for data retention and purging. Destroying an entire backup violates retention policies and prevents the backup from being used to restore data. Encrypting the active database directly (instead of creating shadow encrypted tables) would interfere with (commonly used) incremental backups and introduce additional query overheads. In this paper we have shown how a framework using cryptographic erasure is able to facilitate compliance with data purging requirements in relational database backups.

Our approach does not change the active tables and maintains support for incremental backups while providing an intuitive method for data curators to define purge policies. This framework balances multiple overlapping policies and maintains database integrity constraints (checking policy definitions for entity and referential integrity). We demonstrate that cryptographic erasure supports the ability to destroy individual values at the desired granularity across all existing backups.

Overall, our framework provides a clear foundation for how organizations can implement purging into their backup processes without disrupting the organization's business continuity processes. This is also accomplished without adding any restrictions to existing databases. Our purging framework is able to guarantee purging compliance while being easily integrated into existing databases.

References

1. Ataullah, A.A., Aboulnaga, A., Tompa, F.W.: Records retention in relational database systems. In: Proceedings of the 17th ACM Conference on Information and Knowledge Management, pp. 873–882 (2008)
2. Boneh, D., Lipton, R.J.: A revocable backup system. In: USENIX Security Symposium, pp. 91–96 (1996)
3. Citus Data: pg_cron, https://github.com/citusdata/pg_cron
4. Congress, U.S.: 28 u.s. code §1732 (1948)
5. Dudjak, M., Lukić, I., Köhler, M.: Survey of database backup management. In: 27th International Scientific and Professional Conference"Organization and Maintenance Technology (2017)
6. European Parliament: Regulation (EU) 2016/679 of the European parliament and of the council (2020). https://gdpr.eu/tag/gdpr/
7. International Data Sanitization Consortium: Data sanitization terminology and definitions, September 2017. https://www.datasanitization.org/data-sanitization-terminology/
8. IRS: How long should i keep records? https://www.irs.gov/businesses/small-businesses-self-employed/how-long-should-i-keep-records
9. Kamara, S., Lauter, K.: Cryptographic cloud storage. In: Sion, R., Curtmola, R., Dietrich, S., Kiayias, A., Miret, J.M., Sako, K., Sebé, F. (eds.) FC 2010. LNCS, vol. 6054, pp. 136–149. Springer, Heidelberg (2010). https://doi.org/10.1007/978-3-642-14992-4_13
10. Lenard, B., Rasin, A., Scope, N., Wagner, J.: What is lurking in your backups? In: Jøsang, A., Futcher, L., Hagen, J. (eds.) SEC 2021. IAICT, vol. 625, pp. 401–415. Springer, Cham (2021). https://doi.org/10.1007/978-3-030-78120-0_26

11. Office of the Attorney General: California consumer privacy act (CCPA), July 2020. https://oag.ca.gov/privacy/ccpa
12. Reardon, J., Basin, D., Capkun, S.: SoK Secure data deletion. In: 2013 IEEE Symposium on Security and Privacy, pp. 301–315. IEEE (2013)
13. Bringas, P.G., Hameurlain, A., Quirchmayr, G. (eds.): DEXA 2010. LNCS, vol. 6261. Springer, Heidelberg (2010). https://doi.org/10.1007/978-3-642-15364-8

Information Retrieval

Improving Quality of Ensemble Technique for Categorical Data Clustering Using Granule Computing

Rahmah Brnawy[✉] and Nematollaah Shiri

Concordia University, Montréal, Canada
r_brnawy@encs.concordia.ca, shiri@cse.concordia.ca

Abstract. Ensemble clustering is a popular technique to improve quality of clustering. It aims to combine multiple base clusterings (partitions) into a single robust, stable and accurate clustering. Existing ensemble techniques focused more on numerical data due to the presence of well-defined similarity measures for such data. For categorical data, however, lack of objective similarity measures poses challenges to successful clustering such data due to inherent complexity such as uncertainty and overlapping among the clusters. In this work, we use granule computing theories and methodologies to overcome the challenges. In particular, we use granule knowledge to develop an asymmetric measure and leverage measure to minimize the risk of selecting and combing redundant partitions. Using Rough Set theory, we also propose a novel refinement similarity matrix to handle uncertain objects in overlapped areas. The results of our numerous experiments on real-life benchmark datasets illustrate that our proposed techniques significantly outperform existing techniques for categorical data. Furthermore, the proposed techniques yield improved clustering quality also when applied to numerical data.

Keywords: Asymmetric measure · Categorical data · Granule computing · Ensemble clustering · Rough sets · Overlapping clustering

1 Introduction

Ensemble clustering (EC) was proposed to deal with the challenge of selecting a desired clustering algorithm for a given data [13,18,27]. In ensemble clustering, instead of generating a clustering solution, EC first generates a number of diverse, base clusterings of the input dataset and then integrates them into a single, consensus clustering. The ultimate goal is to improve accuracy and robustness of what a single clustering algorithm can achieve. Several approaches and techniques have been proposed to achieve this goal. However, a few solution techniques were particularly proposed for categorical data. Based on the problems focused on, those solutions may be classified into two directions, described as follows.

© Springer Nature Switzerland AG 2021
C. Strauss et al. (Eds.): DEXA 2021, LNCS 12923, pp. 261–272, 2021.
https://doi.org/10.1007/978-3-030-86472-9_24

The first direction includes works which focused on the generation phase. It was observed that not every generated base clustering solution would be beneficial for producing the final clustering result [2,9,27]. Hence, they proposed selecting a smaller subset of the input clusterings which performed equally or better than the ensemble of the full set. Generally speaking, a selection strategy relies on two core elements: selection criteria and selection search method. The former is used to evaluate the quality of partitions (in a clustering), whereas the later is used to select suitable partitions. Toward this goal, Fern and Lin [9] proposed three heuristics search methods which consider both diversity and quality of the ensemble members. Of the three proposed heuristics, the Cluster And Select (CAS) method was empirically demonstrated to achieve the best overall performance. APMM and its extensions [1,2] adopted the normalize mutual information as a selection criterion. Huang et al. [15] proposed the ECI selection criterion using information entropy. Recently, authors in [27] considered a collection of arbitrarily validity criteria as diversity and quality measures and proposed the Sum of Internal Validity Indices with Diversity algorithm (or SIVID, for short). This algorithm uses five different internal validity indices for categorical data.

The aforementioned selection algorithms point to a common problem, to which we refer as the risk of redundancy. The absence of the class label makes it difficult to measure the level of diversity among base clusterings. Therefore, in the related literature, it is assumed that the diversity can be applied using different generative mechanism. However, it is still very likely to have clusterings that are the same or very similar to some other, which create redundancy among base clustering [5]. Of course one can argue to solve redundancy by removing such partitions manually. However, this cannot be achieved in general by most of the existing selection search methods. This is mainly because most selection criteria utilize clustering validation index (CVI) to assign quality scores to base clusterings. In some cases, CVI assigns the same quality score to partitions with different clustering schemes. The authors in [19] refer to this problem as the context meaning problem [19]. As a result, the performance of the existing selection search methods might get affected negatively and similar or very divers partitions get selected, since it is difficult to decide whether those partitions are actually the same or because they were mistakenly assigned the same quality score. Therefore, a new selection criteria which is independent from any CVI is required to overcome the context meaning problem and minimize the risk of redundancy among selected base clusterings. Accordingly, redundant partitions can be identified and removed, and hence selecting only those that have novel contents and patterns about the data.

The second direction of related ensemble research is in the integration phase, where information in the generated base clusterings is summarized in order to build the final clustering result. The co-association matrix (co-matrix, for short) is the most commonly used summarizing technique [16,18,23]. Each entry in a co-matrix indicates the number of times the corresponding pair of objects are assigned to the same clusters. Several co-matrix refinement methods have been proposed to address different problems from different perspectives. For example, the authors in [15,28] observed that in the same base clusterings, pairs of objects with different distances could have different weights in capturing similarity evidences. To avoid this problem, they replaced occurrence frequencies with occurrence probabilities. Inspired by the Adaboost algorithm, Ren et al. [20] proposed a new index in which a larger weight is assigned to objects that are hard to be clustered. Recently, Li et al. [18] distinguish among the objects in base clusters and classified them into stable and unstable objects. Stable objects are used to build the co-matrix and produce the pre-discovered data structures and gradually identify unstable objects to obtain the final clustering result. These refinement matrix techniques were mainly designed and employed over numerical data, and assumed that the clusters are disjoint [5]. Therefore, a new refinement technique is needed, when the assumption is not valid, to handle categorical data and overlapping clusters.

In this research, we adopt concepts and techniques from GrC to address the aforementioned problems to improve the performance of ensemble clustering. For this, we first propose a new selection criteria, called *Information Granulation Quality Scores* (IGQS), which is independent of any external validity index or prior knowledge, and consider the intra-structure of clusters. In particular, information granulation concept is applied. We will then introduce a novel refinement co-matrix, called *Fuzzy Rough Refinement CO_ matrix* (FRRCO), to determine the degree of similarity of objects in the overlapping area. To this end, we employ the principles of hybrid fuzzy-rough. Note that both proposed techniques are independent from each other, but both aim to improve the final quality clustering.

The rest of this paper is organized as follows. Section 2 provides a background and notations. Section 3 introduces our proposed techniques. The experiments and results are presented in Sect. 4. Concluding remarks and future research are provided in Sect. 5.

2   Background and Notations

2.1   Information Granulation

Granule Knowledge GK is used to represent discernibility, that is the ability to distinguish, and provides a strong measure of intra-granule uncertainty [25]. There are extensive studies on the measurement of Information granularity [26]. In this research, we adopt and extend a measure based on the cardinality of granularity, defined as follows [14].

Definition 1 *(Information Granulation (IG)). A cluster ensemble information system S for a cluster C is defined as $S = (U, \Pi, C)$, where $U = \{x_1, x_2, \cdots, x_n\}$ is the set of categorical objects and $\Pi = \{\pi_1, \cdots, \pi_P\}$ is a set of base clustering solutions. Each base clustering $\pi_i \in \Pi$ divides U into a set of clusters, also called granules, denoted by $U/\pi_i = \{C_1^i, C_2^i, \cdots, C_k^i\}$, where k is the number of granules defined by π_i. The IG degree of $C_i \in \pi_i$ with respect to π_j is defined as:*

$$IG(C_i|\pi_j) = \sum_{k=1}^{|\pi_j|} \frac{|G_{ik}|^2}{|C_i|^2} \qquad (1)$$

Note that IG is a bounded measure $1/|C_i| \leq IG(C_i|\pi_j) \leq 1$. The maximum value 1 is attained if cluster C_i has just one granule, and the minimum value is $1/|C_i|$, if it has k granules.

2.2 Hybridization of Fuzzy Sets and Rough Sets

Fuzzy sets and rough sets are different theories developed to deal with uncertainty in data and information [7,8,13]. They have both been adopted and deployed in clustering algorithms to model and process the uncertainty in the data from different perspectives [8,13]. In fuzzy clustering, each object x is assigned to a cluster C with a membership value μ in the range of $[0,1]$, indicating the strength with which x belongs to C. In rough set based clustering, each cluster is represented with two regions, called the *lower* and *boundary* regions. An object x can belong to a cluster's lower region ($\underline{C}$) or to several clusters' boundary regions ($\overline{C}$). Objects that are in the boundary regions can be thought of as being situated in the overlapping portions of two or more clusters. Recently, it has been observed that combining the fuzzy and rough set theories can enhance information processing [7,22]. The standard definitions for the fuzzy lower and upper approximations are as follows [22]:

$$\mu\underline{(C)}(x) = \inf_{y \in U} I(sim(x,y), \mu_C(y)) \qquad (2)$$

$$\mu\overline{(C)}(x) = \sup_{y \in U} \tau(sim(x,y), \mu_C(y)) \qquad (3)$$

Here, I and t-norm(τ) are fuzzy operators, $sim(x,y)$ is the fuzzy similarity relation, which measures the degree to which objects x and y are similar, and $\mu_C(y)$ is the membership value of object y in cluster C. For details, interested readers can refer to [24].

3 The Proposed Techniques

Let us denote the following notations: $U = \{x_1, x_2, ..., x_n\}$ be a set of n objects described by A attributes, and let $\Pi = \{\pi_1, ..., \pi_P\}$ be an ensemble library with P base clusterings. Each base clustering $\pi_i \in \Pi$ is a set of clusters $\pi_i = \{C_1^i, C_2^i, ..., C_K^i\}$, where K is the number of clusters in π_i.

3.1 Information Granulation Selection Criteria

To evaluate the quality of a partition among a collection of base clusterings, we measure the uncertainty of the clusters each include. Intuitively, the uncertainty of a cluster C^i reflects how the objects in C^i are clustered in the ensemble of

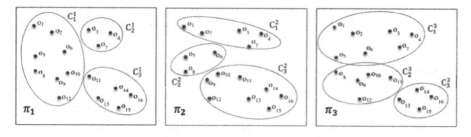

Fig. 1. An ensemble example with three partitions [15].

multiple base clusterings. Generally, a cluster has two states w.r.t other partitions: it is either kept unchanged or decomposed into multiple, refined clusters. The first case indicates maximum certainty degree of a cluster, whereas the second one indicates less certainty degree. To implement this idea in our work, we adopt the axiom definition of information granulation (IG) given in Eq. 1. To illustrate the IG degree, consider the cluster C_1^1 in Fig. 1. As can be seen, C_1^1 is distributed over three clusters in π_2 and over two cluster in π_3. Then, the certainty degree of C_1^1 is expected to be different in these two clusterings, that is, $IG(C_1^1|\pi_3) \neq IG(C_1^1|\pi_2)$. To be precise, using the Eq. 1, we obtain $IG(C_1^1|\pi_2) = 0.34$, and $IG(C_1^1|\pi_3) = 0.52$. Despite being popular and useful, this measure of uncertainty may cause confusion in some cases. For example, consider cluster C_2^1 in partition π_1. It can be seen that cluster C_2^1 does not appear as one granule G in neither partitions π_2 and π_3. However, it appears with other objects, to which we refer as Granule Extended Elements (GEE). In both partitions, the size of the GEE is different, and hence we expect their certainty levels to be different as well. However, according to formula (1), we have that $IG(C_2^1|\pi_2) = IG(C_2^1|\pi_3) = 1$. Ignoring the GEE might cause context meaning problems, mentioned in Sect. 1. To avoid such problem, in this work, we introduce the measure *complement granule information* to determine the purity degree of a cluster w.r.t other partitions. This measure is formulated by redefining Eq. (1) as follows:

$$EIG(C_i|\pi_j) = \frac{1}{|C_i|^2} \sum_{k=1}^{|\pi_j|} |G_{ik}|^2 * (1 - \frac{|GEE|^2}{|C_k|^2}) \tag{4}$$

where $G_{ik} = C_i \cap C_k$ and GEE is the difference between C_i and C_k. Considering again the cluster C_2^1, we obtain $EIG(C_2^1|\pi_2) = 0.84$ and $IG(C_2^1|\pi_3) = 0.6735$. That is, *EIG* reaches its maximum certainty, that is 1.0 when it appears as one pure cluster; and as the number of GEE increases, *EIG* decreases as an indication of low uncertainty degree. Given this, we define our quality measure as the total information granulation score of partition π_i over all base partitions. This can be calculated by summing the knowledge granulation of *EIG* for each base partition. Formally,

$$IGQS(\pi_l|\Pi) = \frac{1}{1 - |\Pi|} * \sum_{i=1}^{|\pi_l|} \sum_{j=1, j\neq i}^{|\Pi|} EIG(C_i|\pi_j) \tag{5}$$

The proposed selection criterion is simple and intuitive yet it has the following advantageous properties:

– Scalability: Unlike most of the existing selection criterion, $IGQS$ dose not require the entire dataset to be maintained in the main memory.
– knowledge-free: $IGQS$ is independent from any validation index(internal or external) and user input parameter.
– Asymmetric measure: two or more partitions get the same quality score if and only if they are the same. Hence, depending on the selection search method, they can be manually deleted or grouped together without loosing information.

3.2 Fuzzy Rough Refinement Matrix

Generally, soft clustering provides better results than hard clustering and might discover much information about the real structure of the data. However, the standard co_matrix given below cannot summarize such clusters efficiently [4].

$$co\_matrix(x_i, x_j) = \frac{1}{P} \sum_{\pi=1}^{P} \sum_{m=1}^{|\pi|} Sim(x_i, x_j|C_m) \tag{6}$$

$$Sim(x_i, x_j|C_m) = \begin{cases} 1, & \text{if } x_i, x_j \in C_m \\ 0, & \text{otherwise} \end{cases} \tag{7}$$

This is because the co_matrix is based on binary assignment and ignores the fact that pair of objects might allocate in the overlapping area. In this section, we further take advantage of granule computing (GrC) and propose a novel refinement matrix based on the principle of Fuzzy Rough set theory. Our approach mainly designed for categorical data, but we have test its applicability for numerical data as well. Before going into details, we present the steps of the proposed refinement co_matrix as the following.

1. Generate P soft clustering.
2. Map the fuzzy clustering results into fuzzy rough results using Eqs. 8 and 9.
3. Based on objects region, create the FRRCO matrix using Eq. 10.
4. Apply an integration algorithm over the constructed matrix.

Given the general approach, we next explain the proposed approach (steps 2 and 3) in details. After the soft base clusterings are generated, a mapping function is applied to construct the fuzzy rough regions. In the related literature, there two main approaches to define the fuzzy rough regions. The standard one as given in Eq. 2 and 3, is based on the fuzzy logic operators [17]. The other approach, the recent one, is based on the notation of similarity [13,21,24]. In this paper, we adopt the later approach and use the definition introduced in [13] and [24]. This yields a new fuzzy rough similarity function defined as follows:

$$\mu(\underline{C_j})(x) = Average\{1 - sim(x,y)\}, \ \forall y \notin C_j \tag{8}$$

$$\mu(\overline{C_j})(x) = Average\{sim(x,y)\}, \ \forall y \in C_j \tag{9}$$

where $\mu(\underline{C_j})(x)$ indicates that an object x belongs to the lower region of C_j^L if its average dissimilarities with all the samples from the sample domain belonging to $U - C_j$ is the maximum. Analogously, $\mu(\overline{C_j})(x)$ indicates that an object x belongs to the upper region C_j^B if its average similarities with all the samples from the sample domain belonging to C_j is the minimum. In the above definitions, sim is the similarity function computed based on Jacquard coefficient [10]. Our choice of the Jacquard coefficient is justified for being based on set operations, making it more suitable for describing the properties and structures of categorical data [27]. Once regions constructed, a refinement co-matrix measure, called Fuzzy Rough Refinement CO-matrix (FRRCO) is applied to summarize the soft base clusterings. The proposed FRRCO is defined as:

$$FRRCO(x_i, x_j) = \frac{1}{P} \sum_{\pi=1}^{P} \sum_{m=1}^{|\pi|} sim(x_i, x_j | C_m) \tag{10}$$

$$sim(x_i, x_j | C_m) = \begin{cases} 1 & \text{if } x_i, x_j \in C_m^L \\ certainty(x_i, x_j) & \text{if } x_i, x_j \in C_m^B \\ sim(x_i, x_j) & \text{if } x_i \in C_m^L \text{ and } x_j \in C_m^B \\ 0 & \text{otherwise} \end{cases} \tag{11}$$

where

$$certainty(x_i, x_j) = \begin{cases} 0.5 & \text{if } \mu_{x_i^B}(y) = \mu_{x_j^B}(y) \\ [0.1, 0.4] & \text{otherwise} \end{cases} \tag{12}$$

In FRRCO, we may have one of the following three cases for each pair of objects in a cluster C_m. (i) The pair belong to the lower region $C_m{}^L$, and hence expected to be quite similar and have a significant contribution in creating the cluster. In this case, their corresponding entry in FRRCO would be 1. (ii) The pair belong to the boundary region or the overlapping region C_m^B, so they are expected to be uncertain. In this case, according to RST their corresponding value should be 0.5. However, this uncertainty value is too strict. It assumes that pairs of objects in the boundary region belong to same overlapping area. Obviously, this kind of assumption does not reflect the actual situation. Objects in the boundary regions might come from different classes. Hence, such objects should be assigned less certainty values than the values assigned to objects in the same class labels. To decide whether boundary objects come from the same class label or different ones, we adopt the FKNN (Fuzzy K Nearest neighbor) algorithm as follows:

$$\mu_{x_i^B}(y) = \sum_{j=1}^{K} \mu_{ij} \frac{1}{dis(y_j, x_i^B)^{\frac{2}{\alpha-1}}} \Bigg/ \sum_{j=1}^{K} \frac{1}{dis(y_j, x^B)^{\frac{2}{\alpha-1}}} \tag{13}$$

where $dis(.)$ is the simple matching coefficient, α is the fuzzification parameter. Given this, a pair of objects in the boundary get the value 0.50 if they share the same FKNN. Otherwise, they are assigned a value in the range [0.1, 0.4]. (iii) The pair belong to different regions. In this case, since objects in the lower-region are relatively similar, their corresponding entry in FRRCO is computed as the similarity between the pair of objects.

4   Experiments and Results

In this section, we present the results of our experiments using multiple and popular real-life benchmark datasets from the UCI machine learning repository [3]. These are: Dermatology (Derma), Congressional Voting (Vote), Soybean (Soy), Mushroom (Mush), Breast Cancer (BC), Hayes-Roth (Hayes), Zoo, Balance Scale (BS). A brief description of these datasets is provided in Table 1. In order to evaluate the effectiveness of the proposed techniques, the clustering results obtained are compared with the ground truth provided with the dataset. For this, we used the external criterion NMI, which essentially measures the agreement degree of the clustering results produced by an algorithm and the ground truth. The maximum NMI value is 1. If the clustering result is close to the true class distribution, the NMI value returned is high. For the proposed selection criteria IGQS, we compare the results against SIVID [27], APMM [1], as well as the baseline model (referred to as "Base" hereafter). For the proposed refinement technique FRRCO, we compare its performance against the conventional co_matrix. We also perform statistical significance tests for our results.

Table 1. Descriptions of the datasets.

Datasets	Derma	Vote	Soy	Mush	BC	Hayes	Zoo	BS
No. objects	366	435	47	8124	699	160	101	625
No. attributes	35	16	21	23	9	5	16	5
Classes	6	2	4	2	2	3	7	3

4.1   Evaluation of the Proposed Selection Criteria IGQS

Below we describe the set up of the experiments performed. For generating the base clusterings (partitions), the algorithms performed include k-modes [10], squeezer [12], CLUC^{++} [6], for which we used different number of clusters, initialization, and permutation (only for the squeezer algorithm). In our experiments, we considered P = 40 as the size of the base clustering. The results reported are the average of the results obtained in 30 runs.

We applied two search methods (SM) to select the candidate partitions for the integration. The first SM applied is ranking approach, to which we refer as Rank and select, or RAS for short. In this approach, partitions are ranked according to their quality degree. Based on this ranking, we select the top T partitions to participate in the final phase. The second SM applied is based on clustering partitions, to which we refer as Cluster and Select, or CAS for short. Using CAS, we select K different base clusterings. This is done by first creating a similarity matrix of size $\Pi * \Pi$, which uses IGQS as the similarity criterion. We then apply a typical clustering algorithm to partition the base clustering into K groups. As is customary, we use spectral clustering algorithm in our work. From each group, we then select a partition with the highest quality as a candidate for the integration phase. The candidate selected partitions are summarized using the standard co_matrix. Subsequently, a hierarchical clustering algorithm (Average linkage) is applied over the co_matrix to generate the final clustering result. For both types of search methods, the best sizes for T and K reported are obtained by considering the best result running from 2 to P/2.

Table 2 shows that IGQS always outperforms the base ensemble clustering (Base) and APMM algorithms. Our proposed technique also outperforms SIVID in most cases, except for a few, which are marked in the table with an asterisk. For instance, SIVID outperforms IGQS in clustering the dataset Hays. Our results indicate comparable performance in clustering Soybeans and Mushroom datasets. Even though SIVID performs better than our IGQS in some cases, it is important to note that, unlike IGQS, SIVID requires to maintain the dataset in the main memory which increases both the run time and storage utilization [27].

Overall, the proposed selection criteria along with CAS provided superior clustering results. We attribute the higher accuracy of CAS to the asymmetric and granularity properties which is inherent and built into our proposed IGQS. This can be seen clearly when CAS is adopted, as the properties of IGQS minimize the risk of grouping dissimilar partitions within the same group and successfully prunes redundant clusters in the ensemble. Thus, the level of both diversity and quality is increased, which results in increased quality of the clustering yield. For the case of RAS, we can still see the advantage of the asymmetric and granularity properties. Since RAS is based on ranking, without loosing reliability, redundant base clusterings can be manually deleted. However, cluster that are just slightly different still cannot be removed, which can cause some level of redundancy. This explains why the clustering results with RAS in many cases is lower than CAS.

Table 2. NMI values for the compared algorithms using average linkage.

Datasets	SM	Derma	Vote	Soy	Mush	BC	Hayes	Zoo	BS
Base	Full	0.687	0.7542	0.8809	0.7156	0.6775	0.0046	0.7735	0.0306
IGQS	RAS	0.794	0.8442	**1.0**	0.7457	0.7422	0.0116	**0.8767**	0.0567
	CAS	**0.818**	0.7935	**1.0**	**0.8566**	**0.7491**	0.0942	0.8568	**0.1901**
SIVID	RAS	0.6995	**0.9095***	**1.0**	0.7662	0.6785	**0.2160***	0.8433	0.0307
APMM	CAS	0.6718	0.7697	0.9194	0.7175	0.6792	0.0121	0.7869	0.0252

4.2   Evaluation of the Proposed Refinement Techniques

In what follows we first specify the settings of the proposed refinement techniques, FRRCO and then present the evaluation results. We run fuzzy k-modes algorithms with different settings (initialization, iteration, fuzzification). We also applied the standard k-Modes over random subspace to create overlapping clusters [11]. We set the number of KNN in the range [2–5] with different certainty coefficient values and the best result is reported. For the integration phase, we applied the fuzzy c-means [10] over the created matrices, and then compared the results against the ground truth using NMI. Table 3 presents the performance of the FRRCO technique. The value of the best performance for each data set is shown in bold. From the table, we can clearly see superiority of the proposed techniques over the standard co_matrix. In general, the proposed techniques successfully improved the accuracy of the final clustering results. We attribute this success to the hybrid granularity theories. Simply, our techniques performed intra-clustering by dividing clusters into regions, when necessary. This helped discover some of actual relationships of objects in clusters, which in turn increased the reliability of co_matrix. Note that the proposed refinement matrix was applied to a well-known numerical data (with suitable settings) and compared against WOEC (Weighted Objects Ensemble Clustering) algorithm [20]. The FRRCO outperformed WOEC on five out of eight datasets. We omitted the experiment and the results due to the space limitation.

Table 3. The performance of FRRCO technique over categorical data.

Datasets	Derma	Vote	Soy	Mush	BC	Hayes	Zoo	BS
co_matrix	0.37215	0.4923	0.51779	0.5079	0.5285	0.0098	0.7111	0.0201
FRRCO	**0.5113**	**0.5154**	**0.7212**	**0.777**	**0.7931**	**0.0879**	**0.7805**	**0.0861**

4.3   Statistical Significance Tests

As showed in the previous section, our both proposed techniques resulted in improved performance of the ensemble clustering process. To determine whether the improved performance obtained using different algorithms was by chance or not, and also to decide which technique performed better, we applied some statistical significance tests, for which we used a non-parametric statistical test, called Wilcoxon rank sum test [27]. As in related literature, we set the significance level at 5%. Table 4 reports the p-values obtained by Wilcoxon rank sum test. All

Table 4. The p-values between the proposed techniques and existing algorithms

Algorithms	Derma	Vote	Soy	Mush	BC	Hayes	Zoo	BS
IGQS vs SIVID	$4.25e{-}2$	**0.017^{*}**	0.00	0.017	0.033	**$5.71e{-}4^{*}$**	$8.44e{-}3$	$4.29e{-}2$
IGQS vs APMM	$4.44e{-}3$	$8.88e{-}2$	$1.69e{-}2$	$6.34e{-}2$	$2.34e{-}4$	$1.89e{-}2$	$3.11e{-}2$	$1.03e{-}5$
FRRCO vs matrix	$1.11e{-}8$	$4.30e{-}5$	$4.94e{-}9$	$1.43e{-}2$	$1.50e{-}3$	$6.72e{-}2$	$1.38e{-}7$	$3.16e{-}4$

the p-values reported in the table are less than 0.05 (5% significance level). This is a strong evidence against the null hypothesis, which indicates that the better NMI values produced by the proposed techniques are statistically significant and not occurred by chance. Except for a few cases marked with an asterisk, SIVID performed significantly better than IGQS.

5   Conclusions and Future Work

In this paper, we adopted the concept of granule computing GrC and proposed two powerful techniques, which improve the performance of cluster ensemble techniques for categorical data. The IGQS was proposed to address the problems in selecting a subset of base clusterings. Unlike most of the existing selection criteria, IGQS dose not require a prior knowledge of the data set and has no bias towards the characteristic of the dataset. The second technique FRRCO was proposed to address the uncertainty caused by overlapping clustering, which improved the reliability of the co_matrix. The results of our extensive experiments using the real-life benchmark data showed superiority of both techniques of ensemble clustering as compared to the base techniques as well as the existing ones. The statistical significant tests performed clearly indicated that the performance improvement measured are both significant and not by chance. As for the future work, we plan to investigate ways to extend IGQS to soft clustering in order to develop a fuzzy ensemble clustering framework for categorical data. We will also study other cluster ensemble approaches on GrC, for instance, stacking and blending.

Acknowledgements. Rahmah's research was supported by a scholarship from the government of Saudi Arabia.

References

1. Abbasi, S., Nejatian, S., Parvin, H., Rezaie, V., Bagherifard, K.: Clustering ensemble selection considering quality and diversity. Artif. Intell. Rev. **52**(2), 1311–1340 (2018). https://doi.org/10.1007/s10462-018-9642-2
2. Alizadeh, H., Minaei-Bidgoli, B., Parvin, H.: Cluster ensemble selection based on a new cluster stability measure. Intell. Data Anal. **18**(3), 389–408 (2014)
3. Asuncion, A., Newman, D.: UCI machine learning repository (2007)
4. Bagherinia, A., Minaei-Bidgoli, B., Hossinzadeh, M., Parvin, H.: Elite fuzzy clustering ensemble based on clustering diversity and quality measures. Appl. Intell. **49**(5), 1724–1747 (2018). https://doi.org/10.1007/s10489-018-1332-x
5. Boongoen, T., Iam-On, N.: Cluster ensembles: a survey of approaches with recent extensions and applications. Comput. Sci. Rev. **28**, 1–25 (2018)
6. Brnawy, R., Shiri, N.: K-mixed prototypes: a clustering algorithm for relational data with mixed attribute types. In: Proceedings of the 34th ACM/SIGAPP Symposium on Applied Computing (SAC), pp. 542–545 (2019)
7. Chen, J., Mi, J., Lin, Y.: A graph approach for fuzzy-rough feature selection. Fuzzy Sets Syst. **391**, 96–116 (2020)

8. Duan, Q., Yang, Y.L., Li, Y.: Rough k-modes clustering algorithm based on entropy. IAENG Int. J. Comput. Sci. **44**(1), 13–18 (2017)
9. Fern, X.Z., Lin, W.: Cluster ensemble selection. ASA Data Sci. J. **3**(1), 128–141 (2008)
10. Gan, G., Ma, C., Wu, J.: Data clustering: theory, algorithms, and applications, vol. 20. SIAM (2007)
11. He, X., Feng, J., Konte, B., Mai, S.T., Plant, C.: Relevant overlapping subspace clusters on categorical data. In: Proceedings of the 20th ACM SIGKDD International Conference on Knowledge Discovery and Data Mining, pp. 213–222 (2014)
12. He, Z., Xu, X., Deng, S.: Squeezer: an efficient algorithm for clustering categorical data. J. Comput. Sci. Technol. **17**(5), 611–624 (2002)
13. Hu, J., Li, T., Luo, C., Fujita, H., Yang, Y.: Incremental fuzzy cluster ensemble learning based on rough set theory. Knowl.-Based Syst. **132**, 144–155 (2017)
14. Hu, J., Li, T., Wang, H., Fujita, H.: Hierarchical cluster ensemble model based on knowledge granulation. Knowl.-Based Syst. **91**, 179–188 (2016)
15. Huang, D., Wang, C.D., Lai, J.H.: Locally weighted ensemble clustering. IEEE Trans. Cybern. **48**(5), 1460–1473 (2018)
16. Iam-On, N., Boongoen, T., Garrett, S., Price, C.: A link-based cluster ensemble approach for categorical data clustering. IEEE Trans. Knowl. Data Eng. **24**(3), 413–425 (2012)
17. Jensen, R., Jensen, R., Shen, Q.: New approaches to fuzzy-rough feature new approaches to fuzzy-rough feature selection (2), 1–17 (2017)
18. Li, F., Qian, Y., Wang, J., Dang, C., Jing, L.: Clustering ensemble based on sample's stability. Artif. Intell. **273**, 37–55 (2019)
19. Li, F., Qian, Y., Wang, J., Dang, C., Liu, B.: Cluster's quality evaluation and selective clustering ensemble. ACM Trans. Knowl. Disc. Data (TKDD) **12**(5), 1–27 (2018)
20. Ren, Y., Domeniconi, C., Zhang, G., Yu, G.: Weighted-object ensemble clustering: methods and analysis. Knowl. Inf. Syst. **51**(2), 661–689 (2016). https://doi.org/10.1007/s10115-016-0988-y
21. Saha, I., Sarkar, J.P., Maulik, U.: Ensemble based rough fuzzy clustering for categorical data. Knowl.-Based Syst. **77**, 114–127 (2015)
22. Sheeja, T., Kuriakose, A.S.: A novel feature selection method using fuzzy rough sets. Comput. Ind. **97**, 111–116 (2018)
23. VegaPons, S., RuizShulcloper, J.: A survey of clustering ensemble algorithms. Int. J. Pattern Recognit. Artif. Intell. **25**(03), 337–372 (2011)
24. Wang, C., Huang, Y., Shao, M., Fan, X.: Fuzzy rough set-based attribute reduction using distance measures. Knowl.-Based Syst. **164**, 205–212 (2019)
25. Yao, Y.: Three-way decision and granular computing. Int. J. Approx. Reason. **103**, 107–123 (2018)
26. Zhao, J., Zhang, Z., Han, C., Zhou, Z.: Complement information entropy for uncertainty measure in fuzzy rough set and its applications. Soft. Comput. **19**(7), 1997–2010 (2014). https://doi.org/10.1007/s00500-014-1387-5
27. Zhao, X., Liang, J., Dang, C.: Clustering ensemble selection for categorical data based on internal validity indices. Pattern Recogn. **69**, 150–168 (2017)
28. Zhong, C., Yue, X., Zhang, Z., Lei, J.: A clustering ensemble: two-level-refined co-association matrix with path-based transformation. Pattern Recogn. **48**(8), 2699–2709 (2015)

Online Optimized Product Quantization for Dynamic Database Using SVD-Updating

Kota Yukawa[1]($\boxtimes$) and Toshiyuki Amagasa[2] 🆔

[1] Degree Program in Computer Science, University of Tsukuba, Tsukuba, Japan
yukawa@kde.cs.tsukuba.ac.jp
[2] Center for Computational Sciences, University of Tsukuba, Tsukuba, Japan
amagasa@cs.tsukuba.ac.jp

Abstract. Approximate nearest neighbor (ANN) search allows us to perform similarity search over massive vectors with less memory and computation. Optimized Product Quantization (OPQ) is one of the state-of-the-art methods for ANN where data vectors are represented as combinations of codewords by taking into account the data distribution. However, it suffers from degradation in accuracy when the database is frequently updated with incoming data whose distribution is different. An existing work, Online OPQ, addressed this problem, but the computational cost is high because it requires to perform of costly singular value decompositions for updating the codewords. To this problem, we propose a method for updating the rotation matrix using SVD-Updating, which can dynamically update the singular matrix using low-rank approximation. Using SVD-Updating, instead of performing multiple singular value decompositions on a high-rank matrix, we can update the rotation matrix by performing only one singular value decomposition on a low-rank matrix. In the experiments, we prove that the proposed method shows a better trade-off between update time and retrieval accuracy than the comparative methods.

Keywords: ANN search · Product quantization · Online PQ

1 Introduction

The nearest neighbor (NN) search is a fundamental operation over a database whereby one can retrieve similar vectors for a given query vector. Besides, k-nearest neighbor (kNN) search outputs not only the nearest but also k ranked nearest neighbors. They are frequently used in a broad spectrum of applications, such as multimedia search over images, audio, or videos, and inside of machine learning algorithms as well.

The computational complexity of the kNN search increases according to the dimensionality and the database's size being processed. More specifically, it requires (ND) where D and N represent the dimensionality and the number of data, respectively. So, it is important to improve efficiency when dealing with high-dimensional data at scale. It should be noticed that, in a kNN search, we do not need the exact distances among vectors if the relative distances among vectors are well-preserved. To take advantage of

© Springer Nature Switzerland AG 2021
C. Strauss et al. (Eds.): DEXA 2021, LNCS 12923, pp. 273–284, 2021.
https://doi.org/10.1007/978-3-030-86472-9_25

this property, the approximate nearest neighbor (ANN) search is often used. The idea is to use approximated vectors so that we can quickly process kNN queries and save memory space as well.

The Product Quantization (PQ) [1] is one of the well-known methods for ANN search. In PQ, we divide each vector into some subvectors. Then, for the subvectors of the same subspace, we apply k-means clustering to quantize them. As a result, we can represent each vector as a combination of codewords corresponding to the cluster centroid, thereby significantly reducing data volume. Moreover, PQ has been shown to reduce the quantization error compared to conventional vector quantization because each vector is represented as a product of subvectors. PQ shows a good trade-off between accuracy/efficiency and memory cost. To improve the method, many variants have been published in recent years. The optimized PQ (OPQ) [2] introduces a vector space rotation as a pre-transformation to minimize the quantization error. OPQ can be done by mutually optimizing the optimal codebook and rotation matrix and successfully showed better accuracy than the original PQ.

Meanwhile, the number of real-time information sources has been drastically increasing due to the proliferation of network-attached devices. Therefore, there has been a growing demand for processing dynamic data, such as data streams [3, 4]. The kNN search on such dynamic databases has been used in many applications, such as video search, and we can consider to apply PQ/OPQ to such data to get the benefits as discussed above. However, it gives rise to another problem - the codebook optimized for the existing data becomes obsolete gradually as new data arrives, resulting in search accuracy degradation. Online PQ [5] proposed dynamically updating the codebook for new data and showed efficiency in updating the codebook while maintaining the search accuracy. Besides, Online Optimized PQ [6] proposed dynamically updating the rotation matrix to adapt Optimized PQ to time series data. However, it requires costly singular value decomposition (SVD) when updating the rotation matrix, leading to a long time to update [6].

In this paper, we address the problem of improving the efficiency of Online Optimized PQ. More specifically, we update the codebook and rotation matrix by applying the singular value decomposition updating method called SVD-Updating [7], thereby reducing the updating cost of Online Optimized PQs. In SVD-Updating, the computational cost is reduced using the singular value decomposition of a low-rank approximation matrix instead of the singular value decomposition on a full-rank matrix. In our experiments, we use a total of three datasets, text and image. We prove that the proposed method shows a good trade-off between the two methods, maintaining almost the same accuracy with less model update time than existing methods.

2 Preliminaries

2.1 Approximate k-Nearest Neighbor Search

Let us assume a set of vectors $X = \{x_0, \ldots, x_{N-1}\} \in \mathbb{R}^D$. Given a query vector $q \in \mathbb{R}^D$, the nearest neighbor (NN) search over X is defined as:

$$x* = argmin_{x \in X} dist(x, q),$$

where $dist(\cdot)$ is a distance measure. In the approximate NN (ANN) search, an approximated distance measure is used instead of the original one. Besides, in the k-NN (k-ANN) search, top k nearest vectors are retrieved.

2.2 Vector Quantization, Product Quantization, and Its Variants

The idea of *vector quantization* is to approximate a vectors by a short representation called *codeword,* thereby reducing the space and time required to process them. More concretely, the distance between the query and data is approximated by the distance between the codewords corresponding to the query and the data. In the basic vector quantization [8], vectors $X = \{x_0, \ldots, x_{N-1}\} \in \mathbb{R}^D$ are clustered using the k-means method. Each clustered vector is approximated as a cluster centroid $C \in c_i (0 \leq i \leq k-1)$. The cluster centroid c_i is called a codeword, and the set of cluster centroids C is called a codebook. The objective function of the codebook is to minimize the sum of squared errors, i.e.,

$$\min_{C} \sum_{X} \|x - c_i(x)\|^2 \tag{1}$$

The *product quantization* (PQ) [1] was proposed as an improved method. Each vector is divided into M subvectors of equal length:

$$x = \langle x^1 \ldots x^M \rangle \tag{2}$$

Thus, any vector can be expressed as an element in the cartesian product of M sets of subvectors. Then, for each set of subvectors, we apply VQ to compress the subvectors. As a result, we can represent a vector in terms of M sub-codewords. The objective function for the codebook optimization is as follows: x^m

$$\min_{C^1,\ldots,C^M} \sum_{X} \sum_{m=1}^{M} \|x^m - c_i^m(x^m)\|^2, \tag{3}$$
$$s.t \; C = C^1 \times \ldots \times C^M$$

where x^m represents the m-th subvector of x, and $c_i^m(x^m)$ represents the i-th codeword corresponding to x^m. It has been proven that the quantization error of PQ is smaller than that of VQ, because each vector is represented as a Cartesian product of sub-codebooks.

The *optimized product quantization* (OPQ) [2] was proposed to improve PQ. The idea is to reduce the quantization error by applying rotation to the dataset. To this end, they use a rotation matrix R in addition to the original PQ. The objective function for the rotation matrix and codebook optimization is:

$$\min_{R,C^1,\ldots,C^M} \sum_{X} \sum_{m=1}^{M} \|x^m - c_i^m(x^m)\|^2, \tag{4}$$
$$s.t \; c \in C = \{c | Rc = C^1 \times \ldots \times C^M, R^T R = I\}$$

The optimal rotation matrix can be obtained by solving the orthogonal procrustes problem [9, 10]. Therefore, the optimal rotation matrix can be obtained by using singular value decomposition as follows:

$$SVD\left(XC(X)^T\right) = USV^T,$$

$$R = VU^T \tag{5}$$

To optimizes the rotation matrix and the codebook, they apply alternative optimization over the codebook and the rotation matrix.

2.3 Online Product Quantization for Dynamic Data

When dealing with dynamic data where new vectors are continually added, the performance of PQ and OPQ gradually degrades, because the codebook is optimized to minimized for the set of vectors when the codebook is created while the distribution of incoming vectors may differ from the past dataset. To this problem, the *online product quantization* (online PQ) [5] retrains the codebook only by using the new vectors, thereby making it possible to adapt the codebook to the change in vector distribution. The objective function is:

$$\min_{C^1, \ldots, C^M} \sum_X \sum_{m=1}^{M} \| x^{t,m} - c_i^{t,m}(x^{t,m}) \|^2, \tag{6}$$
$$s.t \ C = C^1 \times \ldots \times C^M$$

where $c_i^{t,m}(x^{t,m})$ is the nearest neighbor codeword corresponding to the input data $x^{t,m}$ in the m-th subspace of the t-th data.

The *online optimized PQ* [6] applied optimized PQ to dynamic data. To this end, they solve SVD to optimize the rotation matrix when a new vector is added. As we can see, it is expensive to solve SVD against high-dimensional data.

2.4 Other Methods Based on PQ

So far, there have been various works to improve PQ [1]. One approach is to use pre-transformation [2, 11] where the vectors are rotated using a rotation matrix to reduce the quantization error and to improve the search accuracy. [12, 13] attempt to represent a sub-vectors in terms of sum of codewords from multiple codebooks, thereby improving the search accuracy. In *additive quantization* [14] and *composite quantization* [15], PQ quantization is generalized and formulated. Some methods use generalized pre-quantization algorithms to improve encoding speed [16], to use supervised data [17], to use multimodal models [18], or to add online learning algorithms for stream data [5]. Besides, PQ is applied to neural networks, such as CNN, to consistently minimize codebook and network losses [19, 20]. However, there have been no work that allows us to apply optimized PQ to dynamic data in an efficient way.

3 Proposed Method

As we observed, optimized PQ presents a good performance than ordinary PQ, while it requires expensive SVD to optimize rotation matrix. To adapt it to dynamic data, we proposed to apply SVD-Updating that allow us to solve SVD against dynamic data in an efficient way. Basically, we employ the approach proposed in online PQ [5] and extend it by introducing rotation to reduce the quantization error, where the codebooks and the rotation matrix are alternatively optimized. In the following, we describe the details of the proposed method.

3.1 Updating Codebook

Let us recall that the objective function is defined as:

$$\min_{R,C^{t,1},...,C^{t,M}} \sum_{X} \sum_{m=1}^{M} \|x^{t,m} - c_i^{t,m}(x^{t,m})\|^2, \tag{7}$$
$$s.t\; C^t = \{R^t c^t \in C^t = C^{t,1} \times \ldots \times C^{t,M}\}$$

To optimize it, we need to compute $c_k^{t+1,m} = c_k^{t,m} + \frac{1}{n_k^m}(\hat{x}^{t,m} - c_k^{t,m})$, where $x^{t,m}$ is the m-th subvector of the t-th input data, $c_k^{t,m}$ is the codeword corresponding to $x^{t,m}$, n_k^m is the number of vectors in the cluster corresponding to the codeword $c_k^{t,m}$, and $\hat{x}^{t,m} = x^{t,m}R^t$. Intuitively, we modify the position of each codeword so that the corresponding cluster centroid is located at the center of vectors including the new vectors. Notice that this can be computed only using new vectors. Hence, it does not depend on the size of existing dataset.

3.2 Update Rotation Matrix Using SVD-Updating

Similarly, when new vectors are added, the existing rotation matrix is not optimal for the data with the new data. The optimal rotation matrix can be obtained by solving the orthogonal Procrustes problem, which can be solved by SVD. Suppose that the optimal rotation matrix is obtained as in Eq. (8) where X is the input vector and Y is the vector generated from the codebook.

$$SVD(XY^T) = USV^T$$

$$R = VU^T \tag{8}$$

In this work we employ SVD-Updating [7] to solve the problem without decomposing whole matrix. Besides, it allows us to approximate the vectors by low-rank singular value decomposition, thereby saving the computation. The original SVD-Updating was proposed to address the problem of recommender systems where the matrix represent user-item relationship. In the following, we show how we can adapt it to optimize the rotation matrix in OPQ.

When a new vector X' is added, we want to solve the following formula without recalculating the entire matrix.

$$W = A_k + X'Y'^T\# \tag{9}$$

$$SVD(W) = U_W S_W V_W^T \tag{10}$$

$$R_W = V_W U_W^T \tag{11}$$

where A_k is obtained by a low-rank approximation of $A = XY^T$, i.e.,

$$SVD(A) \approx SVD(A_k) = U_k S_k V_k^T \tag{12}$$

From Eqs. (9) and (12), we get:

$$W = U_k S_k V_k^T + X'Y' \tag{13}$$

$$\therefore U_k^T W V_k = S_k + U_k^T X'Y'^T V_k \# \tag{14}$$

Then,

$$Q = S_k + U_k^T X'Y'V_k \tag{15}$$

From Eqs. (10), (14) and (15), we get:

$$U_k^T U_W S_W V_W^T V_k = U_Q S_Q V_Q^T \tag{16}$$

$$\therefore U_W = U_k U_Q, \tag{17}$$

$$\therefore S_W = S_k, \tag{18}$$

$$\therefore V_W = V_k V_Q \tag{19}$$

In this way, the U_W, S_W, and V_W can be obtained by Equations (17), (18), and (19). Thus, the new rotation matrix R_W is calculated as follows:

$$R_W = V_W U_W^T \tag{20}$$

$$\therefore R_W = V_k V_Q U_Q^T \# U_k^T. \tag{21}$$

$$(Q = S_k + U_k^T X'Y'^T, SVD(Q) = U_Q S_Q V_Q^T)$$

In other words, the new rotation matrix R_W in Equation (21) only needs to compute SVD of the $(k \times k)$ matrix Q, instead of computing the SVD of $(r \times r)$ matrix W in Equation (13). The overall algorithm of the proposed *online OPQ with SVD-Updating* is shown in Algorithm 1.

3.3 Orthogonality

The *folding-in* method [21, 22] is used to update the SVD instead of SVD-Updating, because the computational cost of folding-in is smaller than that of SVD-Updating. However, the orthogonality of the updated singular vectors is not maintained when using folding-in. Therefore, it is inappropriate to use it for updating the rotation matrix. On the other hand, SVD-Updating guarantees orthogonality of the updated singular vectors. Therefore, the updated singular vectors U_W and V_W become orthonormal matrices. The vector $R_W = V_W U_W^T$ calculated from these vectors is a product of orthonormal matrices and R_W is also an orthonormal matrix. For this reason, R_W calculated by SVD-Updating is suitable for a rotation matrix.

Algorithm 1. Online OPQ using SVD-Updating

Input : streaming datasets $\{x\} \in \mathbb{R}^{N \times D}$,
      rotation matrix R,
      value of low rank approximation l,
      initialize $M \times K$ sub-codebook $c_{1,1}^0, \ldots, c_{m,k}^0, \ldots, c_{M,K}^0$ using the data by OPQ,
      initialize $M \times K$ counters $n_{1,1}^0, \ldots, n_{m,k}^0, \ldots, n_{M,K}^0$ that indicate the number of
      initial data assigned to the $c_{m,k}^0$.

Output : rotation matrix R^t

1: $[U_l^0, S_l^0, V_l^0] \leftarrow$ LowRankApproximation(U, S, V, l)
2: **for** t $\leftarrow$ 1 to N **do**
3:      rotate the data: $\hat{x}^t = x^t R$
4:      // (i) update the code book
5:      $[c^{t+1}, n^{t+1}] \leftarrow$ updateCodeBook$(c^t, \hat{x}^t, n^t)$
6:      // (ii) update rotation matrix
7:      $Q^t \leftarrow S_l^t + U_l^{t^T} \cdot x^t \cdot c^t (x^t)^T \cdot V_l^t$
8:      $[U_Q, S_Q, V_Q] \leftarrow$ SVD(Q)
9:      $U_l^{t+1} \leftarrow U_l^t U_Q$
10:     $S_l^{t+1} \leftarrow S_Q$
11:     $V_l^{t+1} \leftarrow V_l^t V_Q$
12:     $R^{t+1} \leftarrow V_l^{t+1} U_l^{t+1^T}$
13: **end for**

3.4 Computational Complexity

Time Complexity. The transformation of the input vector x by the rotation matrix R (Line 1 of Algorithm 1) takes $\mathcal{O}(ND^2)$. To update the codebook, online PQ (Line 5) requires $\mathcal{O}(NDK + NM + ND)$. Next, updating the rotation matrix requires $\mathcal{O}(ND^2l^2)$ for the calculation of matrix Q (Line 7), $\mathcal{O}(l^3)$SVD on the $(l \times l)$ matrix (Line 8) requires $\mathcal{O}(l^3)$, and $\mathcal{O}(Dl^2)$ is needed for calculating U_l^{t+1} and V_l^{t+1} (Lines 9 and 11). Finally, the computation of the updated rotation matrix R^{t+1} in the 12th row is $\mathcal{O}(D^2l)$. From the above, the time calculation complexity of the whole algorithm is $\mathcal{O}(DK + NM + ND^2l^2 + l^3 + Dl^2 + D^2l)$.

Space Complexity. The rotation matrix R requires $\mathcal{O}(D^2))$, the codebook requires $\mathcal{O}(KD)$, and the counter $n_{m,k}^t$, which counts the number of data belonging to each codeword for updating the codebook, requires $\mathcal{O}(KM)$. The spatial computation of the matrices Q, U_l^t, S_l^t and V_l^t required to update the rotation matrix is $\mathcal{O}(l^2)$, $\mathcal{O}(Dl)$, $\mathcal{O}(l^2)$ and $\mathcal{O}(Dl)$, respectively. From the above, the spatial computation of the whole algorithm is $(D^2 + KD + KM + 2Dl + 2l^2)$.

4 Experiments

4.1 Experimental Setting

To evaluate the performance of the proposed method, we have conducted a set of experiments. The comparative methods are PQ, optimized PQ, online PQ, and online OPQ. Besides, we tested the case where the model (codebook and rotation matrix) is trained only once in the beginning as the baseline (labeled as "no update"). On the other hand, we tried the best case where the model is retrained with the entire dataset whenever a new batch is added (labeled as "all"). For online PQ, we updated the codebook only with the latest batch. For online OPQ, there are two cases: 1) updates the codebook and rotation matrix with the conventional algorithm which is without SVD-Updating (labeled as "online opq") and 2) the proposed method which is using SVD-Updating (labeled as "online opq(SVD-updating)").

The comparison metrics are Recall@R and update time. Recall@R is the probability that the ground-truth nearest neighbor data is contained within the top R data extracted as approximate nearest neighbors. Update time is the running time required to update the codebook (and rotation matrix) in each model.

We experimented with three different datasets, including text and images. The details of datasets are shown in Table 1. Each dataset is queried with an input batch, and then the model is updated with it. For the text data, the dataset is ordered in chronological order. For image datasets, they are divided into classes (comprising two batches for one class) according to the class label to simulate the change in data distribution over time.

Table 1. Dataset details

Dataset	# Classes	Size	Feature	# Dim
News20 [23]	20	18,845	BERT [24]	768
Caltech-101 [25]	101	9,144	GIST [27]	960
SUN397 [26]	397	108,753	GIST [27]	960

4.2 Effects of Low-Rrank Approximation

We tested the performance with different rank values. The experimental results are shown in Fig. 1. Each value represents the ratio of the approximated rank to the original dimensionality (e.g., 0.8 for 960-dimension means 768-dimension). Besides, we checked the rank of the matrix of the first batch (labeled as "auto") and used the value to approximate the matrix (Table 2). In terms of update time, the lower rank cases performed better due to their reduced amount of computation.

In the meantime, for the recall values, there is no much difference between the full- and low-rank results. This implies that it is possible for us to focus on meaningful dimensions (out of full dimensions) to perform an NN search so that we can maintain the recall while achieving faster computation over low-ranked matrices. In the subsequent experiments, we use "auto" as the low-rank approximation.

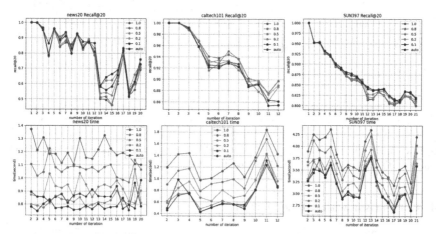

Fig. 1. Comparison results of Recall@20, update time with low rank approximation for each dataset

Table 2. Calculated rank value for each dataset

Dataset	Rank
News20	267
Caltech101	112
SUN397	77

4.3 Comparison with Non-online Methods

The results of the comparison experiments with PQ and OPQ are shown in Fig. 2. Recall shows that the proposed method is better than the "no update" method. This is because the "no update" method trains the model in the first batch and does not update the model. As a result, it does not adapt to the newer batches. The "all" method retrains the model using all the data and thus shows a better recall than others, including the proposed method.

On the other hand, in comparing the update time, the "all" method increases as more batches are input. This is because training is performed on all batches, including the input batches, leading to a longer time for training with increasing data. The OPQ update time increases more than that of PQ because both the codebook and the rotation matrix need to be optimized. On the other hand, in the proposed method, the update time is constant. This is because it can update the model using only the input batch data (and not all data).

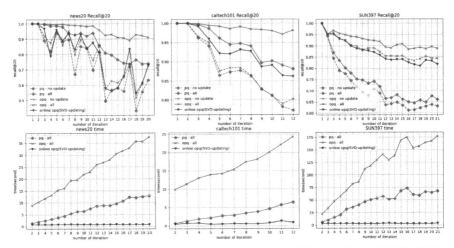

Fig. 2. Comparison results of Recall@20 and update time of PQ(no update), PQ(all), OPQ(no update), OPQ(all) and online OPQ(SVD-Updating) for each dataset.

4.4   Comparison with Online Methods

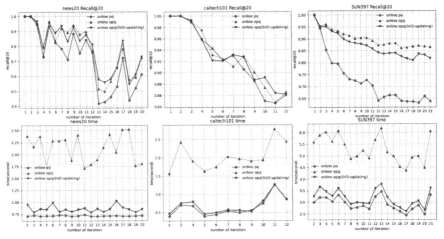

Fig. 3. Comparison results of Recall@20 and update time of online PQ, online OPQ and online OPQ (SVD-Updating) for each dataset

We compared the performance among the online algorithms, i.e., online PQ, online OPQ, and online OPQ (SVD-Updating). The experimental results are shown in Fig. 3. We can observe that the degradation in recall for online OPQ is smaller than that of online PQ. The SUN397 dataset, which contains many data records, showed a significant difference in recall values. In the News20 and caltech101 datasets, the recall values of online OPQ and online OPQ (SVD-Updating) are almost the same, while online OPQ was slightly better than online OPQ (SVD-Updating) on the SUN397 dataset.

In terms of processing time, the proposed scheme was much faster than online OPQ – about 2× faster than online OPQ on the News20 dataset. This is because the size of the matrix to be computed for updating the rotation matrix is smaller because of the efficiency in low-rank approximation with SVD-Updating.

In summary, we can say that the proposed online OP with SVD-Updating achieved a good trade-off between recall and efficiency.

5 Conclusion

In this paper we have proposed an efficient method for approximate NN search over dynamic data based on online OPQ, where the rotation matrix can be updated efficiently using SVD-Updating. It allows us to update the model without conducting SVD over the entire dataset. The experimental results have shown that the proposed method outperformed the previous studies and achieved a better trade-off between recall and efficiency. Our future research includes further improve the performance by employing more sophisticated methods, such as Tree-quantization [13].

Acknowledgment. This work was partly supported by the Project Commissioned by New Energy and Industrial Technology Development Organization (JPNP20006).

References

1. Jégou, H., et al.: Product quantization for nearest neighbor search. IEEE Trans. Pattern Anal. Mach. Intell. **33**(1), 117–128 (2011)
2. Ge, T., et al.: Optimized product quantization. IEEE Trans. Pattern Anal. Mach. Intell. **36**(4), 744–755 (2014)
3. Moffat, A., et al.: Text compression for dynamic document databases. IEEE Trans. Knowl. Data Eng. **9**(2), 302–313 (1997)
4. Dong, A., Bhanu, B.: Concept learning and transplantation for dynamic image databases. In: ICME, pp. 765–768 IEEE Computer Society (2003)
5. Xu, D., et al.: Online product quantization. IEEE Trans. Knowl. Data Eng. **30**(11), 2185–2198 (2018)
6. Liu, C., et al.: Online optimized product quantization. IEEE International Conference on Data Mining (ICDM), pp. 362–371 (2020)
7. Berry, M.W., et al.: Using linear algebra for intelligent information Retrieval. (1994)
8. Gray, R.M., et al.: Quantization. IEEE Trans. Inf. Theor. **44**, 2325–2384 (1998)
9. Sch"onemann, P.: A generalized solution of the orthogonal procrustes problem. Psychometrika **31**(1), 1–10 (1966)
10. Gong, Y., et al.: Iterative quantization: a procrustean approach to learning binary codes. CVPR (2011)
11. Norouzi, M., Fleet, D.J.: Cartesian K-means. In: CVPR, pp. 3017–3024. IEEE Computer Society (2013)
12. Wang, J., et al.: Optimized Cartesian K-means. IEEE Trans. Knowl. Data Eng. **27**(1), 180–192 (2015)
13. Babenko, A., Lempitsky, V.S.: Tree quantization for large-scale similarity search and classification. In: CVPR, pp. 4240–4248. IEEE Computer Society (2015)

14. Babenko, A., Lempitsky, V.S.: Additive quantization for extreme vector compression. In: CVPR, pp. 931–938. IEEE Computer Society (2014)
15. Zhang, T., et al.: Composite quantization for approximate nearest neighbor search. In: ICML, pp. 838–846. JMLR.org (2014)
16. Zhang, T., et al.: Sparse composite quantization. In: CVPR, pp. 4548–4556. IEEE Computer Society (2015)
17. Wang, X., et al.: Supervised quantization for similarity search. In: CVPR, pp. 2018–2026. IEEE Computer Society (2016)
18. Zhang, T., Wang, J.: Collaborative quantization for cross-modal similarity search. In: CVPR, pp. 2036–2045. IEEE Computer Society (2016)
19. Cao, Y., et al.: Deep quantization network for efficient image retrieval. In: Schuurmans, D., Wellman, M.P. (eds.) In: Proceedings of the Thirtieth AAAI Conference on Artificial Intelligence (AAAI), pp. 3457–3463. AAAI Press (2016)
20. Cao, Y., et al.: Deep visual-semantic quantization for efficient image retrieval. In: CVPR, pp. 916–925. IEEE Computer Society (2017)
21. Sarwar, B., et al.: Incremental singular value decomposition algorithms for highly scalable recommender systems. In: Fifth International Conference on Computer and Information Science, pp. 27--28 (2002)
22. Deerwester, S.C., et al.: Indexing by latent semantic analysis. J. Am. Soc. Inf. Sci. **41**(6), 391–407 (1990)
23. Lang, K.: NewsWeeder: learning to filter netnews. In: International Conference on Machine Learning (1995)
24. Devlin, J., et al.: BERT: pre-training of deep bidirectional transformers for language understanding. In: Proceedings of the 2019 Conference of the North American Chapter of the Association for Computational Linguistics: Human Language Technologies, vol. 1 (Long and Short Papers), pp. 4171—4186. Association for Computational Linguistics, Minneapolis (2019)
25. Fei-Fei, L., et al.: Learning generative visual models from few training examples an incremental Bayesian approach tested on 101 object categories. In: Proceedings of the Workshop on Generative-Model Based Vision. Washington, DC (2004)
26. Xiao, J., et al.: SUN database: large-scale scene recognition from abbey to zoo. In: CVPR, pp. 3485–3492. IEEE Computer Society (2010)
27. Friedman, A.: Framing pictures: the role of knowledge in automatized encoding and memory for gist. J. Exp. Psychol. Gen. 108(3), 316–355 (1979)

Querying Collections of Tree-Structured Records in the Presence of Within-Record Referential Constraints

Foto N. Afrati and Matthew Damigos[(⊠)]

National Technical University of Athens, Athens, Greece

Abstract. In this paper, we consider a tree-structured data model used in many commercial databases like Dremel, F1, JSON stores. We define identity and referential constraints within each tree-structured record. The query language is a variant of SQL and flattening is used as an evaluation mechanism. We investigate querying in the presence of these constraints, and point out the challenges that arise from taking them into account during query evaluation.

1 Introduction

Systems that efficiently analyze complex data (e.g., graph, or hierarchical data) are ubiquitous. Such systems include document databases (e.g., MongoDB), or systems combining a tree-structured data model and a columnar storage, such as F1 [15], Dremel/BigQuery [12] and Apache Parquet.

Identity and referential constraints (a.k.a., keys and foreign keys) have been extensively studied in the context of relational databases and, lateron, in the context of XML data model [3,7,9], as well as for graphs [8] and RDF data [4,10]. Recently, key constraints have analyzed for JSON data models [2,14].

In this paper, we consider a theoretical *tree-record data model* for representing collections of tree-structured records. This model is mainly inspired by the Dremel data model [1,12,13], and it applies to document-oriented (i.e., XML, JSON) data stores (e.g., ElasticSearch, MongoDB) and relational databases supporting hierarchical data types (e.g., JSON type in PostgreSQL, MySQL and struct type in Hive). We define identity and referential constraints *within each* tree-structured record (called *within-record constraints*). Unlike relational databases and XML data, where such constraints are used for validating the data, in this work, we take advantage of them to improve query answering. We consider SQL-like query language (such as the one used in Dremel, F1, Apache Drill), and investigate querying in the presence of within-record constraints. We show that there are queries that can be answered only if we use the constraints. To the best of our knowledge this is the first work investigating the problem of querying collections of tree-records with SQL in the presence of such constraints.

Contributions: (1) We define within-record key and foreign key constraints for the tree-structured data model (Sect. 3). (2) We show when these concepts

© Springer Nature Switzerland AG 2021
C. Strauss et al. (Eds.): DEXA 2021, LNCS 12923, pp. 285–291, 2021.
https://doi.org/10.1007/978-3-030-86472-9_26

are well-defined (Sect. 4.2). (3) We show how to use flattening to answer single-table, SQL-like queries (i.e., without joins) in the presence of such constraints (Sect. 4). We also introduce the concept of relative flattening which is a part of the flattened data corresponding to a subtree of the schema.

Related Work: Work on constraints for tree-structured data has been done during the past two decades. Our work, as regards the formalism, is closer to [3,16, 17]. The paper [3] is among the first works on defining constraints on tree-structured data. [14] and [2] focus on the JSON data model and a similar to XPath navigational query language. These works also formalize specification of unique fields and references, they do not define relative keys. Flattening has initially been studied in the context of nested relations and hierarchical model (see [1] and references therein). Dremel [1,12,13], F1 [15] and Drill use flattening to answer SQL-like queries over tree-structured data. Flattening semi-structured data is also investigated in [5,6,11], where the main problem is to translate semi-structured data into multiple relational tables.

2   Defining the Data Model

The tree-record data model considers collections of records (or, *tables*) conforming to a nested, tree-structured schema. A *group type* (or simply *group*) G is a complex data type defined by an ordered list of items (also called *attributes* or *fields*) of unique names which are associated with a data type, either primitive type (e.g., integer, string, etc.) or group type.

The repetition constraint for a field N specifies the number of times that N is repeated within a group and takes one of the following values with the corresponding *annotation*: (1) *required*: N is mandatory, and there is no annotation, (2) *optional*: N is optional (i.e., appears 0 or 1 times) and is labeled by N?, (3) *repeated*: N appears 0 or more times and is labeled by $N*$, (4) *required and repeated*: N appears 1 or more times and is labeled by $N+$. We denote as *repTypes* the set {required, repeated, optional, required and repeated} of repetition types.

A *tree-schema* S of a table T is a tree with labeled nodes such that (1) each non-leaf node (called *intermediate node*) is a group and its children are its attributes, (2) each leaf node is associated with a primitive data type, (3) each node (either intermediate or leaf) is associated with a repetition constraint in *repTypes*, and (4) the root node is labeled by the name T of the table. Each non-required node is called *annotated node*. We de-annotate a node by removing the repetition symbol from its label. $lb(N)$ represents the de-annotated label of a node N in a schema. The path[1] of de-annotated labels between the root and a node N of a schema S is called *reachability path* of N[2].

---

[1] The path between the nodes N_i and N_j is denoted as $N_i.N_{i+1}.....N_j$.

[2] We omit a prefix in the reachability path of a node if we can still identify the node through the remaining path.

Considering a subtree s of S, we denote as *dummy* s the tree constructed from s by de-annotating all the annotated nodes of s and adding to each leaf a single child which is labeled by the $NULL$-value. Let S be a tree-schema and t be a tree that is constructed from S by recursively replacing, from top to down, each subtree s_N rooted at an annotated node $N\sigma$, where σ is an annotation, with (1) either a dummy s_N or k s_N^d-subtrees, if $\sigma = *$, (2) either a dummy s_N or a single s_N^d-subtree, if $\sigma = ?$, (3) k s_N^d-subtree, if $\sigma = +$, where $k \geq 1$ and s_N^d is constructed from s_N by de-annotating only its root. Then, for each non-$NULL$ leaf N of t, we add to N a single child which is labeled by a value of type that matches the primitive type of N. The tree t is a *tree-record* of S. An instance of S, called *tree-instance*, is a multiset of tree-records. The reachability path of a node N in a tree-record is similarly defined as the path from the root of the tree-record to N. We also consider that each node of both tree-schema and tree-record has a unique virtual id (called *node id*).

We now define an instantiation in a multiset rather than in a set notion. Let S be a schema and t be a tree-record in an instance of S. Since each node of S is replaced by one or more nodes in t, there is at least one mapping μ, called *instantiation*, from the node ids of S to the node ids of t, such that ignoring the annotations in S, both the de-annotated labels and the reachability paths of the mapped nodes match. The subtree of t which is rooted at the node $\mu(N)$ is an *instance* of the node N, where N is a node of S. If N is a leaf, an instance of N is the single-value child of $\mu(N)$.

We say that two subtrees s_1, s_2 of a tree-record are *isomorphic* if there is a bijective mapping h from s_1 to s_2 such that the de-annotated labels of the mapped nodes match. We say that s_1 and s_2 of t are equal, denoted $s_1 = s_2$, if they are isomorphic and the ids of the mapped nodes are equal.

3 Within-Record Constraints

In this section, we define identity and referential constraints that hold on each tree-record. Unlike primary keys in the conventional databases, which are unique across all the records of a table, in tree-schemas, we might have fields that uniquely identify other fields (or subtrees) in each record, but not the record itself [3].

To support such type of constraints in a tree-record data model, we define the concept of *identity constraint with respect to a group*. Let S be a tree-schema with root R_o, $\mathcal{D}$ be a tree-instance of S, and N, I be nodes of S such that N is intermediate and I is a descendant of N. Suppose that M is the parent of N, if $N \neq R_o$; otherwise, $M = N = R_o$. An *identity constraint with respect to* N is an expression of the form $I \xrightarrow{1} N$, such that I and all the descendants of I in S are required. We say that $I \xrightarrow{1} N$ *is satisfied* in $\mathcal{D}$ if for each $t \in \mathcal{D}$ and for each instance t_p of M in t, there are not two isomorphic instances of I in t_p. The node I is called *identifier* and N is the *range group* of I.

The identity constraint is similar to the concept of relative key defined for XML documents in [3]. For each $I \xrightarrow{1} N$, we use the symbol # to annotate the identifier I (i.e., $I\#$). We also use a special, dotted edge (I, N), called *identity edge*, to illustrate range group N of I.

We now define the concept of *referential constraint* (or, simply *reference*), which intuitively links the values of two fields. Let I, N and R be nodes of a tree-schema S such that $I \xrightarrow{1} N$, R is not a descendant of N, and I, R have the same data type. A *referential constraint* is an expression of the form $R \xrightarrow{r} I$. A tree-instance $\mathcal{D}$ of S *satisfies* the constraint $R \xrightarrow{r} I$, if for each tree-record $t \in \mathcal{D}$ the following is true: For each instance t_{LCA} of the lowest common ancestor (LCA) of R and N in t, each instance of R in t is isomorphic to an instance of I in t_{LCA}. If I is a leaf, then $R \xrightarrow{r} I$ is called *simple*.

If we have $R \xrightarrow{r} I$, we say that R (called *referrer*) *refers* to I (called *referent*). To represent the constraint $R \xrightarrow{r} I$ in a tree-schema S, we add a special (dashed) edge (I, R), called *reference edge*, which is labeled by r. Let $\mathcal{C}$ be the set of identity and referential constraints over S. Consider now the tree B given by (1) ignoring all the reference and identity edges, and (2) de-annotating the identifier nodes. We say that a collection $\mathcal{D}$ is a tree-instance of the tree-schema S in the presence of $\mathcal{C}$ if $\mathcal{D}$ is a tree-instance of B and $\mathcal{D}$ satisfies all the constraints in $\mathcal{C}$.

4   Querying Tree-Structured Data

In this section, we investigate querying tree-structured tables in the presence of identity and referential constraints. In particular, we use the Select-From-Where-GroupBy expressions used in Dremel [1,12] to query tables defined through a tree-schema. We refer to such a query language as *Tree-SQL*.

A query Q is an expression over a table $\mathcal{T}$ with schema S and tree-instance $\mathcal{D}$ in the presence of within-constraints $\mathcal{C}$, and results a relation (multiset of tuples). We consider *only* simple references in $\mathcal{C}$. Q has the following form, using the conventional SQL syntax: **SELECT** *expr* **FROM** $\mathcal{T}$ [**WHERE** *cond*] [**GROUP BY** *grp*], where *cond*, *expr* and *grp* are defined in terms of the leaves of S. Each leaf node in Q is referred through either its reachability path or a path using reference and identity edges implied by the constraints in $\mathcal{C}$. Intuitively, a query typically determines a mapping from tree-structured data model to relational model. To formally define the query semantics and handle repetition, we use the concept of flattening [1] which is discussed in detail in the next section.

4.1   Flattening Nested Data

In this section, we analyze the flattening operation applied on tree-structured data. Flattening is a mapping applied on a tree-structured table and translates the tree-records of the table to tuples in a relation. By defining such a mapping, the semantics of Tree-SQL is given by the conventional SQL semantics over the *flattened relation* (i.e., the result of the flattening over the table). Initially, we consider a tree-schema without referential and identity constraints. The presence of constraints is discussed in the next section.

Let S be a tree-schema of a table T and $\mathcal{D}$ is an instance of S, such that there is not any reference defined in S. Suppose also that $N_1, \ldots, N_m$ are the leaves of

S. The *flattened relation* of $\mathcal{D}$, denoted $flatten(\mathcal{D})$, is the relation given by the multiset: $\{\{(lb(\mu(N_1)), \ldots, lb(\mu(N_m))) \mid \mu$ is an instantiation of a tuple $t \in \mathcal{D}\}\}$. For each pair (N_i, N_j), $\mu(N_i)$ and $\mu(N_j)$ belong to the same instance of the lowest common ancestor of N_i and N_j in t. Considering now a query Q over S and an instance $\mathcal{D}$ of S, we say that Q is evaluated using *full flattening*, denoted $Q(flatten(\mathcal{D}))$, if $Q(\mathcal{D})$ is given by evaluating Q over the relation $flatten(\mathcal{D})$. The existence of the repeated fields affect the number of tuples in the flattened relation even if these fields are not used by the given query.

To avoid cases where the repetition of a field that is not used in the query has an impact on the query result, we define the concept of *relative flattening*. Let S be a tree-schema of a table T and $\mathcal{D}$ is an instance of S, such that there is not any reference defined in S. Consider also a query Q over T that uses a subset $L = \{N_1, \ldots, N_k\}$ of the set of leaves of S, and the tree-schema S_L constructed from S by removing all the nodes except the ones included in the reachability paths of the leaves in L. Then, we say that a query Q is evaluated using *relative flattening*, denoted $Q(flatten(\mathcal{D}, Q))$, if $Q(\mathcal{D})$ is given by evaluating Q over the relation: $flatten(\mathcal{D}, Q) = \{\{(lb(\mu(N_1)), \ldots, lb(\mu(N_k))) \mid \mu$ is an instantiation from the nodes of S_L to the nodes of $t \in \mathcal{D}\}\}$.

Proposition 1. *Consider a query Q over a tree-schema S such that Q does not apply any aggregation. Then, for every instance $\mathcal{D}$ of S, the following are true: (1) there is a tuple r in $Q(flatten(\mathcal{D}, Q))$ if and only if there is a tuple r in $Q(flatten(\mathcal{D}))$, and (2) $|Q(flatten(\mathcal{D}, Q))| \leq |Q(flatten(\mathcal{D}))|$.*

4.2   Navigating Through References

In the previous section, we ignored the within-record constraints when we explained how to use flattening to answer an SQL-like query. Here, we show how we take advantage of the constraints to extend the query semantics based on the relative flattening. We initially extend the notation of the Tree-SQL as follows. Apart from the reachability paths of the leaves that can be used in *SELECT*, *WHERE* and *GROUP BY* clauses, if there are constraints $R \xrightarrow{r} I$ and $I \xrightarrow{1} G$, we can use paths of the form: $[pathToR].R.[pathToL]$, where the $[pathToR]$ is the reachability path of R, L is a leaf which is a descendant of G, and $[pathToL]$ is the path from G to L. Intuitively, navigating through identity and reference edges, the leaves of G become accessible from R; enabling answering queries that couldn't be defined without using the references.

To formally capture queries using references, we extend the relative flattening presented in the previous section as follows. Let S be a tree-schema of a table T, $\mathcal{D}$ be an instance of S, and $\mathcal{C}$ be a set of identity and referential constraints satisfied in $\mathcal{D}$. Consider also a query Q over T that uses a set of leaves $\mathcal{L} = \{N_1, \ldots, N_k, L_1, \ldots, L_m\}$ of S such that for each L_i in $\mathcal{L}$ there are constraints $R_i \xrightarrow{r} I_i$, $I_i \xrightarrow{1} G_i$ in $\mathcal{C}$, where G_i is an ancestor of L_i. Let also $S_{\mathcal{L}}$ be the tree-schema constructed from S by keeping only the reachability paths of the leaves in $\{N_1, \ldots, N_k, R_1, \ldots, R_m\}$ (without de-annotating any node), and for each i, S_{G_i} be the tree-schema including only the reachability paths of R_i, I_i and the leaves of G_i that are included in $\{L_1, \ldots, L_m\}$. Both $S_{\mathcal{L}}$ and S_{G_i} keep the node ids from S. A query Q is evaluated in the presence of the constraints in $\mathcal{C}$, denoted $Q(flatten(\mathcal{D}, Q, \mathcal{C}))$, if $Q(\mathcal{D})$ is given by evaluating Q over the relation: $flatten(\mathcal{D}, Q, \mathcal{C}) = \{\{(lb(\mu(N_1)), \ldots, lb(\mu(N_k)), lb(\mu_1(L_1)), \ldots, lb(\mu_m(L_m)))|$ μ is an instantiation from the nodes of $S_{\mathcal{L}}$ to the nodes of $t \in \mathcal{D}$, each μ_i is an instantiation from the nodes of S_{G_i} to the nodes of t, for every two L_i, L_j s.t. $G_i = G_j$ and $R_i = R_j$ we have that $\mu_i = \mu_j$, and for each i, we have that $\mu(R_i) = \mu_i(R_i)$ and $lb(\mu_i(R_i)) = lb(\mu_i(I_i)))\}\}$.

We now investigate well-defined references; i.e., whether it is clear which referent is referred by each referrer in a tree-instance. Let S be a tree-schema, and $R \xrightarrow{r} I$, $I \xrightarrow{1} G$ be two constraints over S. Let L be the lowest common ancestor of G and R. Then, we say that the reference $R \xrightarrow{r} I$ is *out-of-range* if there is at least one repeated group on the path from L to G; otherwise, the reference is *within-range*. Intuitively, if a reference is out-of-range, then it might not be clear which is the referent value each referrer value refers to in each tree-record.

Proposition 2. *Let C be a reference $R \xrightarrow{r} I$ over a tree-schema S, and t be a record of a tree-instance $\mathcal{D}$ of S satisfying the constraint. If C is within-range, then for each instance of R in t, there is a single instance of I in t.*

As next steps, we plan to investigate querying tree-schemas having references to intermediate nodes and/or reference cycles. Also, we aim to study flattening when the referrer is defined in the range group of the referent.

References

1. Afrati, F.N., Delorey, D., Pasumansky, M., Ullman, J.D.: Storing and querying tree-structured records in Dremel. In: VLDB Endownment (2014)
2. Bourhis, P., Reutter, J.L., Suárez, F., Vrgoc, D.: JSON: data model, query languages and schema specification. In: PODS (2017)
3. Buneman, P., Davidson, S.B., Fan, W., Hara, C.S., Tan, W.C.: Keys for XML. Comput. Networks **39**(5), 473–487 (2002)
4. Calvanese, D., Fischl, W., Pichler, R., Sallinger, E., Simkus, M.: Capturing relational schemas and functional dependencies in RDFS. In: AAAI (2014)
5. Deutsch, A., Fernández, M.F., Suciu, D.: Storing semistructured data with STORED. In: SIGMOD (1999)
6. DiScala, M., Abadi, D.J.: Automatic generation of normalized relational schemas from nested key-value data. In: SIGMOD (2016)
7. Fan, W.: XML constraints: specification, analysis, and applications. In: DEXA (2005)
8. Fan, W., Lu, P.: Dependencies for graphs. ACM Trans. Database Syst. **44**(2), 1–40 (2019)
9. Fan, W., Siméon, J.: Integrity constraints for XML. J. Comput. Syst. Sci. **66**(1), 254–291 (2003)
10. Lausen, G., Meier, M., Schmidt, M.: SPARQLing constraints for RDF. In: EDBT (2008)

11. Liu, Z.H., Hammerschmidt, B.C., Mcmahon, D.: JSON data management: supporting schema-less development in RDBMS. In: SIGMOD (2014)
12. Melnik, S., et al.: Dremel: interactive analysis of web-scale datasets. In: VLDB Endownment (2010)
13. Melnik, S., et al.: Dremel: a decade of interactive SQL analysis at web scale. In: Proceedings VLDB Endownment (2020)
14. Pezoa, F., Reutter, J.L., Suárez, F., Ugarte, M., Vrgoc, D.: Foundations of JSON schema. In: WWW (2016)
15. Shute, J., et al.: F1: a distributed SQL database that scales. In: VLDB Endownment (2013)
16. Vo, L.T.H., Cao, J., Rahayu, J.W.: Discovering conditional functional dependencies in XML data. In: ADC 2011 (2011)
17. Yu, C., Jagadish, H.V.: XML schema refinement through redundancy detection and normalization. VLDB J. **17**(2), 203–223 (2008)

Dealing with Plethoric Answers
of SPARQL Queries

Louise Parkin[1]([✉]), Brice Chardin[1], Stéphane Jean[2], Allel Hadjali[1],
and Mickaël Baron[1]

[1] ISAE-ENSMA, LIAS, Chasseneuil-du-Poitou, France
`louise.parkin@ensma.fr`
[2] Université de Poitiers, LIAS, Poitiers, France

Abstract. When querying Knowledge Bases (KBs), users are faced with large sets of data, often without knowing their underlying structures. It follows that users may make mistakes when formulating their queries, therefore receiving an unhelpful response. In this paper, we address the plethoric answers problem, the situation where a query produces significantly more results than the user was expecting. We deal with this problem by identifying the parts of the failing query, called *Minimal Failure Inducing Subqueries* (MFIS), that cause plethoric answers. As long as the query contains an MFIS, it will fail to reach a sufficiently low amount of answers. Thanks to these MFIS, interactive and automatic approaches can be set up to help the user reformulate their query. The dual notion of MFIS, *maXimal Succeeding Subqueries* (XSS), is also useful. They are queries with the most parts of the original query that return non plethoric answers. Our goal is to compute MFIS and XSS efficiently, so that they may be used to solve the plethoric answers problem. We propose two algorithms that leverage query and data properties to compute MFIS and XSS. We show experimentally that our algorithms clearly outperform a baseline method on generated queries as well as real user-submitted queries.

1 Introduction

A Knowledge Base (KB) is a collection of entities and facts about them. With the development of the Semantic Web, numerous KBs have been created in academic and industrial areas. A well known example of a KB is *DBpedia* [1]. These KBs store information as RDF triples (subject, predicate and object) and are queried with the *SPARQL* language [2] using triple patterns which are triples containing variables. KBs typically store billions of facts and are often structured using an ontological schema and rules, such as those provided by *RDFS* [3].

A new user querying a KB is often unfamiliar with the KB's structure and the data within it. Thus, *mistakes* or *misconceptions* can manifest in queries, and cause unexpected or unsatisfactory answers. Mistakes refer to the user incorrectly writing their query, for example creating an unwanted Cartesian product by omitting a triple pattern, or misspelling a term. Misconceptions refer to the

C. Strauss et al. (Eds.): DEXA 2021, LNCS 12923, pp. 292–304, 2021.
https://doi.org/10.1007/978-3-030-86472-9_27

difference between a user's view of a KB, and its reality [4]. For example if in a hospital KB, the property *treats* can only link a *Doctor* to a *Patient*, and a user writes a query based on the patients that a *Nurse treats*, they will be frustrated to receive no answers. Alternatively, a user may believe a property *birthPlace* only gives a person's town of birth whereas in the KB *birthPlace* is used for the country, county, town, and address of birth. A query involving *birthPlace* may overwhelm the user by producing four times as many answers as expected. The issue of unexpected answers is a challenge to database usability [5]. There are five unexpected answers problems, each associated with a why-question: no answers (*why-empty*), too few answers (*why-so-few*), too many answers (*why-so-many*), missing expected answers (*why-not*), and unwanted answers (*why-so*). We focus on the too many answers problem, also called the plethoric answers problem, where users struggle to extract useful information from an overwhelming result. A query's results are said to be plethoric when there are more than a threshold K.

Most state of the art methods to deal with plethoric answers rely on ordering results and selecting an adequately sized subset of answers to be returned to the user. These methods are called *top-K* methods. Solutions vary by the way results are ordered, and the extent of user involvement. They guarantee the number of answers will be at most K. Yet, if a query is based on a misconception on the user's part no ordering strategy will solve the underlying problem. In this paper, we claim that the first step to solve the plethoric answers problem should be to understand why a query produces plethoric answers. Our failure causes can be directly provided to users in an effort to educate them in formulating their queries. They can also be used as a basis for automatic or interactive query rewriting, in order to avoid suggesting queries that are known to fail, and thus accelerate the process.

Drawing on work on the empty-answers problem in KBs [6], we propose the basis of a cooperative method to deal with the plethoric answers problem. We provide two notions that can help with query rewriting: the smallest subqueries that cause plethoric answers (MFIS) and the largest subqueries that do not produce plethoric answers (XSS). We propose an algorithm to compute MFIS and XSS, leveraging query properties to avoid executing some subqueries. Improvements based on a data property, i.e. predicate cardinalities, are also discussed. The performance of our algorithms is assessed through experimental evaluation, using queries generated for the *WatDiv* synthetic dataset [7], and user-submitted DBpedia queries from the *Linked SPARQL Queries Dataset* logs [8].

This paper is organized as follows. Section 2 gives a motivating example that will illustrate our proposal throughout the paper. Section 3 details related work. We provide preliminary notions in Sect. 4. Section 5 presents our approaches to calculate MFIS and XSS. Section 6 describes the experimental evaluation of our algorithms. We conclude and introduce future work in Sect. 7.

2   Motivating Example

We consider a simplified hospital KB (Fig. 1a), and a user wanting information on doctors, nurses and patients. Figure 1b shows an example of a user query

subject	predicate	object
d_1	experience	25
d_1	supervises	n_3
d_1	supervises	n_2
d_1	treats	p_1
d_1	treats	p_2
d_2	experience	14
d_2	supervises	n_1
d_2	treats	p_3
n_1	type	SurgicalNurse
n_1	providesCare	p_3
n_2	type	ERNurse
n_2	providesCare	p_2
n_2	providesCare	p_3
n_3	type	ERNurse
n_3	providesCare	p_1
n_3	providesCare	p_2

(a) Knowledge base D

```
Q : SELECT * WHERE {
      ?d treats ?p .            #t1
      ?d experience ?e .        #t2
      ?d supervises ?n .        #t3
      ?n providesCare ?pt .     #t4
      ?n type ERNurse }         #t5
```
(b) Query $Q = t_1 t_2 t_3 t_4 t_5$

?d	?p	?e	?n	?pt
d1	p_1	25	n3	p_1
d1	p_2	25	n3	p_1
d1	p_1	25	n3	p_2
d1	p_2	25	n3	p_2
d1	p_1	25	n2	p_2
d1	p_2	25	n2	p_2
d1	p_1	25	n2	p_3
d1	p_2	25	n2	p_3

(c) Results of Q on D

Fig. 1. A knowledge base, a SPARQL query and its results

defined as a conjunction of triple patterns $Q = t_1 \wedge t_2 \wedge t_3 \wedge t_4 \wedge t_5$ (or $t_1 t_2 t_3 t_4 t_5$ in short). When executing Q on our dummy KB the query produces 8 results (Fig. 1c). On a real KB containing hundreds of doctors and patients, this query would produce thousands of results, which would not be manageable for the user. To illustrate our method, we set a small threshold of plethoric answers $K = 3$, so Q is considered failing as its number of results exceeds this threshold.

A top-K method could be used to reduce the number of results, such as ordering patients alphabetically, but this would only return information about a few patients. Our approach will focus on explaining plethoric answers based on failure causes, so the query can be modified to return fewer answers. In our example, there are three failure causes: $t_1 t_3$, $t_1 t_5$ and t_4. As every subquery of Q containing one of these subqueries fails, each of the failure causes must be resolved in order to reach the desired number of answers. Each failure cause can be interpreted to explain plethoric answers, which will in turn suggest possible corrections.

- $t_1 t_3$ and $t_1 t_5$ indicate that asking for both patients and nurses produces plethoric answers. They suggest splitting the query into two parts, one concerning patients, and the other concerning nurses.
- t_4, indicates that nurses are assigned to a plethoric number of patients. It suggests removing t_4 from the query, or adding some additional conditions.

We also provide the succeeding queries which are not subqueries of any other succeeding query: $t_2 t_3 t_5$ and $t_1 t_2$. They partially meet the user's requirement and can be used as alternative queries, knowing that they have at most K results.

The interpretation of failure causes, and their presentation to a user along with alternative queries in an interactive query refining system is planned for future work, and is not further studied in this paper. We focus here on the efficient computation of the failure causes, for use in a query rewriting system.

3   Related Work

Existing approaches dealing with the plethoric answers problem can be divided into two categories: those focusing on data and those focusing on queries.

Data-oriented methods suppose that the query submitted by the user is correct, and present results in an organised fashion, so that certain information is easily visible. Top-K methods are the most widely used type of data-oriented methods. They order results based on user preferences and return only the top K answers. Ilyas et al. present top-K query processing techniques for relational database systems [9]. Other data-oriented strategies have been proposed for cases where user preferences are unknown. Regret-minimization strategies combine features from top-K and skyline methods [10]. They return a set of K answers which maximizes the minimal satisfaction of any user with any preference function. Finally, grouping methods aggregate results into categories, and show the user the common features of each category [11,12]. If the initial query correctly matches the user's requirement, data-oriented methods can be useful to sort through large result sets. However, if the original query contains an underlying issue, these methods are not appropriate, as they do not attempt to fix the query.

Query-oriented methods modify the user's query so that it returns fewer answers. In the field of fuzzy queries, intensification strategies are used to make patterns present in the user's query more restrictive [13,14]. Alternatively, new patterns are added to the query [15]. They are chosen based on a measure of correlation between predicates so that they are semantically close to the original query and reduce the number of answers. In the field of knowledge graphs, recent work on the why-not and why-so problems – where an expected answer is missing or an unexpected answer appears in the response – can be extended to the plethoric answers problem [16]. Exact algorithms and heuristics are proposed to refine a user's query. A final approach, which is most similar to ours, considers subqueries of the original query, to find the parts with few enough answers [17]. However, this algorithm does not consider failure causes, and uses no inference rules to avoid exploring parts of the subqueries search space. Query-based solutions are more appropriate to address an underlying issue in the original query. However, as none of the existing approaches study the cause of plethoric answers, the query intensification is done blindly. So patterns causing multiple results may be missed or take several attempts to find.

While failure causes have not previously been considered for the plethoric answers problem, they have been used for other unexpected answers problems. For the why-not problem, a divide-and-conquer approach is used, first studying a query's triple patterns and then its SPARQL operators [18]. A failure cause shows users which triple pattern or operator causes an answer to be absent. For the empty answers problem, Godfrey [19] suggested providing users with failure causes (called MFS) and alternate subqueries (called XSS). MFS have subsequently been used in interactive and automatic query relaxation to accelerate the process, by pruning the search space of queries which necessarily fail [20,21]. We propose extending the definitions of MFS and XSS to deal with the plethoric answers problem in the context of RDF KBs.

4   Preliminaries and Problem Statement

We describe the formalism and semantics of RDF and SPARQL necessary for this paper. We use the notations and definitions provided by Pérez et al. [22].

4.1   Basic Notions

Data Model. We consider three pairwise disjoint infinite sets: I the set of IRIs, B the set of blank nodes, and L the set of literals. We denote by T the union $I \cup B \cup L$. An *RDF triple* is a triple (subject, predicate, object) $\in (I \cup B) \times I \times T$.

RDF Queries. Consider V a set of variables disjoint from T. A triple t (subject, predicate, object) $\in (I \cup L \cup V) \times (I \cup V) \times (I \cup L \cup V)$ is a *triple pattern*. We denote by $s(t)$, $p(t)$, $o(t)$, and $var(t)$ the subject, predicate, object and variables of t. *RDF queries* are defined as conjunctions of triple patterns $Q = t_1 \cdots t_n$. The variables of a query are $var(Q) = \bigcup var(t_i)$. We define an order on queries using triple pattern inclusion. Given $Q = t_1 \cdots t_n$, $Q' = t_i \cdots t_j$ is a *subquery* of Q, denoted by $Q' \subseteq Q$, iff $\{i, \cdots, j\} \subseteq \{1, \cdots, n\}$. Then Q is a *superquery* of Q'.

Query Evaluation. A mapping μ from V to T is a partial function $\mu : V \to T$. For a triple pattern t, we denote by $\mu(t)$ the triple obtained by replacing the variables in t according to μ. The domain of μ, $dom(\mu)$, is the subset of V where μ is defined. Two mappings μ_1 and μ_2 are *compatible* if $\forall x \in dom(\mu_1) \cap dom(\mu_2)$, $\mu_1(x) = \mu_2(x)$. The join of two sets of mappings Ω_1 and Ω_2 is: $\Omega_1 \bowtie \Omega_2 = \{\mu_1 \cup \mu_2 \mid \mu_1 \in \Omega_1, \mu_2 \in \Omega_2 \text{ are compatible mappings}\}$. The *evaluation* of a triple pattern t over a KB D, is $[[t]]_D = \{\mu \mid dom(\mu) = var(t) \wedge \mu(t) \in D\}$. The evaluation of a query $Q = t_1 \cdots t_n$ over D is $[[Q]]_D = [[t_1]]_D \bowtie \cdots \bowtie [[t_n]]_D$.

4.2   Notions of MFIS and XSS

In the plethoric answers problem, for a threshold K, a failing subquery of a query Q is a query that returns more than K answers and a succeeding subquery of a query Q is a query that returns at most K answers. We introduce a Boolean property of query failure: $\text{FAIL}_K(Q, D) = |[[Q]]_D| > K$. As the evaluation of the empty query returns one answer mapping no variables, it succeeds (if $K > 0$).

The notion of Minimal Failing Subqueries (MFS) was introduced for the empty answers problem [19]. In that problem, the failure property is monotonic, i.e. if a query fails, its superqueries fail. With this monotony, MFS are the smallest parts of a query that cause failure. In the plethoric answers problem, there is no such monotony. A failing query can have a succeeding superquery: in our example $t_2 t_5$ fails but $t_2 t_3 t_5$ succeeds. We therefore define two new notions.

Definition 1. *A Failure Inducing Subquery (FIS) of a query Q is one of its failing subqueries whose superqueries all fail.*

Definition 2. *A Minimal Failure Inducing Subquery (MFIS) of a query Q is one of its FIS having no subqueries that are FIS.*

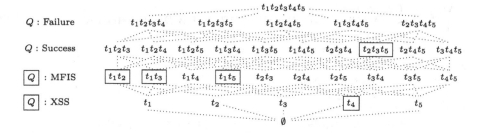

Fig. 2. Lattice of subqueries of $Q = t_1t_2t_3t_4t_5$

If the failure condition is monotonic, MFIS and MFS are equivalent notions. The notion of maXimal Succeeding Subqueries (XSS) was also defined for the empty answers problem. They are the succeeding queries that are most similar to the original query and can be used as alternative queries. The notion of XSS applies to the plethoric answers problem, as it does not require monotony.

Definition 3. *A maXimal Succeeding Subquery (XSS) of a query Q is a succeeding subquery whose superqueries are all FISs.*

Problem Statement. We are concerned with efficiently computing the MFIS and XSS for an RDF query Q and a given threshold K in a KB D.

5 Computing MFIS and XSS

We discuss here MFIS and XSS computation. First, a baseline algorithm is presented, then improved versions are introduced by leveraging various properties.

5.1 Baseline

A baseline approach to calculate all MFIS and XSS is to execute every subquery of the original query. Figure 2 shows the lattice of subqueries of Q from our running example. In this first algorithm, which we call BASE, the lattice is explored in a *Breadth-First* order, so we start by executing the query with the most triple patterns. For a query with n triple patterns, BASE requires $2^n - 1$ query executions (the empty query is not executed), which is time-consuming for queries with many triple patterns. To make the search for MFIS and XSS more efficient, we want to reduce the number of queries to be executed.

5.2 General Properties

A first improvement is to avoid executing queries irrelevant to the search for MFIS and XSS, with a property deduced from the definitions of MFIS and XSS.

Property 1. If a subquery Q' succeeds, and $Q'' \subset Q'$, then Q'' is neither an MFIS nor an XSS.

Var/Full(Q, D, K)

    inputs : A failing query $Q = t_1 \wedge ... \wedge t_n$, a KB D, a threshold K
    outputs: MFIS and XSS of Q

1	mfis $\leftarrow \emptyset$, xss $\leftarrow \emptyset$, fis $\leftarrow \emptyset$, queryStatus $\leftarrow \emptyset$, list $\leftarrow \{Q\}$
2	**while** list $\neq \emptyset$ **do**
3	$Q' \leftarrow$ first query of list in BFS ordering
4	list $\leftarrow$ list $- \{Q'\}$
5	parents_fis $\leftarrow$ true
6	**foreach** $t \in$ triplePatterns$(Q) -$ triplePatterns(Q') **do**
7	parents_fis $\leftarrow$ parents_fis $\wedge ((Q' \wedge t) \in$ fis$)$
8	**if** parents_fis **then**
9	**if** $Q' \notin$ queryStatus **then**
10	queryStatus $[Q'] \leftarrow \mathrm{FAIL}_K(Q', D)$
11	**if** queryStatus $[Q']$ **then** // if Q' fails
12	fis $\leftarrow$ fis $\cup \{Q'\}$
13	mfis $\leftarrow$ mfis $-$ superQueries(Q')
14	mfis $\leftarrow$ mfis $\cup \{Q'\}$
15	**foreach** $t \in$ triplePatterns(Q') **do**
16	**if** $(Q' - t) \notin$ list **then**
17	list $\leftarrow$ list $\cup (Q' - t)$
18	**if** $var(Q' - t) = var(Q')$ **then**
19	queryStatus $[Q' - t] \leftarrow$ true
20	**else if** $\mathrm{card}_{\max}(p(t), D) = 1 \wedge s(t) \in var(Q' - t)$ **then**
21	queryStatus $[Q' - t] \leftarrow$ true
22	**else** // Q' is successful, and therefore an XSS
23	xss $\leftarrow$ xss $\cup \{Q'\}$
24	**return** mfis, xss

Algorithm 1: Enumerate the MFIS and XSS of a query Q

When using this property to avoid query executions, query success or failure is not known over the whole lattice but partial knowledge is sufficient here. Next, we consider deduction rules that predict query failures without executing them. As we run a breadth-first-search, a query is studied after all its superqueries, so we can leverage properties deducing the failure of a query from the failure of its superqueries.

Property 2. Given a query Q and triple pattern t, if $var(Q \wedge t) = var(Q)$ then $Q \wedge t$ fails $\Rightarrow Q$ fails.

Property 2 states that, if removing a triple pattern from a query does not remove any variables, then the number of answers cannot decrease. In our example, adding t_5 (?n type ERNurse) to a query containing variable n adds a constraint on n, and so cannot increase the number of answers.

Adding Properties 1 and 2 to BASE creates an improved algorithm, VAR, shown in Algorithm 1 (lines 20 and 21 do not apply, they are used in the next version). The main data structures are a list (list) of subqueries to evaluate and a map (queryStatus) storing the result of their evaluations: failure or success (lines 9–10). From the list, we consider queries from the lattice in a breadth-first order (lines 1–4). According to Property 1, every direct superquery of Q' has to be

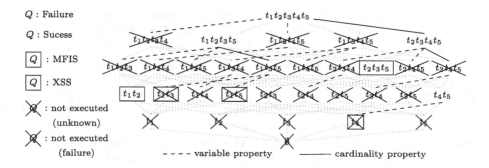

Fig. 3. Lattice of subqueries of Q with FULL algorithm

an FIS for further consideration (lines 5–8). On query failure, the sets fis and mfis are updated (lines 5–8): the subquery Q' being considered replaces its direct superqueries in the mfis set as those can no longer be minimal (lines 12–14). We then consider every direct subquery of Q' (lines 15–21) for future evaluation (line 17), checking if Property 2 is applicable (line 18) to predict its failure without executing it. If, instead, Q' succeeded, it is added to the xss set (line 23).

5.3 Cardinality-Based Property

The last improvement leverages a property involving both the query and the data. We consider triple patterns that add a piece of information to each answer but do not overall change the number of answers. In the running example if t_2 (?d experience ?e) is removed from a query, as each person has at most one *experience*, each answer will lose a piece of information, but no two answers will become identical. So the number of answers will not decrease. We focus on predicates with maximum cardinality 1, of which any subject has at most one occurrence.

Definition 4. *The global maximum cardinality of a predicate p in a dataset D is [23]:* $\mathrm{card}_{\max}(p, D) = \max\limits_{s|\exists\ p,o:(s,p,o)\in D} |\{(s,p,o) \mid (s,p,o) \in D\}|$

Property 3. Given a query Q, and a triple pattern t with a fixed predicate $p(t)$, if $\mathrm{card}_{\max}(p(t), D) = 1$ and $s(t) \in var(Q)$ then $Q \wedge t$ fails $\Rightarrow Q$ fails.

The complete algorithm, FULL, is created by adding Property 3 to VAR (lines 20–21).

Figure 3 shows the lattice of subqueries of the query from the motivating example, and the subqueries avoided by FULL. For example, t_2 has maximum cardinality 1, the subject of t_2 (d) is one of the variables of $t_1t_3t_4t_5$, and $t_1t_2t_3t_4t_5$ fails. We deduce that $t_1t_3t_4t_5$ fails using Property 3. In our example, FULL only executes 6 queries (rather than 31 for BASE and 9 for VAR).

Other cardinality definitions, like Class Cardinality [23], consider a smaller set of subjects than global cardinalities so may provide more precise values.

But these are query specific so a single cardinality value does not hold over the whole lattice. Cardinalities would need to be calculated at each query evaluation. Extending our approach with other cardinalities is a perspective for future work.

The deduction rule based on variables (Property 2) is applicable to any query in any dataset as it relies only on information contained within the original query. However, the cardinality-based condition (Property 3) requires additional information: if cardinalities are not enforced by the schema, they must be calculated for all predicates. For frequently modified KBs, cardinalities would need to be updated each time the data is changed. This requires KB administrators to provide updated cardinality values or users to regularly query the KB to obtain cardinalities, which is costly. So we provide two versions of our algorithm: the FULL version with all optimization properties, and the variable-only version, VAR.

6   Experimental Evaluation

Hardware. Our experiments were run on a Ubuntu Server LTS system with an Intel(R) Xeon(R) Gold 5118 CPU @ 2.30 GHz and 32 GB RAM. The results presented are the average of five consecutive runs of the algorithms. To prevent a cold start effect, a preliminary run is performed but not included in the results.

Algorithms. The BASE, VAR and FULL algorithms are implemented[1] in Oracle Java 1.8 64 bits and run using the Jena TDB triplestore. Cardinalities are precomputed for the FULL algorithm. We set the threshold for plethoric answers $K = 100$, as it is the default limit used by the DBpedia SPARQL endpoint. As Cartesian products are costly to execute, queries containing Cartesian products are split into connected parts so that in each part every triple pattern shares a variable with at least one other triple pattern, and so that separate parts share no variables. Each part is then executed separately, the number of answers of the original query being the product of the number of results of each part.

Synthetic Dataset and Queries. We used a dataset of 11M triples, generated with the WatDiv benchmark. We have considered 21 queries with 4 to 12 triple patterns. There are 7 star queries (all triple patterns have the same subject), 7 chain queries (subject of triple pattern $j + 1$ is the object of triple pattern j) and 7 composite queries (any other configuration). All chain queries and some star and composite queries are based on the WatDiv test cases (IL-1-10, F1, F2, F4, C2, C3). We added new composite and star queries to have varied characteristics.

Real Dataset and Queries. We used the 3.9 version of the English DBpedia dataset, which contains 812M triples. Queries come from the LSQ project [8] which recorded user-submitted queries to DBpedia (version 3.5.1) between April 30 and June 20, 2010. Some minor adaptations were made as some original URIs were not compatible with version 3.9 of DBpedia. We have used 9 star or composite queries, containing 4 to 10 triple patterns.

---

[1] Our implementation is available at https://forge.lias-lab.fr/projects/tma4kb with a tutorial to reproduce experiments.

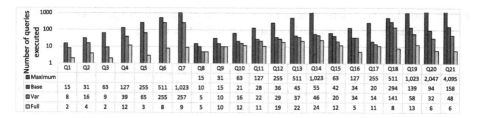

Fig. 4. # Executed queries Watdiv 11M triples

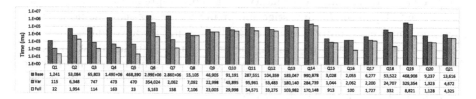

Fig. 5. Execution time Watdiv 11M triples

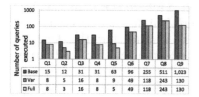

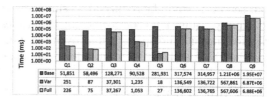

Fig. 6. # Executed queries DBpedia   **Fig. 7.** Execution time DBpedia

6.1 Results

Synthetic Dataset and Queries. First we study the performance of the three algorithms on a WatDiv generated dataset of 11M triples. The number of executed queries and the execution time for each algorithm are given in Figs. 4 and 5.

For all the queries tested, we verify that VAR and FULL execute at most as many queries as BASE. This is a guarantee of the properties presented in Sect. 5. The improvement is less notable for chain (Q8 to Q14) and composite (Q15 to Q21) queries. Indeed, these queries have subqueries containing Cartesian products. Since we execute Cartesian products by separating them into connected parts, queries that have a succeeding superquery can be executed as part of a Cartesian product. The number of queries executed by the BASE algorithm if Cartesian products were executed directly, $2^n - 1$, is given in the *Maximum* bars on Fig. 4. Overall VAR and FULL execute respectively 46% and 75% fewer queries than BASE. The execution times follow the same general trend. FULL is faster than VAR, itself faster than BASE. The improvement in execution time is smaller than the improvement in query executions. Indeed, all query executions are not equal, and the executions avoided can have short execution times. Overall VAR saves 65% of the BASE execution time, and FULL saves 83%.

Real Dataset and Queries. We show the number of executed queries and the run-time of each algorithm in Figs. 6 and 7. As in the WatDiv experiments, VAR and FULL execute at most as many queries as BASE. However, we notice that FULL rarely executes fewer queries than VAR. This can be explained by cardinalities in DBpedia. Few of the predicates used have maximum cardinality 1, so the cardinality property cannot be applied to them. Only 0.67% of predicates in DBpedia have maximum cardinality 1, but 66% of predicates have maximum cardinality 2. Upon further investigation, many predicates that should in theory have maximum cardinality 1, such as BirthDate, in fact have maximum cardinality 2. This can be due to errors in the data or uncertain information [24,25]. Using a curated dataset would likely improve the benefit of the cardinality-based pruning. Consequently, VAR and FULL have very similar execution times, each saving around 78% of the baseline time.

Some queries, like Q9, have long execution times (over an hour). The execution time depends heavily on the time the triplestore takes to answer a query on our server. It could be reduced by using a distributed solution or optimizing how the query evaluation is performed by the triplestore. This has not yet been investigated and is a prospect of improvement. Our experiments show that for most queries, VAR and FULL run in a few seconds, despite using a centralized server.

7   Conclusion

In this paper, we have addressed the plethoric answers problem in the context of RDF queries. We have identified that none of the approaches proposed in the literature try to identify why the user query produced plethoric answers. Yet, several approaches developed for other unsatisfactory answers problems have shown that the first step in a query adjustment process designed to meet the user expectation should be understanding why the query failed.

Our goal was to fill this gap. We have first shown that the notions defined for other unsatisfactory answers problem are too restrictive for our context and defined a more general notion, named MFIS. Starting from a baseline method to calculate all MFIS and XSS of a failing query, we have proposed improvements based on query and data properties, to reduce the number of queries that need to be executed, and therefore reduce the run-time of our algorithms. Experiments using both synthetic and real data show that our optimized algorithms offer a significant improvement. VAR saves 71% of the baseline time and can be used for any query, and FULL saves 82% of the baseline time but requires additional information: predicate cardinalities.

The next step will be to use the MFIS and XSS to aid in rewriting queries with plethoric answers. Query modification can be performed entirely by the user (i.e. we provide the MFIS and XSS and the user interprets them to adapt their query), entirely automatically, or with an interactive approach, where the user is guided through changes applied to their query.

References

1. Lehmann, J., et al.: DBpedia - a large-scale, multilingual knowledge base extracted from Wikipedia. Semantic Web **6**(2), 167–195 (2015)
2. Harris, S., Seaborne, A.: Sparql 1.1 query language. W3C Recommendation (2013)
3. Brickley, D., Guha, R.: RDF schema 1.1. W3C Recommendation (2014)
4. Webber, B.L., Mays, E.: Varieties of user misconceptions: detection and correction. In: IJCAI 1983, vol. 2, pp. 650–652 (1983)
5. Jagadish, H.V., et al.: Making database systems usable. In: SIGMOD 2007, pp. 13–24 (2007)
6. Fokou, G., Jean, S., Hadjali, A., Baron, M.: Cooperative techniques for SPARQL query relaxation in RDF databases. In: Gandon, F., Sabou, M., Sack, H., d'Amato, C., Cudré-Mauroux, P., Zimmermann, A. (eds.) ESWC 2015. LNCS, vol. 9088, pp. 237–252. Springer, Cham (2015). https://doi.org/10.1007/978-3-319-18818-8_15
7. Aluç, G., Hartig, O., Özsu, M.T., Daudjee, K.: Diversified stress testing of RDF data management systems. In: Mika, P., et al. (eds.) ISWC 2014. LNCS, vol. 8796, pp. 197–212. Springer, Cham (2014). https://doi.org/10.1007/978-3-319-11964-9_13
8. Saleem, M., Ali, M.I., Hogan, A., Mehmood, Q., Ngomo, A.-C.N.: LSQ: the linked SPARQL queries dataset. In: Arenas, M., et al. (eds.) ISWC 2015. LNCS, vol. 9367, pp. 261–269. Springer, Cham (2015). https://doi.org/10.1007/978-3-319-25010-6_15
9. Ilyas, I.F., Beskales, G., Soliman, M.A.: A survey of top-k query processing techniques in relational database systems. ACM Comput. Surv. **40**(4), 1–58 (2008)
10. Xie, M., Wong, R.C.W., Peng, P., Tsotras, V.J.: Being happy with the least: achieving α-happiness with minimum number of tuples. In: ICDE 2020, pp. 1009–1020. IEEE (2020)
11. Chaudhuri, S., Das, G., Hristidis, V., Weikum, G.: Probabilistic ranking of database query results. In: VLDB 2004, vol. 30, pp. 888–899 (2004)
12. Ozawa, J., Yamada, K.: Discovery of global knowledge in a database for cooperative answering. In: IEEE 1995, vol. 2, pp. 849–854 (1995)
13. Bosc, P., HadjAli, A., Pivert, O.: About overabundant answers to flexible queries. In: IPMU 2006, vol. 6, pp. 2221–2228 (2006)
14. Moises, S.A., Pereira, S.D.L.: Dealing with empty and overabundant answers to flexible queries. J. Data Anal. Inf. Process. **2**(1), 12–18 (2014)
15. Bosc, P., Hadjali, A., Pivert, O., Smits, G.: Une approche fondée sur la corrélation entre prédicats pour le traitement des réponses pléthoriques. In: EGC 2010, pp. 273–284 (2010)
16. Song, Q., Namaki, M.H., Wu, Y.: Answering why-questions for subgraph queries in multi-attributed graphs. In: ICDE 2019, pp. 40–51 (2019)
17. Vasilyeva, E., Thiele, M., Bornhövd, C., Lehner, W.: Answering "why empty?" and "why so many?" queries in graph databases. JCSS **82**(1), 3–22 (2016)
18. Wang, M., Liu, J., Wei, B., Yao, S., Zeng, H., Shi, L.: Answering why-not questions on SPARQL queries. Knowl. Inf. Syst. **58**, 169–208 (2019)
19. Godfrey, P.: Minimization in cooperative response to failing database queries. Int. J. Cooperative Inf. Syst. **6**(2), 95–149 (1997)
20. Fokou, G., Jean, S., Hadjali, A., Baron, M.: RDF query relaxation strategies based on failure causes. In: Sack, H., Blomqvist, E., d'Aquin, M., Ghidini, C., Ponzetto, S.P., Lange, C. (eds.) ESWC 2016. LNCS, vol. 9678, pp. 439–454. Springer, Cham (2016). https://doi.org/10.1007/978-3-319-34129-3_27

21. Jannach, D.: Techniques for fast query relaxation in content-based recommender systems. In: Freksa, C., Kohlhase, M., Schill, K. (eds.) KI 2006. LNCS (LNAI), vol. 4314, pp. 49–63. Springer, Heidelberg (2007). https://doi.org/10.1007/978-3-540-69912-5_5

22. Pérez, J., Arenas, M., Gutierrez, C.: Semantics and complexity of SPARQL. ACM Trans. Database Syst. **34**(3), 1–45 (2009)

23. Dellal, I.: Management and Exploitation of Large and Uncertain Knowledge Bases. Ph.D. thesis, ISAE-ENSMA - Poitiers (2019)

24. Giacometti, A., Markhoff, B., Soulet, A.: Mining significant maximum cardinalities in knowledge bases. In: Ghidini, C., et al. (eds.) ISWC 2019. LNCS, vol. 11778, pp. 182–199. Springer, Cham (2019). https://doi.org/10.1007/978-3-030-30793-6_11

25. Muñoz, E., Nickles, M.: Mining cardinalities from knowledge bases. In: Benslimane, D., Damiani, E., Grosky, W.I., Hameurlain, A., Sheth, A., Wagner, R.R. (eds.) DEXA 2017. LNCS, vol. 10438, pp. 447–462. Springer, Cham (2017). https://doi.org/10.1007/978-3-319-64468-4_34

Prediction and Decision Support

Feature Selection and Software Defect Prediction by Different Ensemble Classifiers

Natalya Shakhovska[1] and Vitaliy Yakovyna[2]

[1] Artificial Intelligence Department, Lviv Polytechnic National University, Lviv, Ukraine
nataliya.b.shakhovska@lpnu.ua
[2] Faculty of Mathematics and Computer Science, University of Warmia and Mazury in Olsztyn, Olsztyn, Poland
yakovyna@matman.uwm.edu.pl

Abstract. Software defect prediction can improve its quality and is actively studied during the last decade. This paper focuses on the improvement of software defect prediction accuracy by proper feature selection techniques and using ensemble classifier. The software code metrics were used to predict the defective modules. JM1 public NASA dataset from PROMISE Software Engineering Repository was used in this study. Boruta, ACE, regsubsets and simple correlation are used for feature selection. The results of selection are formed based on hard voting of all features selectors. A new stacking classifier for software defects prediction is presented in this paper. The stacking classifier for defects prediction algorithm is based on combination of 5 weak classifiers. Random forest algorithm is used to combine the predictions. The obtained prediction accuracy was up to 96.26%.

Keywords: Ensemble of classifiers · Feature selection · Software defect analysis

1 Introduction

A software defect is a bug in the program code that causes the program to crash or produces incorrect results. Most of the defects arise from errors made in the source code of the program or its design. Software defect prediction locates defective module to help developers improve the quality of software. Early detection of software defects reduces development costs and improves software quality and reliability.

Software defect prediction methods are used to either classify modules prone to defects or to predict the number of defects expected in software module. A number of different methods are used to classify and predict defects. They are divided into methods for predicting the expected number of defects that may be found in a software artifact (prediction), as well as methods for predicting whether a given software artifact may contain a defect (classification). Machine learning techniques use algorithms based on statistical and data mining techniques and can be applied to classify and predict defects. These methods are similar to regression methods and use the same input independent variables. The aim of the paper is to develop to new ensemble of classifiers for software defects prediction and frequent patterns mining based on feature selection. The proposed

© Springer Nature Switzerland AG 2021
C. Strauss et al. (Eds.): DEXA 2021, LNCS 12923, pp. 307–313, 2021.
https://doi.org/10.1007/978-3-030-86472-9_28

approach can be used also in other domains. The developed ensemble allows more accurate estimation of software defects than logistic regression and random forest. In addition, the time complexity is lower than for deep learning approach, particularly RRN.

2 Related Works

Datasets for software defects prediction consist of a lot of features. However, the quality of prediction should be improved [1]. That is why many researchers focus on selecting variables or features of the original project that are relevant to the target project. In [2], the neighbor filter method is used to delete instances of the original project, the functions of which are not close enough to the functions of the target project. The choice of functions based on correlation was studied in [3]. Naïve Bayes classifier is used for feature selection and defects prediction in [4]. However, the nature of dataset requires the combination of different methods for features selection.

Wrapper methods rely on feature importance information from some classification or regression methods, and therefore can find deeper patterns in the data than filters. Wrappers can use any classifier that determines the degree of importance of the features. Wrappers are used in [5–7].

The next step is data analysis and classification of software defects. Machine learning and deep learning are widely used for software defects prediction. The new approach that leverages deep learning techniques to predict the number of defects in software systems is developed in [8]. The specially designed deep neural network-based model to predict the number of defects is proposed. A new unsupervised based embedding method SimAST-Token2Vec is designed in [9] to learn meaningful representation for these extracted token vectors. Bi-directional Long Short-Term Memory (BiLSTM) neural network is used to automatically learn semantic features from embedded token vectors. To enhance software reliability, in paper [10], a deep learning-based method called DP-ARNN (defect prediction via attention-based recurrent neural network) is developed. Specifically, DP-ARNN leverages RNN to automatically generate syntactic and semantic features from source code. The attention mechanism to capture crucial features is employed.

3 Methods and Results

3.1 The Dataset

Public dataset from PROMISE Software Engineering Repository [11] was used in this study. The JM1 dataset on software defect prediction was selected. The source of this dataset is NASA and the NASA Metrics Data Program. The dataset contains 10,885 entries (modules) along with 21 code metrics, used as features and listed in [12], and a "TRUE/FALSE" field indicating either the module contains one or more reported defects, used as a target. Among all modules listed in the dataset, 2,107 have reported defects. The dataset is imbalanced.

All calculations were made using RStudio. Data was passed through Data Sampler for balancing (a subset of 50/50 True / False was randomly selected). As a result, from a dataset of 10885 records, 5443 records were selected for which further models were built. The randomized 90% of the dataset were used as a training set, and the rest 10% – as a validation set. One of the obstacles while building software reliability model based on its code metrics is the fact that a lot of metrics highly correlate. Thus, pairwise Pearson correlation coefficient greater than 0.900 was obtained for 20 metrics pairs [12]. The quality of prediction is calculated for whole dataset and for selected features.

3.2 Research Methods: Feature Selection

Collinear features are those that correlate with each other. The presence of these features can adversely affect the learning rate of the model (especially for wrapping methods), reduce interpretability and performance, and contribute to model overfitting. In the presence of correlated features, algorithms based on an assessment of the importance of permutations of a feature may be wrong, since removing one of them will not reduce the quality of the model. Consequently, the importance of the permutation for them may be low and they will all be discarded, regardless of their usefulness. From the set of mutually correlated features, you can remove all but one without significant loss of information. That is why the correlation between features is found and correlation matrix is built. A significant correlation of some metrics was revealed, and the correlation between e and t, as well as between b and v is equal to one. Another 18 pairs of metrics have a Pearson correlation coefficient greater than 0.9.

Boruta is heuristic algorithm for selecting significant features based on the use of Random Forest. The essence of the algorithm is that at each iteration, features are removed whose Z-measure is less than the maximum Z-measure among the added features. To get the Z-measure of a feature, it is necessary to calculate the importance of the feature, obtained using the built-in algorithm in Random Forest, and divide it by the standard deviation of the feature importance. The obtained most significant variables are McCabe's cyclomatic complexity, $v(g)$; McCabe's design complexity, $iv(g)$; Halstead's total operators & operands, n; Halstead's volume, v; Halstead's line count, $lOCode$; Halstead's count of lines of comments, $lOComment$; Halstead's Total operators, $total\_Op$; number of branches in the flow graph, $branchCount$.

ACE (Artificial Contrasts with Ensembles) is another algorithm that can be used for feature selection. The main idea of the ACE algorithm is similar to the idea of the Boruta algorithm – each feature is filled in randomly by shuffling its values. Random Forest is launched on the resulting sample. However, unlike Boruta, it does not remove the found features with the least importance, which can improve the quality of measurements of important features. The most important features found by the ACE algorithm, on the contrary, are removed, which allows the algorithm to find patterns. The ACE issues the deleted signs as a response. The result of ACE for two primary components is the following: lOCode, b, v, t, e, total_Op, n, loc, total_Opnd, branchCount, v(g), d, uniq_Opnd, lOBlank, uniq_Op, lOComment, d.

The R package leaps has a function regsubsets that can be used for best subsets, forward selection and backwards elimination depending on which approach is considered most appropriate for the application under consideration. The most significant variables are the following: loc, e, lOCode, v(g), l, branchCount, n, uniq_Op,uniq_Opnd.

The final results of selection are forming based on hard voting of all features selectors. So, we divide features by following groups based on the results of feature selection algorithms:

- important variables (v.g, n, v, lOCode, branchCount),
- tentative variables (loc, l, d, i, e, uniq_Op,uniq_Opnd, total_Op, lO_Comment),
- rejected variables (the rest of features).

3.3   Research Methods: Classification

The research methodology is organize as following:

1. We divide the available sample into training and testing. For training, we will select 70% of the original sample.
2. We train six models for important variables, for important and tentative variables and for while dataset respectively (LDA – linear discriminant analysis, CART – classification tree, SVM – support vector machine with a radial kernel, RF – random forest, KNN – nearest neighbors, NNET – perceptron with one hidden layer with 15 neurons and Gaussian activation function).
3. We carry out resampling of the obtained models, perform a statistical analysis of two generated model quality metrics (Accuracy and J. Cohen's Kappa index) on the training set, rank the models and estimate the confidence intervals of the criteria used.

The statistical analysis for whole dataset is given below (Table 1).

Table 1. The Accuracy comparison

Classifier	Accuracy		
	Important variables	Important and tentative variables	Whole features
CART	0.833	0.666	0.616
LDA	0.857	0.800	0.646
SVM	0.902	0.800	0.732
KNN	0.833	0.714	0.663
RF	0.900	0.833	0.733
NNET	0.833	0.800	0.715

As result, the best accuracy is found for dataset consists of important variables.

3.4 Research Methods: The Ensemble of Classifiers

The three most popular methods for combining the predictions from different models are:

- Boosting. Building multiple models each of which learns to fix the prediction errors of a prior model in the chain.
- Bagging. Building multiple models from different subsamples of the training dataset.
- Stacking. Building multiple models and supervisor model that learns how to best combine the predictions of the primary models.

The ensemble will be built for selected important features. Two Boosting algorithms (c5.0 and Stochastic Gradient Boosting, SGB) are used for comparison. The accuracy and Kappa values for these algorithms are listed in Table 2.

Table 2. Prediction accuracy and Kappa index for c5.0 and SGB algorithms.

Algorithm	Min.	1st quartile	Median	Mean	3rd quartile	Max.
c5.0 accuracy	0.4	0.6666667	0.6666667	0.6825397	0.8000000	0.8333333
SGB accuracy	0.4	0.6666667	0.8000000	0.7450794	0.8511905	1.0000000
c5.0 Kappa	−0.4	0	0.0000000	0.03848297	0.0000000	0.5714286
SGB Kappa	−0.4	0	0.4227273	0.32427044	0.5840336	1.0000000

The next, bagged CART and Random Forest (RF) algorithms were used. Accuracy and Kappa index for these algorithms are listed in Table 3.

Table 3. Prediction accuracy and Kappa index for bagged CART and RF algorithms.

Algorithm	Min.	1st quartile	Median	Mean	3rd quartile	Max.
CART accuracy	0.3333333	0.6666667	0.6666667	0.7050794	0.8000000	1
RF accuracy	0.3333333	0.6666667	0.7142857	0.7330159	0.8333333	1
CART Kappa	−0.5	0	0	0.1663473	0.4000000	1
RF Kappa	−0.5	0	0	0.2033032	0.5714286	1

For stacking, let us first look at creating 5 sub-models for the dataset, namely: LDA, CART, GLM, kNN, SVM. The RF is used to combine the predictions. 100 trees are built for RF. Stacking results – accuracy and Kappa values along with their standard deviations (SD) – for RF with 1053 samples using 5 predictors and 2 classes ('bad', and 'good') are presented in the Table 4. Here mtry is the number of variables available for splitting at each tree node.

Accuracy was used to select the optimal model using the largest value. The final value used for the model was $mtry = 2$.

Table 4. Stacking results for 100 trees RF with 1053 samples.

mtry	Accuracy	Kappa	Accuracy SD	Kappa SD
2	0.9626439	0.9179410	0.01777927	0.03936882
3	0.9623205	0.9172689	0.01858314	0.04115226
5	0.9591459	0.9106736	0.01938769	0.04260672

4   Conclusions and Future Work

This paper presents the results of feature selection and software defect prediction using ensemble classifier. The data preprocessing allows us to increase the quality of analysis. Boruta, ACE, regsubsets and simple correlation are used for feature selection. The results of selection are formed based on hard voting of all features selectors. A stacking classifier for defects prediction algorithm based on combination of 5 poor classifiers was developed in this paper. Random forest algorithm is used to combine the predictions. Using this ensemble allow us to increase the defect prediction accuracy up to 96.26%. Therefore, the proposed algorithm shows the higher accuracy comparing to our previous results published in [12]. The reported in [12] prediction accuracy was about 86% when applied to the whole dataset.

The further research will include the development of hierarchical classifier.

References

1. Deep Singh, P., Chug, A.: Software defect prediction analysis using machine learning algorithms. In: 2017 7th International Conference on Cloud Computing, Data Science & Engineering – Confluence, pp. 775–781, IEEE, Noida, India (2017)
2. Turhan, B., Menzies, T., Bener, A.B., Di Stefano, J.: On the relative value of cross-company and within-company data for defect prediction. Empir. Softw. Eng. 14(5), 540–578 (2009)
3. Yu, Q., Jiang, S., Qian, J.: Which is more important for cross-project defect prediction: instance or feature? In: Proceedings of the International Conference on Software Analysis, Testing and Evolution (SATE), pp. 90–95, IEEE, Kunming, China (2016)
4. Sfenrianto, Purnamasari, I., Bahaweres, R.B.: Naive Bayes classifier algorithm and particle swarm optimization for classification of cross selling. In: 4th Intern. Conf. on Cyber and IT Service Management, pp. 1–4 (2016)
5. Mafarja, M., Mirjalili, S.: Whale optimization approaches for wrapper feature selection. Appl. Soft Comput. 62, 441–453 (2018)
6. El Aboudi, N., Benhlima, L.: Review on wrapper feature selection approaches. In: 2016 Intern. Conf. on Engineering & MIS (ICEMIS), pp. 1–5, IEEE, Agadir, Morocco (2016)
7. Naik, N., Mohan, B.: Optimal feature selection of technical indicator and stock prediction using machine learning technique. In: Somani, A.K., Ramakrishna, S., Chaudhary, A., Choudhary, C., Agarwal, B. (eds.) ICETCE 2019. CCIS, vol. 985, pp. 261–268. Springer, Singapore (2019). https://doi.org/10.1007/978-981-13-8300-7_22
8. Qiao, L., Li, X., Umer, Q., Guo, P.: Deep learning based software defect prediction. Neurocomputing 385, 100–110 (2020)
9. Chen, D., Chen, X., Li, H., Xie, J., Mu, Y.: DeepCPDP: deep learning based cross-project defect prediction. IEEE Access 7, 184832–184848 (2019)

10. Fan, G., Diao, X., Huiqun, Y., Yang, K., Chen, L.: Software defect prediction via attention-based recurrent neural network. Sci. Program. **2019**, 1–14 (2019). https://doi.org/10.1155/2019/6230953
11. Sayyad Shirabad, J., Menzies, T.J.: The PROMISE Repository of Software Engineering Databases. School of Information Technology and Engineering, University of Ottawa, Canada. http://promise.site.uottawa.ca/SERepository (2005)
12. Shakhovska, N., Yakovyna, V., Kryvinska, N.: An Improved Software Defect Prediction Algorithm Using Self-organizing Maps Combined with Hierarchical Clustering and Data Preprocessing. In: Hartmann, S., Küng, J., Kotsis, G., Tjoa, A.M., Khalil, I. (eds.) DEXA 2020. LNCS, vol. 12391, pp. 414–424. Springer, Cham (2020). https://doi.org/10.1007/978-3-030-59003-1_27

Traffic Flow Prediction Through the Fusion of Spatial-Temporal Data and Points of Interest

Wanzhi Xiao, Li Kuang[(✉)], and Ying An[(✉)]

School of Computer Science and Engineering, Central South University, Changsha, China
{wanzhixiao,kuangli,anying}@csu.edu.cn

Abstract. Urban traffic flow prediction plays an important role in alleviating traffic congestion and helping manage public safety. However, the existing solutions to traffic flow prediction still fail to fuse various heterogeneous information effectively, such as historical traffic flow and the semantics of locations, and also fail to capture the traffic flow correlation in the spatial dimension at different levels. Therefore, in this paper, we propose a new traffic flow prediction method to predict the inflow and outflow of each region in the city. Specifically, we first fuse the historical traffic flow, POI, and timestamps data crosswise through a series of gating layers to learn the spatial-temporal interactions. Second, we propose a hierarchical adaptive graph convolution network, which generates multiple learnable adjacency matrices at different network layers to capture the flow correlations in the spatial dimension at different levels. Finally, we take the prediction of inflow, outflow, and overall flow as three tasks and design multi-task learning to improve the prediction performance. Experiments on a real traffic dataset show that our method outperforms the state-of-the-art solutions.

Keywords: Traffic flow prediction · Spatial-temporal data · Graph convolution

1 Introduction

As an essential part of intelligent transportation systems, traffic flow prediction aims to predict the future traffic flow (inflow and outflow) of each region in the city based on historical observation data. Timely and accurate traffic flow prediction can provide people with suggestions of travel route planning and help management departments better allocate road resources and deploy police, thereby reducing traffic congestion and maximizing traffic capacity.

At the same time, traffic flow prediction is also very challenging. Researchers mainly focus on how to model the spatial and temporal correlation to make more accurate predictions. The development of mainstream can be divided into two phases: 1) The researchers in the early time treat the whole city's traffic flow as a pixel map that changes over time [1] and use convolutional neural networks (CNN) to predict city-wide crowd flow. But CNN is limited to model Euclidean spatial relationship, and it is difficult to capture long-distance spatial correlation. 2) Recent studies model non-Euclidean spatial correlations on urban road networks [2, 3], combining graph neural networks (GNN)

© Springer Nature Switzerland AG 2021
C. Strauss et al. (Eds.): DEXA 2021, LNCS 12923, pp. 314–327, 2021.
https://doi.org/10.1007/978-3-030-86472-9_29

and recurrent neural networks [4] to make traffic flow prediction. Although spatial and temporal correlation has been captured in previous studies, three important challenges have not been well solved:

1. First, traffic flow prediction is affected by multiple factors, such as historical traffic flow, time, and points of interest (POI) [1], and the existing methods can not learn the spatial-temporal interactions well. Time-related data are continuous variables, while POI data are static, discrete variables. It is difficult to achieve ideal prediction performance by simply concatenating [5, 6] these data at each time step.
2. Second, most current approaches can not effectively capture spatial correlations at different levels. For example, traffic congestion may increase the local impact, while regions with commuting activities (e.g., residential region to an office region) increase the effects of global-level traffic transition. Most GNN-based method construct a fixed adjacent matrix based on distance [7, 8] or historical flow similarity [9] between regions. However, a fixed adjacent matrix can not reflect the real relationship between regions because of the dynamically changing traffic transition, and also fails to capture spatial correlations at different levels.
3. In addition, most previous studies predict inflow and outflow in a specific region as one task [1, 4], resulting in a decrease in prediction performance. Inflow and outflow are two traffic series in time dimensions, which are different in a specific period. For example, the office area has a large inflow and a small outflow in the morning rush hour, but the situation may be the opposite in the residential area. Predicting these two types flow as one task by using convolution neural network [5] or recurrent neural network [10], ignoring the difference between traffic series.

To address the above three challenges, we propose a new traffic flow prediction framework to predict the traffic flow in the city. First, we designed a gated fusion network to fuse traffic flow, time, and POI to learn the spatial-temporal interactions. A gated fusion network consists of a series of gating layers, which can select relevant features and suppress unnecessary input components. Then, we proposed a hierarchical adaptive graph convolution network in the spatial dimension by learning different adjacency matrices at different network layers to better capture multi-level spatial correlation. After learning the high-level spatial-temporal features, we treat inflow, outflow, and the overall flow(both inflow and outflow) predictions as three tasks and use multi-task learning to improve the prediction performance. In general, the contributions of this paper are as follows:

1. We proposed an effective data fusion method and designed a gated fusion network to fuse spatial-temporal data to learn their spatial-temporal interactions.
2. We designed a hierarchical adaptive graph convolution network to capture the multilevel spatial correlation by learning different adjacency matrices at different network layers.
3. We performed multi-task learning in the prediction stage. Inflow, outflow, and overall flow are predicted as three tasks to further improve prediction performance.

4. Experiments on a public dataset prove the result of our method exceeds all baselines. We have released the code of our method, which can be available at https://github. com/css518/HSTGNN.

The organization of this paper is as follows: Sect. 2 introduces the related work. Section 3 gives the relevant definitions and formalization of the problem. Section 4 describes the details of our proposed method. Section 5 shows the experiments. Section 6 summarizes the whole paper.

2   Related Work

2.1   Convolution Neural Network-Based Traffic Prediction Method

In recent years, deep learning has become the mainstream method to solve spatial-temporal prediction problems. Convolution neural networks have been widely used in the traffic prediction field [1, 11]. Zhang et al. [1] divided the whole city into a grid map and proposed a spatial-temporal residual network for grid-based urban crowd flow prediction. Kuang et al. [12], Guo et al. [13] employed 3D convolutional neural networks to predict traffic flow to further improve the prediction performance. However, due to the invariance of convolution kernel size, it is difficult to capture a different range of spatial dependence for different inputs. Besides, the CNN-based method can only capture Euclidean spatial correlation on the grid data, which may not be in line with the real space division, and is also limited in multi-step prediction.

2.2   Graph Neural Network-Based Traffic Prediction Method

To find traffic prediction methods in non-Euclidean space, researchers have begun to apply graph neural network (GNN) to traffic prediction in recent years. DCRNN [7], STGCN [2] builds a graph structure based on the distance of road sensors. But a fixed adjacent matrix can not reflect real spatial dependence between nodes. For this reason, STMGCN [3] use multiple data-driven adjacent matrices to capture spatial correlations from different perspectives. To capture the dynamics change relationships between nodes, GMAN [8] leverage attention mechanism to adjust the weight of edges between nodes. However, these methods still require fixed adjacent matrix. GraphWaveNet [5], AGCRN [14] noticed that fixed graph structure could not reflect the true spatial dependence, so they use a learnable adjacency matrix to capture dynamic spatial correlations between nodes. However, using one adjacent matrix as the input of graph convolution can hardly capture hierarchical spatial correlation in different levels. In the fusion of multi-source spatial-temporal data, previous studies [5, 15] directly concatenate traffic data and timestamp data for speed prediction. These fusion methods using concatenation can not effectively capture the interactions of multi-source spatial-temporal data. several works [16, 17] use data profiling techniques to fuse spatial-temporal data, but they still not consider data heterogeneity. Different from these works, we learn spatial-temporal interactions by a series of gating layers and learn multi-level spatial correlation by learning different matrices matrix at different network layers.

3   Preliminaries

In this section, we formalize the problem, and our goal is to predict the future multi-step traffic flow of each region in the city. For clarity, first, define the symbol.

Definition 1. Historical traffic sequence $X = \{X_1, X_2, ..., X_T\} \in R^{T \times N \times C}$, where R is the vector dimension, T is the number of time slice and C denotes the number of features, $X_T \in R^{N \times C}$ represents the traffic flow of N regions at time T.

Definition 2. Traffic graph $G = (V, A, \Gamma, \xi)$, $V = \{v_1, v_2, ..., v_N\}$, is a set of nodes, $|V| = N$, is a node on the graph, representing a region of the city, $A = \{a_{i,j}\} \in R^{N \times N}$ is the adjacency matrix and $a_{i,j}$ represents the relation between node i and node j. $\Gamma = \{t_1, t_2, ..., t_T\} \in R^{T \times C}$ denotes time embedding, t_T is one-hot embedding encoded by time of day, day of week. $\xi = \{\varepsilon_1, \varepsilon_2, ..., \varepsilon_N\} \in R^{N \times C}$, indicating geospatial information of N nodes, $\varepsilon_N \in R^{1 \times N}$ is POI information on node N.

Problem Definition. Given previous P steps traffic sequence $[X_{t-P+1}, X_{t-P+2}, \ldots X_t]$, traffic graph G, learn a function f to predict traffic flow in next T steps:

$$[X_{t+1}, X_{t+2}, \cdots, X_{t+T}] = f\left([X_{t-p}, X_{t-p+1}, \cdots, X_t], G\right)$$

4   Method

Figure 1 shows the framework of our method, which mainly includes three steps: data preparation, model learning, and model prediction. First, three gated fusion networks are used to cross-fuse traffic flow X, timestamps Γ, and POI ξ to capture their interactions. Then, the first branch of the model, hierarchical adaptive graph convolution network, is used to capture the spatial feature O_s The second branch uses gated one-dimensional convolution to capture the time feature O_t, the third branch uses two-layer feedforward neural network to capture the mixed spatial-temporal feature O_f, and then performs summation fusion of O_s, O_t and O_f. Finally, multi-task learning is used to map the output.

4.1   Gated Fusion Network

How to effectively fuse spatial-temporal data to better learn their interactions? From the perspective of data, traffic flow, as the main feature, is a continuous variable, geospatial information is a discrete variable, and timestamp is a one-hot encoding variable. These three types of data have different internal distributions, and it is difficult to control the degree of non-linearity between the input variables and the predicted target. Therefore, we designed a gated fusion network (Fig. 2) to fuse the heterogeneous data. Specifically, we take traffic flow as the main feature, POI, and timestamps as the auxiliary feature. To capture the spatial-temporal interaction, we use three gated fusion networks to cross fuse traffic flow and timestamps, traffic flow and POI, and the three features. Formally, the gated fusion network is defined as follows:

$$GFN(a, c) = LayerNorm(a + GLU(h_1)) \tag{1}$$

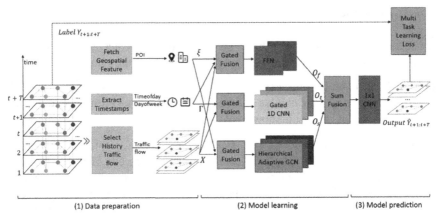

Fig. 1. Framework of the proposed method.

$$h_1 = W_1 h_2 + b_1 \tag{2}$$

$$h_2 = ELU(W_2[a, c] + b_2) \tag{3}$$

where *GFN* is gated fusion network, a is the main feature, c is the auxiliary feature, W and b are parameters. First, a, c are concatenated, and then high-level feature h_2 is learned by the fully connected layer. The Exponential Linear Unit activation function (ELU) is employed to further increase the nonlinearity. After ELU, The Gated Linear Unit (GLU) [18] is used to suppress any unnecessary components in a given input, and selectively retain and forget some features:

$$GLU(x) = \sigma(W * x + b) \odot (V * x + b) \tag{4}$$

where $\sigma(\cdot)$ is the sigmoid activation function, which can determine how much information flows to the next layer. $*$ represents the convolution operation, W, V and b are trainable parameters, $\odot$ denotes the element-wise Hadamard product. Residual connection and layer normalization is added after GLU, which is used to make the training process more stable.

4.2 Hierarchical Adaptive Graph Convolution

In capturing the spatial correlation of traffic flow, GCN [19] is widely used in traffic prediction. GCN can capture non-European spatial dependence. GCN updates its representation by aggregating the feature of neighboring nodes, which is formalized as:

$$Z^{(l+1)} = \sigma\left(AZ^{(l)}W^{(l)} + b^{(l)}\right) \tag{5}$$

where $Z^{(l)} \in R^{N \times C}$ is the node feature of the layer l, σ is the sigmoid activation function, $W^{(l)}$ and $b^{(l)}$ are parameters in layer l, $A \in R^{N \times N}$ represents the adjacency matrix.

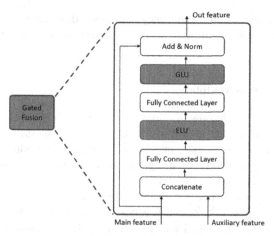

Fig. 2. Gated fusion network

In order to capture the dynamics and hierarchical spatial correlation of traffic flow between regions, we designed a hierarchical adaptive graph convolution. We use two trainable parameters E_1, E_2 to construct adjacent matrix. E_1, $E_2 \in R^{N \times C}$ represent the embedding of the source node and the target node respectively, and C is the embedded dimension. Then we use fully connected layer to map E_1 and E_2, so that there are different adjacency matrices in different layers:

$$E_1^{(l+1)} = E_1^{(l)} W^{(l)} + b^{(l)} \tag{6}$$

$$E_2^{(l+1)} = E_2^{(l)} W^{(l)} + b^{(l)} \tag{7}$$

where $E_1^{(l)}$, $E_2^{(l)}$ is the source node embedding and target node embedding of the layer l. $W^{(l)}$ and $b^{(l)}$ are mapping parameters. Then, the adjacency matrix can be defined as:

$$A^{(l)} = SoftMax\left(ReLU\left(E_1^{(l)} E_2^{(l)T}\right)\right) \tag{8}$$

where $A^{(l)} \in R^{N \times N}$ is the adjacency matrix of layer l. Then, the graph convolution process is implemented on the constructed adjacency matrix. We use diffusion graph convolution [7] to implement the graph convolution, which using a limited number of k steps to simulate the diffusion process of the graph signal. Given the adaptive adjacency matrix $A^{(l)}$ of the layer l, and perform a limited k steps diffusion process on it, which can be formalized as:

$$Z^{(l+1)} = \sum_{i=1}^{K} (A^{(l)})^i Z^{(l)} W_i^{(l)} \tag{9}$$

where $(A^{(l)})^i$ is the power series of $A^{(l)}$, representing the probability of diffusion on the graph, $Z^{(l)}$, and $Z^{(l+1)}$ represent the node feature of layer l and layer $l + 1$ respectively, and $W_i^{(l)}$ is the parameter of graph convolution. We capture spatial correlation from different levels by using different adjacency matrices at different network layers, which can also prevent the problem of node feature over smoothing due to the deepening of network layers.

4.3 Gated Convolution Module

In the time dimension, traffic flow is a time series. The traffic flow at a certain moment in a location is similar to the traffic flow at that moment in the previous few moments and yesterday, which is trendy and periodic [1]. Therefore, we add timestamps as the order information of time series and use gated one-dimensional convolution to capture the temporal correlation in the traffic series.

Specifically, we perform one-hot encoding of time of day and day of week to get the time embedding $\Gamma \in R^{N \times C}$, and use it as the auxiliary feature, the traffic flow $X \in R^{T \times N \times C}$ as the main feature. Then, input them to the gated fusion network Eq. (1), and outputs the fused feature $X \in R^{T \times N \times C_1}$:

$$\Gamma = OnehotEncoding(timeofday, dayofweek) \tag{10}$$

$$X = GRN(X, \Gamma) \tag{11}$$

Then X is input into gated convolution module (Fig. 3), we use a $1 \times k$ convolution kernel with a dilation factor to convolve in the time dimension of X. The gated convolution module takes X as input and has two branches, which can be denoted as:

$$X_t^{(l+1)} = LayerNorm(\sigma(\Theta_1 * X_t^{(l)} + b_1)) \odot LayerNorm(g(\Theta_2 * X_t^{(l)} + b_2)) \tag{12}$$

where $X_t^{(l+1)} \in R^{(T-k+1) \times N \times C}$ is the output of the gated convolution module, Θ_1 and Θ_2 are convolution kernel, b_1 and b_2 are bias, $\odot$ denotes the dot product. g($\cdot$) is the *tanh* activation function of the output part, and $\sigma(\cdot)$ is the sigmoid function, which controls how much information can flow to the next layer. To make the training more stable, we add layer normalization after the activation function.

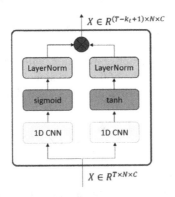

Fig. 3. Gated convolution module.

Then, we sum the final output O_t of the gated convolution module, the spatial feature O_s output by the hierarchical graph convolution, and the feature mixed spatial-temporal feature O_f output by the two-layer feedforward neural network to obtain the high-level spatial-temporal feature Y:

$$Y = O_s + O_t + O_f \tag{13}$$

4.4 Multi-task Learning Loss

After obtaining high-level spatial-temporal features, we design multi-task learning on the prediction stage. We define three types of flow prediction in each region, namely inflow, outflow, and overall flow (both inflow and outflow), as three prediction tasks. Multi-task learning can strengthen the learning ability between related tasks. We use three convolution networks to map high-level features $Y \in R^{T \times N \times C}$ to three types of flow, and our goal is to predict the overall flow.

$$\widehat{Y}_{all} = \Theta_1 * Y + b_1 \tag{14}$$

$$\widehat{Y}_{in} = \Theta_2 * Y + b_2 \tag{15}$$

$$\widehat{Y}_{out} = \Theta_3 * Y + b_3 \tag{16}$$

where Θ and b are parameters, $\widehat{Y}_{all} \in R^{T \times C \times 2}$, $\widehat{Y}_{in}, \widehat{Y}_{out} \in R^{T \times C \times 1}$, are the predicted overall flow, inflow, outflow respectively, after defining three related tasks, we use mean absolute error to calculate the loss of each prediction flow and then jointly optimize them.

$$L = \sum_{i=1}^{N} \sum_{j=1}^{T} \lambda_{all} \left| y_{i,j}^{all} - \widehat{y}_{i,j}^{all} \right| + \lambda_{in} \left| y_{i,j}^{in} - \widehat{y}_{i,j}^{in} \right| + \lambda_{out} \left| y_{i,j}^{out} - \widehat{y}_{i,j}^{out} \right| \tag{17}$$

where N is the number of nodes, T is the number of time steps to be predicted, $\lambda_{all}, \lambda_{in}, \lambda_{out}$ are three hyper-parameters used to adjust the importance of tasks.

5 Experiments

In this section, we conducted experiments on a real traffic dataset. We aim to answer the following research questions:

RQ1: How does our proposed method compare to the state-of-the-art traffic prediction model?

RQ2: How do the components designed in the model contribute to the final prediction performance?

RQ3: How does the model parameter setting affect the prediction performance?

RQ4: How do different data fusion methods affect the experimental results?

5.1 Experimental Settings

Task Descriptions
We use a public dataset named TaxiBJ [4] for the traffic flow prediction task. The specific information of the dataset is as Table 1. This dataset divides Beijing city into 1024 grids, representing 1024 regions of the city, and the area of each grid is 1 km^2. Each grid serves as a node in the graph. It counts the hourly in and out traffic flow of each grid. The POI information corresponding to the grid is used as geospatial data. The traffic flow of the past 12 h is used to predict in and out traffic flow of each region in the next 3 h. The training set, the validation set, and the test set are divided at a ratio of 8:1:1, and our settings are the same as those in [4]. We use mean absolute error (MAE) and root mean square error (RMSE) as evaluation metrics.

Table 1. Dataset description

Dataset	TaxiBJ
Prediction target	Inflow and outflow
Time interval	2015/2/1–2015/6/2
Time slot granularity	1 h
Timestamps	3600
Nodes	1024
Geospatial features	POI

Parameter Settings
We use the PyTorch framework to implement our model and conducted experiments on a computer equipped with an Intel(R) Xeon(R) E5 CPU and two NVIDIA Titan XP GPU cards. The detailed parameters of the model are as follows: (1) The dimension of the hidden unit is set to 32, and the model has four layers of network (2) We set the learning rate of the model to 0.001. (3) The weights of multi-task learning loss λ_{all}, λ_{in}, λ_{out} are set to 0.5, 0.25, 0.25 respectively (4) The number of training epochs is 200, and batch size is 32. Besides, we use the Adam optimizer and perform gradient clipping to train the model.

5.2 Performance Results (RQ1)

Table 2 shows the comparison between our model and nine baselines. Each model has been trained five times, and the results are in the form of mean ± standard deviation. We compared MAE and RMSE, including predicting the next 1-h, next 2-h, next 3-h, and overall performance. Our predict performance exceeds all baselines. The overall MAE and RMSE dropped by 4.2% and 6.9%, respectively.

From Table 2, we can observe that the two statistical models of HA and ARIMA perform the worst. GBRT, as a commonly used machine learning model, has relatively good accuracy in regression prediction. ST-ResNet and STDN, two convolution-based deep learning methods, are not effective in predicting long-term traffic flow, because these two methods do not capture long-term temporal dependence. DCRNN and Graph-WaveNet are two methods based on graph neural networks, which have a better result than convolution methods. but these two model fails to capture the spatial dependence on the multi-levels. ST-MetaNet and ST-MetaNet + use meta learning to generate model parameters to learn spatial-temporal correlation, the result is further improved. But they still do not consider the hierarchical relationship of traffic flow and did not filter the noise data of meta-information. Our method considers the hierarchical spatial dependence, and effectively fused spatial-temporal data, retains important features. In addition, our method employs multi-task learning and achieves a relatively stable performance.

Table 2. Performance comparison of all methods on TaxiBJ dataset.

Models	Overall		1 h		2 h		3 h	
	MAE	RMSE	MAE	RMSE	MAE	RMSE	MAE	RMSE
HA	26.2 ± 0.00	56.5 ± 0.00	26.2 ± 0.00	56.5 ± 0.00	26.2 ± 0.00	56.5 ± 0.00	26.2 ± 0.00	56.5 ± 0.00
ARIMA	40.0 ± 0.00	86.8 ± 0.00	27.1 ± 0.00	58.3 ± 0.00	41.2 ± 0.00	77.0 ± 0.00	51.8 ± 0.00	108.0 ± 0.00
GBRT	28.8 ± 0.04	60.9 ± 0.15	22.3 ± 0.01	47.7 ± 0.01	29.9 ± 0.06	62.7 ± 0.14	34.3 ± 0.11	70.5 ± 0.23
ST-ResNet [1]	18.7 ± 0.53	36.1 ± 0.59	16.8 ± 0.50	31.9 ± 0.69	18.9 ± 0.57	36.4 ± 0.71	20.3 ± 0.52	39.5 ± 0.46
STDN [11]	23.4 ± 2.49	43.2 ± 2.95	21.5 ± 3.24	37.4 ± 3.34	24.7 ± 3.43	44.4 ± 3.71	24.1 ± 1.27	46.9 ± 3.11
DCRNN [7]	17.8 ± 0.13	36.1 ± 0.15	15.8 ± 0.05	32.3 ± 0.08	18.2 ± 0.15	36.9 ± 0.17	19.4 ± 0.19	38.9 ± 0.24
GraphWaveNet [5]	17.1 ± 0.06	35.0 ± 0.14	15.2 ± 0.08	31.1 ± 0.27	17.4 ± 0.09	35.7 ± 0.30	18.6 ± 0.11	37.8 ± 0.25
ST-MetaNet [4]	16.7 ± 0.13	33.6 ± 0.15	14.8 ± 0.05	29.6 ± 0.08	17.1 ± 0.15	34.3 ± 0.17	18.2 ± 0.19	36.5 ± 0.24
ST-MetaNet + [10]	16.5 ± 0.16	33.2 ± 0.35	14.7 ± 0.18	29.7 ± 0.40	16.9 ± 0.16	33.9 ± 0.35	17.8 ± 0.17	35.8 ± 0.53
HSTGNN (ours)	15.8 ± 0.15	30.9 ± 0.13	14.6 ± 0.05	28.7 ± 0.15	16.0 ± 0.12	31.4 ± 0.25	16.9 ± 0.11	32.8 ± 0.23

5.3 Ablation Experiments (RQ2)

In this section, we conduct ablation experiments to verify the effectiveness of each component of our model. For comparison, we consider the following model variants.

1. gated fusion: Remove the gated fusion network and use concatenate fusion instead.
2. adaptive gcn: Remove the hierarchical adaptive graph convolution, and use a single adaptive adjacent matrix for graph convolution instead.
3. multi-task learning: Remove two sub-task of predict inflow, outflow, only keep the main task to predict the overall flow(both inflow and outflow).
4. with all: Keep all components.

Table 3. Ablation experiment

Index	Removed componet	MAE			RMSE		
		1 h	2 h	3 h	1 h	2 h	3 h
1	Gated fusion	14.68	16.24	17.25	29.00	32.36	34.48
2	Adaptive gcn	14.87	16.52	17.30	29.27	32.85	34.48
3	Multi-task learning	15.20	16.06	16.82	29.48	31.61	33.00
4	With-all	14.65	16.01	16.90	28.75	31.41	32.85

From Table 3 We can observe that when all the components are added, the prediction result of the model is better, and the performance is more stable.

Effectiveness of gated fusion network: the prediction loss increases after remove gated fusion network, especially in the medium and long-term predictions. This shows the gating mechanism is improved non-linearity and suppresses unnecessary components in the input and preserves important features.

Effectiveness of hierarchical adaptive graph convolution: This component maintains different adjacency matrices in different layers, and adjusts the weight of the edges between nodes with the training process, adapts to the traffic flow that changes with time, and can capture the spatial dynamic dependencies of nodes at different layers in the traffic graph.

Effectiveness of multi-task learning: multi-task learning serves as implicit data enhancement, enhancing the collaborative learning of related tasks and improving overall performance.

5.4 Evaluation on Framework Settings (RQ3)

In this section, we verify the influence of hyper-parameters on model prediction. including the number of hidden units (hidden size) of the network and the number of network layers. Except for the changed parameters, other parameters are consistent with the experimental settings of Sect. 5.1.

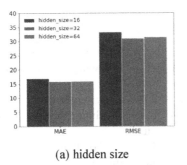
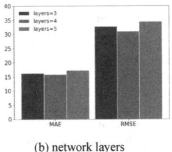

(a) hidden size   (b) network layers

Fig. 4. The influence of different hyper-parameters on overall prediction results

It can be seen from Fig. 4(a) that our model is generally insensitive to hidden size and performs best when the hidden size is 32. At the same time, the performance for hidden size of 16 is slightly worse, because the hidden units are not enough, which leads to insufficient data fitting. As shown in Fig. 4(b), when the number of network layers is 5, too deep network layers lead to over-fitting, which reduces the prediction result, and when the number of network layers is 3, its performance is not as good as the setting of net-work layers of 4, this is because shallower network layers can lead to insufficient fitting.

5.5 Evaluation on Spatial-Temporal Fusion Methods (RQ4)

To further verify the effectiveness of the gated fusion network, we conducted a comparative experiment on the TaxiBJ dataset.

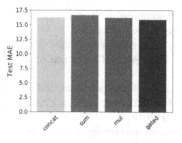
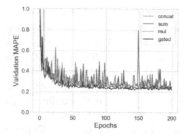

(a) gated fusion vs other fusion   (b) the curves of validation MAPE

Fig. 5. Performance of different fusion methods.

We consider four data fusion methods: (1) concat: concatenate traffic flow, time embedding, and poi, respectively. (2) sum: use sum fusion to fuse spatial-temporal data. (3) mul: fuse spatial-temporal data by dot multiplication. (4) gated: gated fusion network. Figure 5(a) shows the overall MAE of four spatial-temporal data fusion methods on the test set. Compared to other data fusion methods, gated fusion network increases the ability of feature selection by adding GLU gates, which can suppress some unnecessary features and retain important features. MAPE $\in [0, +8]$ is selected to observe the convergence. Figure 5(b) shows the convergence of these fusion methods on the validation set. In general, the gating fusion method is superior to other methods in stability and convergence, which benefits from the activation ability of the gating mechanism.

6 Conclusion and Future Work

This paper proposed a new deep learning-based method for traffic flow prediction. We designed an effective gated fusion network to fuse spatial-temporal data. Gated fusion network can suppress some unnecessary features and learn spatial-temporal interactions. To capture the hierarchical dependence in space, we designed a hierarchical adaptive graph convolution using function mapping to learn different adjacency matrices on different layers, which conform to the spatial hierarchical and dynamic correlations of traffic flow. Furthermore, we design three traffic flow prediction tasks: overall flow, inflow, and outflow prediction, and then use multi-task learning for data enhancement. Experiments on a public dataset show that our method exceeds all baselines.

Acknowledgements. Supported by the National Natural Science Foundation of China (No. 61772560), Natural Science Foundation of Hunan Province (No. 2019JJ40388).

References

1. Zhang, J., Zheng, Y., Qi, D.: Deep spatio-temporal residual networks for citywide crowd flows prediction. In: Proceedings of the AAAI Conference on Artificial Intelligence, vol. 31, no. 1 (2017)
2. Yu, B., Yin, H., Zhu, Z.: Spatio-temporal graph convolutional networks: a deep learning framework for traffic forecasting. arXiv preprint arXiv:1709.04875 (2017)
3. Geng, X., Li, Y., Wang, L., et al.: Spatiotemporal multi-graph convolution network for ride-hailing demand forecasting. In: Proceedings of the AAAI Conference on Artificial Intelligence, vol. 33, no. 01, pp. 3656–3663 (2019)
4. Pan, Z., Liang, Y., Wang, W., et al.: Urban traffic prediction from spatio-temporal data using deep meta learning. In: Proceedings of the 25th ACM SIGKDD International Conference on Knowledge Discovery & Data Mining, pp. 1720–1730 (2019)
5. Wu, Z., Pan, S., Long, G., et al:. Graph WaveNet for deep spatial-temporal graph modeling. arXiv preprint arXiv:1906.00121 (2019)
6. Guo, S., Lin, Y., Feng, N., et al.: Attention based spatial-temporal graph convolutional networks for traffic flow forecasting. In: Proceedings of the AAAI Conference on Artificial Intelligence, vol. 33, no. 01, pp. 922–929 (2019)
7. Li, Y., Yu, R., Shahabi, C., et al.: Diffusion convolutional recurrent neural network: data-driven traffic forecasting. arXiv preprint arXiv:1707.01926 (2017)
8. Zheng, C., Fan, X., Wang, C., et al.: GMAN: a graph multi-attention network for traffic prediction. In: Proceedings of the AAAI Conference on Artificial Intelligence, vol. 34, nol. 01, pp. 1234–1241 (2020)
9. Bai, L., Yao, L., Kanhere, S., et al.: STG2Seq: spatial-temporal graph to sequence model for multi-step passenger demand forecasting. arXiv preprint arXiv:1905.10069 (2019)
10. Pan, Z., Zhang, W., Liang, Y., et al.: Spatio-temporal meta learning for urban traffic prediction. IEEE Trans. Knowl. Data Eng. (2020)
11. Yao, H., Tang, X., Wei, H., et al.: Revisiting spatial-temporal similarity: a deep learning framework for traffic prediction. In: Proceedings of the AAAI Conference on Artificial Intelligence, vol. 33, no. 01, pp. 5668–5675 (2019)
12. Kuang, L., Yan, X., Tan, X., et al.: Predicting taxi demand based on 3D convolutional neural network and multi-task learning. Remote Sens. **11**(11), 1265 (2019)
13. Guo, S., Lin, Y., Li, S., et al.: Deep spatial–temporal 3D convolutional neural networks for traffic data forecasting. IEEE Trans. Intell. Transp. Syst. **20**(10), 3913–3926 (2019)
14. Bai, L., Yao, L., Li, C., et al.: Adaptive graph convolutional recurrent network for traffic forecasting. arXiv preprint arXiv:2007.02842 (2020)
15. Wang, Y., Yin, H., Chen, H., et al.: Origin-destination matrix prediction via graph convolution: a new perspective of passenger demand modeling. In: Proceedings of the 25th ACM SIGKDD International Conference on Knowledge Discovery & Data Mining, pp. 1227–1235 (2019)

16. Mohamed Mostafa, M.A.R., Vucetic, M., Stojkovic, N., et al.: Fuzzy functional dependencies as a method of choice for fusion of AIS and OTHR data. Sensors **19**(23), 5166 (2019)
17. Caruccio, L., Cirillo, S.: Incremental discovery of imprecise functional dependencies. J. Data Inf. Qual. (JDIQ) **12**(4), 1–25 (2020)
18. Dauphin, Y.N., Fan, A., Auli, M., et al.: Language modeling with gated convolutional networks. In: International Conference on Machine Learning, pp. 933–941. PMLR (2017)
19. Kipf, T.N., Welling, M.: Semi-supervised classification with graph convolutional networks. arXiv preprint arXiv:1609.02907 (2016)

Predicting Student Performance in Experiential Education

Lejia Lin[1], Leonard Wee Liat Tan[2], Nicole Hui Lin Kan[1], Ooi Kiang Tan[3], Chun Chau Sze[2], and Wilson Wen Bin Goh[2(✉)]

[1] School of Computer Science and Engineering, Nanyang Technological University, Singapore, Singapore
`LINL0031@e.ntu.edu.sg`
[2] School of Biological Sciences, Nanyang Technological University, Singapore, Singapore
`wilsongoh@ntu.edu.sg`
[3] College of Engineering, Nanyang Technological University, Singapore, Singapore

Abstract. Experiential learning is a key development area of artificial intelligence in education (AIEd). It aims to provide learners with intuitive environments for autonomous knowledge formation and discovery through interactive experiences. However, experiential learning in AIEd faces two main challenges. Firstly, measuring learning performances in unstructured and informal educational settings is difficult. Secondly, providing frequent or timely feedback on student performance is inefficient. To address these issues, this paper explores using natural language processing (NLP) and the tool for the automatic analysis of cohesion (TAACO) features as indicators of student performance in an experiential learning course. Both NLP and TAACO features were tested on a baseline CART decision tree (DT) machine learning (ML) model with and without a grade population distribution mask to predict student final scores at the end of the course. Our results show that (1), the use of a distribution specific Gaussian mask significantly increases prediction accuracy of the CART DT. (2), NLP and TAACO features provide high information value for ML prediction tasks. (3), the CART DT is able to accurately classify learner grade scores against human assessments.

Keywords: Natural language processing · Statistics · Scientific method · Decision trees · Education · Experiential learning · Data mining

1 Introduction

Background and Motivation. Experiential learning (EL) [10] is a method of teaching instruction which differs from traditional classroom pedagogical contexts that are highly structured and organized [10]. EL relies on interactions with the environment as instructive feedback components to build learner

© Springer Nature Switzerland AG 2021
C. Strauss et al. (Eds.): DEXA 2021, LNCS 12923, pp. 328–334, 2021.
https://doi.org/10.1007/978-3-030-86472-9_30

knowledge [2]. The key principle of EL implementation is to build an educational construct around student centric objectives [6] in order to create an efficient pedagogical framework [10]. It leverages on a learner's transformation of experience from reflective thought processes through interactions with pedagogical environments. EL processes goes through four essential cognitive mental stages for a learner's knowledge to develop. They are: The concrete experiences, the reflective observations, the abstract conceptualizations and the active experimentations. The authors in [13] show, through real-life implementations that the four mental stages of EL [9], are important to build an augmentation of contextual information as "new knowledge" [10]. EL [10], is generally not favoured because of difficulties in measuring learning progress and outcomes from its highly unstructured, informal approaches. However, recent methods like [7,8] and [12] offer solutions (e.g. rules-based, label and instance balancing, etc.) to tackle performance evaluation biases strategically. The motivation of this study is driven by the following key research questions. Firstly, how do we assume adequate data distributions to more accurately model predictions of open problems (like learner performance) in EL? Secondly, how do different features like NLP and TAACO provide value to learning semantic information for inference tasks in EL? Thirdly, how do ML models perform in predicting learning outcomes? In this paper, we address these key concerns with the use of text cohesion and readability features as critical measures for learning performance in predicting students' grades.

Contribution and Paper Outline. We address these questions with a focused analysis using an experiential learning programme as a case study. Our model addresses the mental challenge of learning from experience by providing the physical construct through Nanyang Technological University's campus wide experiential learning programme, the "deeper experiential engagement project" (DEEP)[1]. As data used in our study, DEEP journals were collected as reflections of progressive affective states of students, towards experiential learning activities over a ten-week period. Motivated strategies for learning questionnaire (MLSQ) architectures [5] are used to grade, label and rank the essays based on learner strategies and motivation [5]. We first extracted text cohesion features using python spaCy's[2] and TAACO[3] toolkit packages. Next, we introduce a baseline decision tree (DT) algorithm framework with NLP (see footnote 2) and TAACO (see footnote 3) cohesion features to tackle the problem of classifying student performance, by estimating feature classifiers on a Gaussian mask which is skewed to the generalized grade distribution of the experiential learning course. Our model accepts as inputs, key relational feature states f_i of i states in student journal essays a_j of j essays from a 10 week learning reflections report blog, to determine the likelihood of classification τ_{ij} over their learning experience as indicative grade scores of their performance. To eliminate class imbalances when training the CART DT, we run the MLSQ graded essays over

---

[1] https://gohwils.github.io/biodatascience/deep_programme.html.
[2] https://spacy.io/.
[3] https://www.linguisticanalysistools.org/taaco.html.

a synthetic minority over-sampling technique (SMOTE) [1], to pre-process data prior to the DT model training process. Our scientific contributions are three-fold: Firstly, we design a new Gaussian mask to skew feature classifications of data based on experiential course grade distributions for the DT. Secondly, our method optimizes training of the DT for F_ϵ, which is used to predict student's learning performance from classified grade scores. Thirdly, we demonstrate the efficacy of using NLP and TAACO features with ML models to accurately predict classifications of student results in our experimentation. The remaining part of the paper is organized as follows: We begin with a brief introduction to key concepts, theories and preliminaries of our proposed model in Sect. 2. Then, we summarize the methods and models we have developed for profiling and predicting student grade outcomes from experiential learning courses in Sect. 3. Finally, we present our experiment implementations, results and discussions in Sect. 4.

2   Preliminaries, Key Concepts and Theories

Preliminaries. We formulate our model problem within the following statement: Given a set of features F_ϵ of ϵ learning assessment samples where $\epsilon \in 1, ..., \eta$, our objective is to predict the likelihood of a student having acquired an assessed grade Θ_ϵ as an indicative measure of their effectiveness in knowledge acquisition from EL activities. An important assumption that this study makes is that the distribution of grades Φ is Gaussian, with a random mean μ over the sample size η, of the student essay dataset χ in ϵ.

Text Cohesion Features. We define cohesion features f_ϵ as important, highly correlated affective semantics expressed in student reflective journal essays. They measure linguistic connectivity across textual phrases, sentences and paragraphs as logical relationships between cognitive learning processes. f_ϵ are indicative measures of student mental states and are used as predictors to their final experiential course grading classifications. In our study, we extract cohesion features f_ϵ as detected occurances of word pair lexicons and thematic structure.

Decision Tree Models. Decision tree models are organized decision structures that accept input features as decision classes and splits them based on a set of apriori classifier rules. As a rules based approach, DT models which generate actual DT architectures are trained from an iterative process that leverages on greedy optimizers which seek to maximize the data classification accuracy. A standard DT model can be described mathematically as:

$$g(T) = \gamma\omega(T) - \delta\beta(T) \tag{1}$$

where $g(T)$ is the DT optimization, which depends on the decision accuracy $\omega(T)$ and tree size $\beta(T)$. Futhermore, γ and δ are accuracy and size (depth) weights respectively. Our model addresses the problem of misclassification and accuracy by profiling a Gaussian distribution score mask to establish decision rule likelihoods at bifurcating non-leaf nodes of the DT. A diagram of our model architecture is shown in Fig. 1.

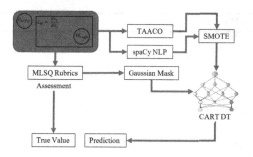

Fig. 1. Overview of the CART DT architecture

3   Model and Methods

Our approach evaluates data from DEEP courses that have been designed from theories of experiential learning [2]. For this study, we used data obtained from BS8101[4] which is concerned with creative project development on food. To begin, we apply an asymmetric Gaussian distribution mask across our available journal grade score classes to approximate the best performing feature/s which will then be used as the best attribute for splitting progressive decision nodes [4,11]. Our skewed probability density Gaussian grade score distribution is given as:

$$\Phi = \frac{p(\Lambda)P(\gamma\Lambda)}{P(\mu)} \tag{2}$$

where γ is the skew factor, μ is the distribution mean and Λ is the assessment grade score. Next, the DT decision branch bifurcation process in our model is determined mathematically as:

$$h : \{L_1^m, L_2^m, ..., L_k^m\} \rightarrow \{Z_1, Z_2, ..., Z_k\} \tag{3}$$

where h is the classifier mapping fit to the training data attribute L_i in m dimensions of data class attributes $z_m \in \{Z_1, Z_2, ..., Z_k\}$. We define W to be the words in the document D of an aggregation of reflective journal essays of data elements $d_m \in D$. Where $d_m = \lim_{m \to k}(|l_m|, z_m)$ and l_i denote the element attributes of the data attribute vector L_i. Algorithm 1 is used to select the best performing decision pathway predictions of student grade classifiers. The Gini Index [13] is given by the method in Algorithm 2.

4   Experiments, Results and Discussion

The experiments were done on two datasets and feature sets using our DT design. Essentially, we run our CART DT design with both NLP and TAACO features. The NLP feature scores were extracted from the essays using the spaCy (see

---

[4] https://dr.ntu.edu.sg/handle/10356/78232?mode=full.

Algorithm 1: CART DT

Result: Decision Tree of Best Fit
initialization;
Training Dataset = D;
Decision Node = H_j;
Decision Depth = j;
Error Function = erf;
Standard Probability Density Function = $\psi(x)$;
Cumulative Density Function = $\Psi(x)$;
Skew Rate = γ ;
Skew-Gaussian Probability Density Function = $\Omega(x)$;
while $j < max - depth$ **do**
  if *Decision Node = H_0 (root node)*;
  then
      Training Data $D_0 = D$;
      Test $(d_m \in z_m)$;
      if $(z_m \in Z_k = 1)$;
      then
          Classifier: $z_m = Z_k$ (leaf node);
          Calculate-Gini();
      else
          Split: $(L)L_i^m$ (left) and $(R)L_i^m$ (right) for each l_i^m;
          For all: $L_i^m = (L)L_i^m \bigcup (R)L_i^m$;
          Such that: $\Omega(L_i^m) = \frac{2}{\sqrt{2\pi}} \exp^{\frac{((R)L_i^m)^2}{2}} \times \frac{1}{2}[1 + erf(\frac{L_i^m}{\sqrt{2}})] = 2\psi(L_i^m)\Psi(\gamma L_i^m)$;
          And $\lim_{m \to k}[\Omega(L_i^m) \to \Omega(z_m)]$;
          Calculate-Gini();
      end
  else
  | repeat;
  end
end

Algorithm 2: Calculate-Gini()

Result: $Gini(D_q)$
Fraction of data elements from $D_q = P_q L_i^m$;
Calculate: $P_{L,q}L_i^m = \frac{(L)L_{i,q}^m}{L_{i,q}^m}$;
And $P_{R,q}L_i^m = 1 - P_{L,q}(L_i^m)$;
Where for class $z_m \in Z_1, Z_2, ..., Z_k$ $P_{L,q}L_i^m \to P_{Z_m,L,q}L_i^m$;
And $P_{R,q}L_i^m \to P_{Z_m,R,q}(L_i^m)$;
For all $P_{Z_m,q}(L_i^m) = P_{Z_m,L,q}(L_i^m) \bigcup P_{Z_m,R,q}(L_i^m)$;
$Gini(D_q) = 1 - \sum_{z_m=Z_1}^{Z_k}(P_{Z_m,q})^2$

---

footnote 2) for python toolkit. While TAACO (see footnote 3) feature scores were extracted from the python add-in module. Details are tabulated in Table 1 and 2 respectively[5].

Results. The tests were run across the baselines and our CART DT algorithm with and without the Gaussian distribution mask Φ. Both NLP and TAACO features were chosen for Kaggle and DEEP datasets after performing regressive correlation analysis. The F1-score is used to measure the accuracy of predictions that is contributed by both precision and recall as important metrics of classifier performance. Finally, during the experimentation, the full datasets obtained from Kaggle and DEEP were partitioned as 80% training and 20% testing data allocations. 40-fold validation [14] was performed over the CART DT across the Mean Absolute Error (MAE) [3] measurement of each run. A summary of our findings are given in Table 3 and 4.

---

[5] https://www.kaggle.com.

Table 1. Statistics of datasets

Dataset	Source	#Essays	Size	Avg. text len
Kaggle[a]	Kaggle Hewlett Foundation	1778	250 MB	250 words
DEEP[b]	NTU's DEEP Programme	37	1.2 MB	5000 words

[a] https://www.kaggle.com/c/asap-aes
[b] https://www.sbs.ntu.edu.sg/Faculty/Sze_chun_chau/Pages/Sze-chun-chau.aspx

Table 2. Baseline models

Baseline	Feature	# Of Features (DEEP)	# Of Features (KAGGLE)
CART DT	NLP Only	10	15
CART DT	TAACO Only	14	23
CART DT	NLP and TAACO	27	10

Table 3. Summary of the CART DT test and train accuracy

Kaggle without Φ					Kaggle with Φ		
NLP	TAACO	Depth	Train accuracy	Test accuracy	Depth	Train accuracy	Test accuracy
Y	N	2	79.5%	82.2%	2	83.2%	78.5%
N	Y	1	93.8%	89.2%	12	100.0%	71.0%
Y	Y	2	80.2%	78.7%	2	81.6%	78.5%
SMOTE'ed Kaggle /wo Φ					SMOTE'ed Kaggle /w Φ		
NLP	TAACO	Depth	Train accuracy	Test accuracy	Depth	Train accuracy	Test accuracy
Y	N	8	93.6%	80.7%	10	95.4%	86.3%
N	Y	8	93.6%	81.2%	18	99.8%	79.6%
Y	Y	12	99.0%	78.9%	8	92.4%	85.1%
DEEP /wo Φ					DEEP /w Φ		
NLP	TAACO	Depth	Train accuracy	Test accuracy	Depth	Train accuracy	Test accuracy
Y	N	9	100.0%	82.4%	14	100.0%	81.25%
N	Y	16	100.0%	76.5%	5	100.0%	81.25%
Y	Y	5	100.0%	82.4%	8	100.0%	93.75%

Table 4. Summary of K-fold cross validated MAE scores

	Grades /wo Φ			Grades /w Φ		
	NLP	TAACO	NLP+TAACO	NLP	TAACO	NLP+TAACO
Kaggle	0.208	0.105	0.208	0.199	0.356	0.195
Kaggle SMOTE'ed	0.193	0.191	0.182	0.135	0.256	0.141
DEEP	0.175	0.279	0.196	0.237	0.113	0.075

Discussion. From the results in Table 3, we can see that: Firstly, there are no obvious trends of TAACO outperforming or underperforming as compared to NLP features. Secondly, when both NLP and TAACO features are used, it generally gives better performance. Thirdly, when TAACO features are added to models which were originally trained in NLP only, an improvement in accuracy is only seen in DEEP models while Kaggle models see a slight drop in performance. Finally, DEEP datasets performs poorly without the Gaussian mask. In conclusion, our results show considerably good accuracies and performance of the CART DT design when both TAACO and NLP features are used together in the classification of student's final grades at the end of the course. There is potential in using machine learning (ML) models for classification and prediction tasks in experiential education and learning. Future work will leverage on results obtained in this study to include various other baseline models such as random forests (RFs), support vector machines (SVMs), Naive Bayes, long short term memory (LSTM), etc. The other direction which we will carry on our research work will also be to verify and test on the generalizability of the ML models for future DEEP journals for the BS8101 and other DEEP experiential learning courses in NTU.

Acknowledgement. CC Sze, WWB Goh and OK Tan acknowledge support from an ACE grant, NTU.

References

1. Chawla, N.V., Bowyer, K.W., Hall, L.O., Kegelmeyer, W.P.: SMOTE: synthetic minority over-sampling technique. J. Artif. Intell. Res. **16**, 321–357 (2002)
2. Bourgault du Coudray, C.: Theory and praxis in experiential education: some insights from gestalt therapy. J. Experiential Educ. **43**(2), 156–170 (2020)
3. De Myttenaere, A., Golden, B., Le Grand, B., Rossi, F.: Mean absolute percentage error for regression models. Neurocomputing **192**, 38–48 (2016)
4. Domingos, P., Hulten, G.: Mining high-speed data streams, pp. 71–80 (2000)
5. Duncan, T.G., McKeachie, W.J.: The making of the motivated strategies for learning questionnaire. Educ. Psychol. **40**(2), 117–128 (2005)
6. Gulz, A., Londos, L., Haake, M.: Preschoolers' understanding of a teachable agent-based game in early mathematics as reflected in their gaze behaviors-an experimental study. Int. J. Artif. Intell. Educ. **30**(1), 1–36 (2020)
7. Jiang, W., Pardos, Z.A.: Towards equity and algorithmic fairness in student grade prediction. arXiv preprint arXiv:2105.06604 (2021)
8. Karthikeyan, R., Satheesbabu, S., Gokulakrishnan, P.: Machine learning based student performance analysis system. Inf. Technol. Ind. **9**(1), 1174–1181 (2021)
9. Kolb, D.: The process of experiential. Adult Continuing Educ. Teach. Learn. Res. **4**, 159 (2003)
10. Kolb, D.A., Boyatzis, R.E., Mainemelis, C., et al.: Experiential learning theory: previous research and new directions. Perspect. Think. Learn. Cogn. Styles **1**(8), 227–247 (2001)
11. Rutkowski, L., Jaworski, M., Pietruczuk, L., Duda, P.: Decision trees for mining data streams based on the gaussian approximation. IEEE Trans. Knowl. Data Eng. **26**(1), 108–119 (2013)
12. Seno-Alday, S., Budde-Sung, A.: Teaching a while measuring b: cultural bias in assessing student performance. J. Int. Educ. Bus. (2021)
13. Sharma, H., Kumar, S.: A survey on decision tree algorithms of classification in data mining. Int. J. Sci. Res. (IJSR) **5**(4), 2094–2097 (2016)
14. Wong, T.T.: Performance evaluation of classification algorithms by k-fold and leave-one-out cross validation. Pattern Recogn. **48**(9), 2839–2846 (2015)

Log-Based Anomaly Detection with Multi-Head Scaled Dot-Product Attention Mechanism

Qingfeng Du[✉], Liang Zhao, Jincheng Xu, Yongqi Han, and Shuangli Zhang

School of Software Engineering, Tongji University, Shanghai, China
{du_cloud,1931525,xujincheng,2011438,1931551}@tongji.edu.cn

Abstract. Anomaly detection is one of the key technologies to ensure the performance and reliability of software systems. Because of the rich information provided by logs, log-based anomaly detection approaches have attracted great interest nowadays. However, it's time-consuming to check the large amount of logs manually due to the ever-increasing scale and complexity of the system. In this work, we propose a log-based automated anomaly detection approach called *LogAttention*, which embeds log patterns into semantic vectors and subsequently uses a self-attention based neural network to detect anomalies in the log pattern sequences. LogAttention has the ability to capture contextual and semantic information in the log patterns and to attend far more long-range dependencies in the log pattern sequence. We evaluate LogAttention on two publicly available log datasets, and the experimental results demonstrate that our proposed approach can achieve better results compared to the existing baselines.

Keywords: Anomaly detection · Log analysis · Log data · Attention · Deep learning

1 Introduction

Anomaly detection is an essential task to ensure the reliability and the performance of software systems, which aims to detect anomalies in time to avoid further failures and losses. As the system gets increasingly complex and error-prone, there is an urgent need for the effective approaches of anomaly detection.

Anomaly detection approaches can be divided into several categories from different views. Log-based anomaly detection approaches have attracted great interest in recent years because of the rich information in logs. System logs can accurately describe the system status and events. Therefore, they can provide the necessary support for system diagnosing. Logs are usually viewed as plain texts with two essential components: constant parts and variable parts [7]. In this paper, we refer to the constant parts as *log patterns*.

With the ever-increasing scale and complexity of software systems, it's not easy to manually detect anomalies from logs [16]. Up to now, many automatic

© Springer Nature Switzerland AG 2021
C. Strauss et al. (Eds.): DEXA 2021, LNCS 12923, pp. 335–347, 2021.
https://doi.org/10.1007/978-3-030-86472-9_31

approaches [1,10] have been proposed to ease the pressure of manual analysis. Most of them need to extract log pattern sequences from raw logs. According to how log pattern sequences are vectorized, these approaches can be further classified into two categories, including log pattern counter based approaches and deep learning based approaches [12]. Log pattern counter based approaches suffer from the dependency on the parsing accuracy. In recent years, deep learning based approaches using Long Short-Term Memory (LSTM) have enjoyed tremendous success [5,12,16]. Instead of using the log pattern count vector, they extract sequential features hidden in the log sequence with recurrent structures. Nonetheless, it is difficult for LSTM to accurately learn the dependency when the path between two interconnected log patterns becomes too long. These problems pose major obstacles for the development of automatic approaches.

In order to overcome these obstacles, we propose a log-based automatic anomaly detection approach called *LogAttention*, which has the ability to learn the semantic information and long-range dependencies effectively in real-world logs. Firstly, LogAttention leverages Drain [6] to extract log patterns from raw logs. Then LogAttention generates semantic vectors for log patterns by aggregating multiple word vectors based on TF-IDF scores. The semantic vectors can reduce the dependency on parsing accuracy. The core part of LogAttention is *multi-head scaled dot-product attention* [14], which takes the sequence of log pattern vectors as input and subsequently detects the possible anomalies. Multi-head scaled dot-product attention is able to learn long-range dependencies in the log pattern sequence and each head can be trained in parallel.

In this work, we evaluate LogAttention on two publicly available datasets which collect logs from Hadoop system and Blue Gene/L supercomputer respectively. The overall experimental results show that our proposed approach can effectively detect anomalies in real systems compared with existing baselines. In addition, we use different window lengths to slice the logs of BGL and evaluate LogAttention's learning ability of long-range dependencies. The results show that LogAttention performs even better when detecting anomalies in the longer log sequences. Finally, we evaluate the effectiveness of the proposed vectorization method.

Our main contributions are listed below:

1. We propose a new vectorization method for log patterns, the multiple word vectors of which are aggregated based on TF-IDF to distinguish the different levels of word importance for anomaly detection.
2. We propose LogAttention, an end-to-end approach using multi-head scaled dot-product attention to detect anomalies effectively.
3. We perform extensive experiments to evaluate LogAttention on two well-established datasets.

2   Related Works

Detecting anomalies in logs can be divided into four steps: log collection, log parsing, feature extraction, and anomaly detection [7]. In this work, we concentrate on log parsing and anomaly detection.

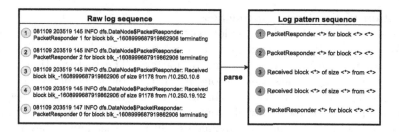

Fig. 1. An example of the raw log sequence and the extracted log patterns.

2.1 Log Parsing

Each log contains a constant part and a variable part [16]. Log parsing extracts log patterns (or the constant parts) from the raw logs. Figure 1 shows the example of the raw log sequence consisting of 5 consecutive logs and the extracted log patterns. In [17], 13 approaches on 16 datasets from various systems are comprehensively evaluated. The online approach Drain [6] shows the best performance and stability among all the considered approaches. In order to speed up the parsing process, fixed depth tree is adopted to encode parsing rules in Drain.

2.2 Log-Based Anomaly Detection

We classify existing log-based anomaly detection approaches into three groups:(1) Supervised machine learning approaches; (2) Unsupervised machine learning approaches; (3) Zero-positive machine learning approaches.

Supervised Machine Learning Approaches

A lot of supervised machine learning approaches have been applied to detect anomalies in literature. SVM (Support Vector Machines) is adopted to detect anomalies in [10]. In [16], an empirical study has been conducted to show the instability of logs in practice, and subsequently LogRobust is proposed to detect anomalies in unstable logs.

Unsupervised Machine Learning Approaches

In [15], PCA (Principle Component Analysis) is used to detect anomalies. In [11], Invariant Mining is used to detect anomalies. LogLens[3] can automatically detect anomalies with no (or minimal) knowledge of the target system.

Zero-Positive Machine Learning Approaches

Zero-positive refers to the approaches which only require normal data for anomaly detection [4]. DeepLog [5] uses LSTM to learn patterns from normal logs. It detects anomalies when logs deviates from normal distribution. LogAnomaly [12] detects sequential and quantitative anomalies at the same time. In [1], LSTM is used to learn temporal correlations and detect performance anomalies in logs.

Limitation of Existing Approaches

For those approaches based on the log pattern counter such as [10,11,15], they transform a log pattern sequence into a count vector in which each dimension represents the occurrence of a certain log pattern. The vectorization method poses the following threats. On the one hand, the count vector won't change even if the log sequence is inversed. In other words, the vector is not sensitive to the order of log sequence, while some anomalies are just manifested in the order which are called sequential anomalies. On the other hand, they are very sensitive to the results of log parsing, because the wrong log pattern will be assigned another index, which may lead to unexpected behaviors.

For those approaches based on deep learning such as [5,12,16], they can overcome the drawbacks mentioned above because they do not rely on count vectors. Besides, semantic vectorization of log patterns proposed in [16] and [12] can solve the problem of unseen log patterns. However, these approaches still suffer from the following threat: it's difficult for LSTM to accurately learn the dependencies in the longer log sequences.

3   LogAttention Design

3.1   Overview

To overcome the drawbacks of existing approaches, we propose LogAttention to effectively detect anomalies in diverse systems. The design of LogAttention has been shown in Fig. 2. In the log parsing step, we use Drain [6] to extract log patterns. After that, log patterns are transformed into semantic vectors to reduce the influence of log parsing errors. Semantic vector sequences will be fed into an multi-head scaled dot-product attention based neural network to detect whether the log sequence is anomalous or not. Each log pattern vector can attend to all the other log pattern vectors in the input sequence. In this way, anomaly features of log sequence will be automatically learned from contextual information.

3.2   Log Parsing

In log parsing, constant parts are separated from raw logs and variable parts are masked with $<*>$ [7]. For example, raw log "Received block blk_-1608999687919 of size 91178 from /10.250.10.6" can be interpreted as "Received block $<*>$ of size $<*>$ from $<*>$", which is a valid log pattern. In this work, we directly introduce Drain [6] to extract these log patterns for its performance and stability. Similar to existing log parsing approaches, Drain can not achieve 100% accuracy on any datasets, let alone incomplete logs caused by accidents. Parsing errors may result in missing necessary words in constant parts or adding redundant words into constant parts. For example, "Received block blk_-1608999687919 of size 91178 from /10.250.10.6" may be transformed into "Received block $<*>$ of size $<*>$ from /10.250.10.6" wrongly because there are too many logs reporting data from this IP address and consequently the IP address seems to be a constant part. In the following section, we will show how our vectorization method can mitigate the influence of such errors.

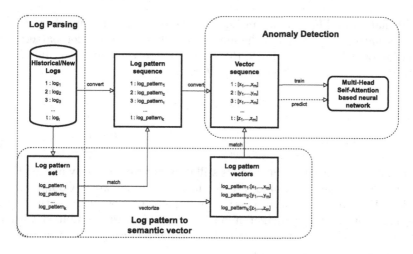

Fig. 2. The framework of LogAttention

3.3   Log Pattern Vectorization

Inspired by semantic vectorization [16], we view the log patterns as natural language and build log pattern semantic vectors to get rid of the dependency on parsing accuracy as much as possible. Different from Template2Vec [12], pre-trained word vectors from FastText [9] are directly used to represent the words in the log pattern. Then, TF-IDF [13] is used to distinguish the different levels of word importance and aggregate these word vectors. Finally, the log pattern vector is calculated by aggregating all the weighted word vectors. In this way, the influence of parsing errors can be minimized because few different words will make little change to the log pattern vector. As shown in Eq. (1), w_i is the TF-IDF score of each word and v_{word_i} is the corresponding word vector.

$$v_{logpattern} = \frac{1}{N} \sum_{i=1}^{N} w_i v_{word_i} \tag{1}$$

The main drawback of TF-IDF in [16] is that IDF is calculated in the log pattern set, where $IDF(word_i) = \frac{L}{L_{word_i}}$, L is the number of all log patterns and L_{word_i} is the number of log patterns containing the target word. In other words, [16] uses the whole log pattern set as the corpus, which is inappropriate. Consider the logs from Hadoop system. There are 48 log patterns extracted by Drain and the word "Exception" occurs in 24 log patterns, which means "Exception" isn't important at all. Obviously, it is not reasonable. Actually, most log patterns containing "Exception" occurs only few times in raw logs, which means "Exception" is very important according to the IDF score in raw logs. To overcome this drawback, we consider the importance of different words in raw logs to better represent the log patterns. In our TF-IDF method, IDF takes the occurrences of log patterns into consideration and TF keeps the same.

The calculation of TF is shown in Eq. (2) where L_{word} is the number of $word_i$ in the log pattern and L_{total} is the number of all words in the log pattern. The calculation of IDF is shown in Eq. (3) where L_{logs} is the number of lines in raw logs and occ_j is the occurrence of $logpattern_j$.

$$TF(word_i) = \frac{L_{word}}{L_{total}} \tag{2}$$

$$IDF(word_i) = \frac{L_{logs}}{\sum_{j=1}^{N} y_j \cdot occ_j}, y_j = \begin{cases} 0 & word_i \notin logpattern_j \\ 1 & word_i \in logpattern_j \end{cases} \tag{3}$$

It is noteworthy that our method is still different from directly using TF-IDF in the raw logs because variables are not included in our work. Compared with the TF-IDF used in log pattern set, our implementation reduces the impact of log patterns with the smaller number of occurrences. Combined with the aggregation of word vectors, our vectorization method can reduce the impact of parsing errors as much as possible.

3.4 Log Anomaly Detection

After the step of log pattern vectorization, each log sequence $S = [log_1, \ldots, log_t]$ can be transformed into a matrix $M = [v_{logpattern_1}, v_{logpattern_2}, \cdots, v_{logpattern_k}]$. Taking M as input, LogAttention adopts a self-attention based neural network to detect whether an anomaly is hidden in S.

Self-attention mechanism is a variant of the attention mechanism in which Query(Q), Key(K) and Value(V) are obtained from the same input. [14] proposes a specific self-attention called multi-head scaled dot-product attention and compares it with recurrent and convolutional layers. The results show that self-attention layers have the better ability to learn long-range dependencies and benefit from the better efficiency. Inspired by these characteristics, we adopt multi-head self-attention layers instead of LSTM to overcome its possible drawbacks in previous work [5,12,16].

The main structure of our proposed model has been shown in Fig. 3. Transformed from M, the input matrix $X \in R^{t \times h}$ can be represented as follows:

$$X = dropout(M^{\top} W^h + PosEmb) \tag{4}$$

$$PosEmb = W^p \tag{5}$$

where $W^h \in R^{m \times h}$ and m is the dimension of original log pattern vector. We use the learned positional embeddings $W^p \in R^{t \times h}$ to represent the absolute position of logs as shown in Eq. (5).

For the ith head, three linear transformations will produce Q_i, K_i and V_i using X:

$$Q_i = XW_i^Q, K_i = XW_i^K, V_i = XW_i^V \tag{6}$$

where $W_i^Q, W_i^K, W_i^V \in R^{h \times d_{head}}$ and $d_{head} = h/heads$.

With Q_i, K_i and V_i in the i_{th} head, scaled dot-product attention mechanism will calculate the attention scores. Then each log pattern embedding will attend

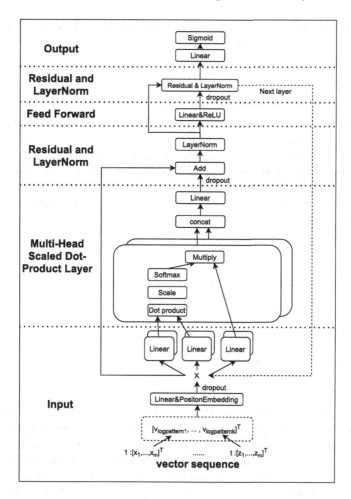

Fig. 3. Multi-head scaled dot-product based neural network

to all the other log patterns within the sequence according to the attention scores. The matrix calculation of scaled dot-product has been shown below:

$$head_i = softmax(\frac{Q_i K_i^{\top}}{\sqrt{d_{head}}})V_i \tag{7}$$

The result of all heads will be concatenated together to produce a new matrix for the next layer by a linear transformation:

$$X_o = (head_1 \oplus head_2 \oplus \cdots \oplus head_i)W^o \tag{8}$$

where $\oplus$ is the operator of concatenation and $W^o \in R^{h \times h}$.

As shown in Eq. (9), two kinds of regularizations will be applied before X^o is fed to the feed-forward layer, including layer normalization and residual dropout.

As shown in Eq. (10), the result of the current loop is then fed to the next loop to perform the same calculation as X.

$$X_o = LN(X + dropout(X_o)) \tag{9}$$

$$X_i = f_{above}(X_{i-1}) \tag{10}$$

After several loops, we get the output $X_{final} \in R^{t \times h}$. We directly use last line of X_{final} denoted as $y^\top$ to calculate the final anomaly probability for the following two reasons: (1) Each element in the sequence has already contained the information of other elements, so has the last one. (2) The last line of X_{final} corresponds to the last log pattern in the log sequence, which is the latest information of the sequence. Therefore, the final anomaly probability is calculated by $y^\top$ as in Eq. (11).

$$AnomalyProb = \sigma(y^\top W^t + b^t) \tag{11}$$

$$\sigma(x) = \frac{1}{1 + e^{-x}} \tag{12}$$

4   Evaluation

4.1   Datasets and Criteria

Datasets

1. **HDFS**: HDFS represents the Hadoop Distributed File System, which is a distributed Hadoop application [8]. This data set is generated with benchmark workloads and manually labelled as normal or abnormal. Each line of HDFS log contains a unique block ID. Therefore, the operations in logs can be naturally organized by session windows identified by block ID [7]. There are 11,175,629 logs in total and they can be sliced into 575061 sessions, in which 16838 sessions are labelled as anomalous. See [15] for more details. We randomly pick 6000 normal sessions and 6000 abnormal sessions from HDFS as the training set.

2. **BGL**: BGL represents BlueGene/L, which is a supercomputer system at Lawrence Livermore National Labs (LLNL). The logs can be divided into alert messages and non-alert messages, which are identified by alert category tags [8]. There are 4,747,963 logs in total and 348,460 logs are labelled as anomalous. BGL logs have no identifier like "block ID" for each job session. Therefore, we will use sliding windows to slice logs into a set of log sequences. Log sequence is anomalous as long as it contains an anomalous log. We randomly pick 40000 normal log sequences and 40000 abnormal log sequences from BGL as the training set.

Baselines
We compare LogAttention with Logistic Regression (LR) [2], Invariant Mining (IM) [11], PCA [15] and LogRobust [16], which contains supervised approach,

Table 1. Experimental results on HDFS dataset and BGL dataset

	HDFS			BGL		
	Precision	Recall	F1 score	Precision	Recall	F1 score
LogAttention	0.975	0.998	0.986	0.978	0.992	0.985
LogRobust	0.933	0.968	0.950	0.950	0.998	0.974
LR	0.955	0.911	0.933	1	0.810	0.820
PCA	0.980	0.670	0.790	0.500	0.610	0.550
IM	0.880	0.950	0.910	0.830	0.990	0.910

unsupervised approach, log pattern counter based approach and deep learning based approach. All the parameters have been fine-tuned for the best accuracy.

Evaluation Metrics
Following previous work, we use *precision*, *recall* and *F1 score* as the evaluation metrics. Log-based anomaly detection is a binary classification problem. The output can be divided into 4 parts: TP, TN, FP, FN. TP is the number of abnormal sequences classified accurately. TN is the number of normal sequences classified accurately. If an anomalous sequence is wrongly classified as a normal sequence, then it will be viewed as FN. Precision, recall and F1 score are calculated as follows:

$$Precision = TP/(TP + FP) \tag{13}$$

$$Recall = TP/(TP + FN) \tag{14}$$

$$F1score = (2 * Precision * Recall)/(Precsion + Recall) \tag{15}$$

4.2 Overall Results and Analysis

In this section, we compare LogAttention with 4 baselines approaches on two datasets: HDFS and BGL. The implementation of LogAttention on HDFS has 4 layers and 8 heads whose dimensions are 32. The experimental results have been shown in Table 1. It can be clearly seen that LogAttention achieves the best F1 score (0.986) and precision (0.975) on HDFS. The results of Invariant Mining and LogRobust are highly competitive with LogAttention in recall. However, they achieve much lower precision, which means they generate far more false alarms than LogAttention. On the contrary, PCA achieves a high precision of 0.98 but a low recall of 0.67, which means 33% anomalies are ignored.

The implementation of LogAttention on BGL has 2 layers and 8 heads whose dimensions are 4. The implementation of LogRobust has 2 LSTM layers with 32 neurons. The window length has been set to 200 here. It is worth mentioning that the results from any window length can be used as the overall results of BGL, depending on the expected scale. The experimental results have been shown in

Table 1. LogAttention and LogRobust achieve comparable F1 scores of 0.985 and 0.974 respectively, which are obviously higher than other approaches. LogAttention yields a slight degradation of 0.6% in recall, but surpasses LogRobust in precision by a large margin of 2.8%.

In conclusion, the experimental results show that LogAttention performs better than baselines for anomaly detection.

Table 2. Experimental results on BGL dataset with different window lengths (LogAtt: LogAttention; LogRob: LogRobust)

	Window length = 200		Window length = 350		Window length = 450		Window length = 500	
	LogAtt	LogRob	**LogAtt**	LogRob	**LogAtt**	LogRob	**LogAtt**	LogRob
Precision	**0.978**	0.950	0.977	**0.978**	**0.977**	0.966	**0.989**	0.962
Recall	0.992	**0.998**	**0.999**	0.990	**0.999**	0.998	**0.995**	0.994
F1 score	**0.985**	0.974	**0.987**	0.984	**0.988**	0.981	**0.992**	0.978

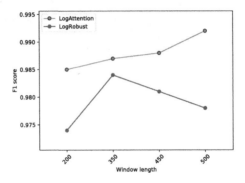

Fig. 4. F1 scores of two approaches with different window lengths

4.3   Experiments on Log Sequences with Different Window Lengths

As described earlier, LogAttention has the ability to learn long-range dependencies in log sequences. To prove this, we evaluate the performance of LogAttention and LogRobust on BGL with different window lengths. LR, PCA and IM will not be considered in this section because they belong to log pattern counter based approaches, which are not sensitive to the parameter of window length and they can't capture sequential patterns in log sequences anyway. The experimental results have been illustrated in Fig. 4, and the more detailed results have been shown in Table 2.

In terms of the F1 score, LogAttention even performs better when the window length becomes larger as shown in Fig. 4. When the window length equals 200, LogAttention achieves the F1 score of 0.985. Then its F1 score increases continuously by 0.002 and 0.003 respectively when the window length becomes 350 and 450. When the window length equals 500, LogAttention achieves the highest F1 score of 0.992. LogRobust achieves the F1 score of 0.974 when the window length equals 200. Then the F1 score increases by 0.01 when the window length increases to 350. In contrast to LogAttention, the F1 score of LogRobust decreases continuously by 0.003 and 0.006 when the window length increases to 450 and 500. Unlike LogAttention, LogRobust achieves its highest F1 score when the window length is 350. Apart from F1 score, we also zoom in on the metrics of precision and recall. The results are reported in Table 2. It can be seen that LogAttention performs better in most cases. It's obvious that LogAttention is not influenced by the increasing window length and can get better results compared to LogRobust.

Table 3. Experimental results of LogAttention with different vectorization methods

	Metrics	**LogAttention**	LogAttention*
HDFS	Precision	**0.975**	0.962
	Recall	0.998	**0.999**
	F1 score	**0.986**	0.980

4.4 Experiments on Log Pattern Vectorization Method

In this section, we compare our vectorization method with [16]. As the baseline, LogAttention is retrained with the vectors generated by the vectorization method proposed in [16], denoted as LogAttention*. We report the results in precision, recall and F1 score in Table 3. As we can see, LogAttention achieves a precision of 0.975 and a F1 score of 0.986, which is much higher than the retrained LogAttention* by 0.013 and 0.006 respectively. As for the recall, LogAttention* surpasses LogAttention by a small margin of 0.001. The difference is almost indistinguishable, so the performance in recall can be considered as comparable for the two methods. In a word, our vectorization method can lead to better results than the one proposed in [16].

5 Conclusion

In recent years, many deep learning based approaches have been proposed to meet the urgent requirement of log-based anomaly detection. Although these approaches demonstrate their superiority, these approaches still have some drawbacks from the LSTM-based models and the vectorization methods. In this paper, we propose an approach, namely *LogAttention*, to detect anomalies effectively. LogAttention adopts the semantic vectorization method based on TF-IDF and the network based on multi-head scaled dot-product attention. We evaluate

LogAttention on two public available datasets with extensive experiments. The experimental results show that LogAttention can effectively detect anomalies in various cases compared to the existing baselines.

Acknowledgment. This work was supported by National Key R&D Program of China (Grant No. 2020YFB2103300).

References

1. Baril, X., Coustié, O., Mothe, J., Teste, O.: Application performance anomaly detection with LSTM on temporal irregularities in logs. In: Proceedings of the 29th ACM International Conference on Information & Knowledge Management, pp. 1961–1964 (2020)
2. Bodik, P., Goldszmidt, M., Fox, A., Woodard, D.B., Andersen, H.: Fingerprinting the datacenter: automated classification of performance crises. In: Proceedings of the 5th European Conference on Computer Systems, pp. 111–124 (2010)
3. Debnath, B., et al.: LogLens: a real-time log analysis system. In: 2018 IEEE 38th International Conference on Distributed Computing Systems (ICDCS), pp. 1052–1062. IEEE (2018)
4. Du, M., Chen, Z., Liu, C., Oak, R., Song, D.: Lifelong anomaly detection through unlearning. In: Proceedings of the 2019 ACM SIGSAC Conference on Computer and Communications Security, pp. 1283–1297 (2019)
5. Du, M., Li, F., Zheng, G., Srikumar, V.: DeepLog: anomaly detection and diagnosis from system logs through deep learning. In: Proceedings of the 2017 ACM SIGSAC Conference on Computer and Communications Security, pp. 1285–1298 (2017)
6. He, P., Zhu, J., Zheng, Z., Lyu, M.R.: Drain: an online log parsing approach with fixed depth tree. In: 2017 IEEE International Conference on Web Services (ICWS), pp. 33–40. IEEE (2017)
7. He, S., Zhu, J., He, P., Lyu, M.R.: Experience report: system log analysis for anomaly detection. In: 2016 IEEE 27th International Symposium on Software Reliability Engineering (ISSRE), pp. 207–218. IEEE (2016)
8. He, S., Zhu, J., He, P., Lyu, M.R.: Loghub: a large collection of system log datasets towards automated log analytics. arXiv preprint arXiv:2008.06448 (2020)
9. Joulin, A., Grave, E., Bojanowski, P., Douze, M., Jégou, H., Mikolov, T.: Fasttext.zip: compressing text classification models. arXiv preprint arXiv:1612.03651 (2016)
10. Liang, Y., Zhang, Y., Xiong, H., Sahoo, R.: Failure prediction in IBM BlueGene/l event logs. In: Seventh IEEE International Conference on Data Mining (ICDM 2007), pp. 583–588. IEEE (2007)
11. Lou, J.G., Fu, Q., Yang, S., Xu, Y., Li, J.: Mining invariants from console logs for system problem detection. In: USENIX Annual Technical Conference, pp. 1–14 (2010)
12. Meng, W., et al.: LogAnomaly: unsupervised detection of sequential and quantitative anomalies in unstructured logs. In: IJCAI, pp. 4739–4745 (2019)
13. Salton, G., Buckley, C.: Term-weighting approaches in automatic text retrieval. Inf. Process. Manage. **24**(5), 513–523 (1988)
14. Vaswani, A., et al.: Attention is all you need. In: Advances in Neural Information Processing Systems, pp. 5998–6008 (2017)
15. Xu, W., Huang, L., Fox, A., Patterson, D., Jordan, M.I.: Detecting large-scale system problems by mining console logs. In: Proceedings of the ACM SIGOPS 22nd Symposium on Operating Systems Principles, pp. 117–132 (2009)

16. Zhang, X., et al.: Robust log-based anomaly detection on unstable log data. In: Proceedings of the 2019 27th ACM Joint Meeting on European Software Engineering Conference and Symposium on the Foundations of Software Engineering, pp. 807–817 (2019)
17. Zhu, J., et al.: Tools and benchmarks for automated log parsing. In: 2019 IEEE/ACM 41st International Conference on Software Engineering: Software Engineering in Practice (ICSE-SEIP), pp. 121–130. IEEE (2019)

Enhancing Scan Matching Algorithms via Genetic Programming for Supporting Big Moving Objects Tracking and Analysis in Emerging Environments

Alfredo Cuzzocrea[1,2], Kristijan Lenac[3], and Enzo Mumolo[4(✉)]

[1] iDEA Lab, University of Calabria, Rende, Italy
`alfredo.cuzzocrea@unical.it`
[2] LORIA, Nancy, France
[3] Faculty of Engineering, Center for Artificial Intelligence and Cybersecurity,
University of Rijeka, Rijeka, Croatia
`klenac@riteh.hr`
[4] University of Trieste, Trieste, Italy
`mumolo@units.it`

Abstract. *Big moving objects* arise as a novel class of big data objects in emerging environments. Here, the main problems are the following: (*i*) tracking, which represents the baseline operation for a plethora of higher-level functionalities, such as detection, classification, and so forth; (*ii*) analysis, which meaningfully marries with *big data analytics scenarios*. In line with these goals, in this paper we propose a novel family of *scan matching algorithms* based on registration, which are enhanced by using a genetic pre-alignment phase based on a novel metrics, fist, and, second, performing a finer alignment using a deterministic approach. Our experimental assessment and analysis confirms the benefits deriving from the proposed novel family of such algorithms.

Keywords: Moving objects · Scan-matching algorithms · Intelligent systems · Genetic optimization

1 Introduction

Nowadays, a great deal of interest is growing around the mobile object tracking problem, especially due to the emerging integration between robotics and *big data applications* (e.g., [11–13]). Following this trend, several mobile object tracking approaches have recently appeared in literature, considering different aspects of the target issue, such as coverage, completeness, effectiveness, efficiency, etc. The category of algorithms that goes under the name of *scan-matching*

A. Cuzzocrea—This research has been made in the context of the Excellence Chair in Computer Engineering – Big Data Management and Analytics at LORIA, Nancy, France.

C. Strauss et al. (Eds.): DEXA 2021, LNCS 12923, pp. 348–360, 2021.
https://doi.org/10.1007/978-3-030-86472-9_32

(e.g., [14–16]) supports mobile objects positioning in indoor environments based on the acquisition of maps of the environment surrounding the target mobile objects. Maps are acquired from two successive points in the objects' path using a range-scanner sensor positioned on mobile objects themselves. The first acquisition is called reference scan and the second actual scan. The actual scan is sometimes also called new scan. By overlapping the maps acquired at two successive positions on the path it is possible to estimate the relative movement of the object between these two positions.

In this paper we describe a family of *scan-matching based registration algorithms* called HGLASM-g which perform scan-matching based on a hybrid approach. First, an approximate pre-alignment of two adjacent maps is performed via a new genetic optimization method called GLASM-g; then a variant of the Iterative Closest Point (ICP) algorithm is applied to pre-aligned maps to obtain the final overlap.

In other proposed genetic scan-matching pre-alignment algorithms, the fitness functions are based on metrics between actual and reference scan points that require to know the correspondence of point pairs and the translation and rotation between the two scans. However, when scan acquisitions include noise, correspondence errors may arise. Moreover, also translation and rotation corrections can lead to errors when they are too large.

In order to overcome such issues, in this paper we propose a novel metric which does not require neither points pair correspondences nor translations and rotation corrections. Indeed, our metric is based on lookup tables built around the reference scan points. The fitness function weights the hits of actual scan points in the lookup table. The genetic pre-alignment then finds the scan with the highest fitness within a search space of given size. This guarantees also the maximum robustness towards both the acquisition errors and the Initial Position Errors (IPE). It is well known that ICP performance depends on the quality of points pair correspondence and on the accuracy of the starting point estimation. We overcome this limitation by choosing the initial guess of ICP via genetic pre-alignment, which makes it close to the true solution. This way, point correspondence, translation and rotation estimations are performed correctly and, as a consequence, iteration failures are reduced. On the other hand, even *adaptive metaphors*, perhaps developed in different contexts (e.g., [21]), can be exploited to this end.

The algorithms described in this paper form a family in the sense that each algorithm is characterized by different values of the target search space size. Each size allows us to solve different registration problems and hence different mobile object tracking scenarios. If the search space size is small, in fact, the algorithm can recover from small errors only, while, if the search space is higher, also higher errors can be recovered. However, computation complexity increases as the search space size gets higher.

A similar hybrid algorithm is described by Martinez *et al.* in [1]. Therefore, we consider the latter algorithm in a comparative approach, and we experimentally show that all the terms of comparison with this algorithm are improved thanks to the hybrid algorithms proposed in this paper. Furthermore, we also show that our approach is able to recover from greater initial positioning and acquisition errors. The key for improvement is the definition of a new metric used for computing the fitness function of the genetic procedure. The proposed target scan-matching algorithm is described for the 2D case, but it can be used in the 3D case as well. Improvements obtained with our proposed algorithm are measured *both* in terms of accuracy and noise robustness. Indeed, the estimation of the initial position of target mobile object often comprises significant errors. For instance, when the mobile object is equipped with a legged or wheeled locomotion, and the initial position is estimated by means of odometric approaches, there may be slippage with respect to the floor, which entails significant errors in the initial position of the object. As a consequence, accuracy of algorithms is seriously affected by IPE.

2   Background Concepts

We are given two sets of bi-dimensional points $A = a_1, \ldots, a_N$ and $B = b_1, \ldots, b_M$, where a_i and b_i are 2×1 column vectors. The two sets A and B are scan descriptions of the environment as seen by a range sensor put on a mobile object from two points P_A and P_B. The first scan represented by the set A is the *reference* scan while B is the *new* scan taken after a movement of the mobile object. If we overlap the two scans, that is by determining the optimum rotation and translation of the set B wrt the set A, an estimation of the movement can be obtained. First of all, points correspondences must be estimated between A and B. A generic correspondence search algorithm takes the points from the two scans, $p_{i_k}, i = 1 \ldots N$ and $p_{j_k}, j = 1 \ldots M$, and establishes a set of k corresponding points pairs (p_{i_k}, p_{j_k}), $k = 0 \ldots K$ where $0 \leq K < M \cdot N$. A straightforward and fast algorithm for establishing point correspondences between two scans simply considers the polar coordinates of the reference and the actual scan points projected in the same coordinate frame of the reference scan. The scan is then traversed with increasing angle and points that belong to the same angle step which are closer than a distance threshold are matched. An example of the correspondence between the points a_i and b_i obtained with such 'polar coordinate' approach is reported in Fig. 1 by the lines connecting the points. This 'polar coordinates' approach has been used in [1] in their hybrid two phase genetic + ICP approach for the genetic phase. However the experiments have shown that using this simple approach in iterative correspondence point algorithms leads to convergence failures and poor performance.

3   Genetic Pre-alignment Phase

Evolutionary Algorithms, such as for example Genetic Algorithms (GA) or Particle Swarm Optimization (PSO) algorithms are heuristic processes inspired by the natural evolution in biological systems. Evolutionary optimization algorithms are

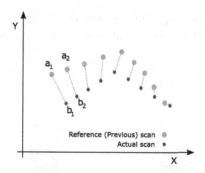

Fig. 1. An example of point pair correspondences.

popular in solving complex or nonlinear problems such as for example optimization or classification. These algorithms are characterized by some key concepts, such as the generation of an initial number of solutions, called the population or swarm of individuals, the calculation of the function to optimize, called fitness, and the generation of a new population until the best individual survives.

The procedures for initial creation and generation of the new population are different in the various versions of the evolutionary algorithms. Since in GA each solution is encoded in binary notation, GA performs basically discrete optimization. PSO instead represents the solutions as particles with position and speed encoded as real variables. All the particles form a swarm, while positions are the real variable to be optimized.

The proposed hybrid algorithm consists in the discrete estimation of translation and rotation by means of an evolutionary algorithm followed by a fine optimization by means of a deterministic algorithm. In has to be remarked that the discrete variables (x, y, ϕ) are related to the discretization of the search space. Sub-optimal values of the variables can be obtained by discrete optimization algorithms and a finer optimization by continuous optimization algorithms. In this paper we perform discrete optimization with Genetic Algorithms and finer optimization with Iterative Closest Point algorithms which are very efficient if the starting point is close to the true value.

4   A Novel Family of Enhanced Hybrid Scan Matching Algorithms

The pre-alignment step is inspired by the algorithm called Genetic Lookup based Algorithm for Scan Matching (GLASM) described by Lenac *et al.* in [3]. The algorithm described in [3] uses a metric based on a binary lookup table.

In the proposed approach we first improve GLASM by using the 8-bit encoding to store the probability density of measurements being close to points of reference scan in the lookup table. We call this improved variant of the existing binary GLASM technique as *GLASM-g*. Then, this improved variant of the existing GLASM technique is combined with MbICP [2]. The new scan B, evaluated starting from the odometric estimation of robot movements, is fed as input to GLASM-g, together with the previous scan, A. The output of GLASM-g x', y', ϕ', is then used as starting point of the MbICP algorithm that compares A and B, thus producing the final output x, y, ϕ.

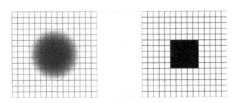

Fig. 2. A detail of a lookup table surrounding an isolated point of the reference scan. Left: Radial, Gradual. Right: Squared, Binary.

In Fig. 2 the difference between the two algorithms is shown: in the GLASM-g version the lookup table is displayed as a gray scale image, while in the GLASM version the lookup table is seen as a black and white image.

While in the GLASM lookup table each cell was either 1 or 0 the GLASM-g table can contain a range of values that model the probability of matching a point using a normal distribution. In the proposed implementation 256 different values are used requiring 1 byte of memory per cell.

The computation of the fitness function works as follows. For each point of the new scan a roto-translation and a discretization are performed to bring the point in the same reference frame of the lookup table and select the corresponding cell. However, once the cell is selected, instead of simply incrementing the fitness with a binary value, the fitness is incremented with a value corresponding to the probability of matching a point that was saved in the lookup table. A representation of this probability is shown in Fig. 2, left panel, with gray levels. In Algorithm 1 we report the pseudo code of the fitness computation used in GLASM-g.

Algorithm 1. GLASM-g Fitness Computation

Input: B // new scan
Output: $fitness$
$fitness = 0$;
for ($each\ point\ p\ of\ B$) **do**
    // roto-translation to lookup reference frame
    $p' = \text{changereferenceframe}(p)$
    $fitness = fitness \cdot lookup(p')$;
return $fitness$;

The fitness function is essential for the proper functioning of genetic optimization algorithms. For each individual and for each execution it is computed only once. Its correct definition is therefore fundamental also from the point of view of computational complexity, given that a fitness function that requires simple calculations translates into a fast execution of the entire algorithm, leading to an exploration of a greater search space with a greater success ratio.

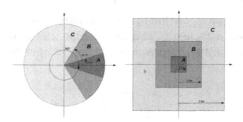

Fig. 3. Search space size of the algorithms *small* (A), *medium* (B), and *large* (C).

The goal of two-dimensional scan matching algorithms is to obtain the path of a mobile object by estimating its relative movements during the path. In other words, we can speak of a space $(translation_X \times translation_Y, rotation)$ within which the scan matching algorithm searches for its estimate.

The individual algorithms withing the family differ in the search space size and the parameters selected for the genetic algorithm. Typically, in order to cover a larger search space, a larger population is used, as well as the number of generations and runs of the genetic algorithm.

In this paper we selected and studied three different algorithms from within the family that we simply call *small*, *medium*, and *large*. The search space size and the combination of genetic parameters of the selected algorithms are depicted in Fig. 3 and listed in Table 1.

Table 1. Search space size of the selected algorithms and the corresponding parameters used for the genetic algorithm

Algorithm	Search space size	Genetic configuration
	$(dX, dY, dRot)$	$(pop \times gen \times runs)$
small (A)	±0.3 m, ±0.3 m, ±17.2°	20 × 6 × 1
medium (B)	±1.0 m, ±1.0 m, ±57.3°	100 × 10 × 1
large (C)	±2.0 m, ±2.0 m, ±180.0°	200 × 12 × 2

The search space of the *small* algorithm is sufficient to recover from small initial errors, while *medium* and *large* are able to correct progressively larger errors, at the cost of computational time. The *large* algorithm is able to recover from arbitrary orientation errors and large translation errors. The search space is obviously centered on the reference scan.

For the refinement step we selected the MbICP algorithm [2] which determines the correspondences between the points of the current scan and the reference scan considering both the rotation and the distance between the points. The algorithm has characteristics of accuracy and low calculation times. MbICP has better characteristics than other iterative scan matching algorithms based on point correspondences, for example [4,5].

Visual examples of the algorithm in operation are shown in Fig. 4. In addition a video file is available attached to this manuscript that illustrates the problem and shows in more detail the proposed algorithms in operation.

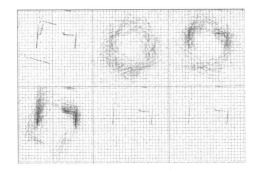

Fig. 4. Visualization of HGLASM-g operation. In the top-left frame the initial position estimate of the new scan is depicted in green and the reference scan in blue. The next three frames show the pre-alignment process depicting the population after 0,5, and 12 generations of the genetic algorithm with the best estimate shown in red. The last two frames show the refinement of the solution using the MbICP algorithm (the first and the last iteration are shown). (Color figure online)

5 Experimental Study

The performance of the family of algorithms described in this paper has been verified through several experiments carried out with a publicly available scan dataset. The dataset used in our experiments was produced by the RAWSEED project [6,7]. The aim of the RAWSEED project was to produce tools to compare the performance of robotic systems. The laser data considered in this dataset was obtained from a Sick sensor. To simplify the experiments, a significant subset of the whole data was randomly extracted across the entire dataset.

The dataset concerns internal and external environments. The movable objects traced in the dataset entered and exited from various environments, and also covered urban environments. The dataset thus provides the opportunity to study the performance of complex navigation algorithms in various types of environments.

It is worth to note moreover that the results reported in this Section have been obtained with a PC equipped with Intel Core 2 Quad Q9550 CPU running at 2,83 GHz.

The assessment of the accuracy obtained by the algorithms requires not only the laser scanning data but also the ground truth data. The Rawseed dataset contains ground truth data, but not for all scans. Furthermore, the reported positioning error is 1 cm, which is not sufficient for our evaluations. To address this inadequacy, the ground truth values were constructed by using a translated and rotated copy of the reference scan as the new scan. The ground truth value is thus known exactly as it corresponds to the position of the reference scan. To prevent the scan from being compared to a copy of itself the scan was split between the two scans with the reference scan consisting of the odd readings in the scan, and the new scan consisting of the even ones.

In all experiments it is important to study how algorithms converge under different conditions. To quantitatively express the robustness of the algorithms with respect to initialization errors, scanning noise and falling in local minima, we defined the Success Ratio. Success Ratio is the percentage of successful comparisons, computed as the number of successful comparisons divided by the total number of comparisons. A success is the case in which the resulting translation and rotation fall within a fixed size ellipsoid around the true values. In all the experiments the pre-established size is 10 cm for distance and $0.57°(0.01$ [rad]) for rotation, respectively. In cases where genetic optimization was used to achieve a coarse pre-alignment of the hybrid algorithm, the predetermined size was 30 cm and $5.73°(0.1$ [rad]) respectively. This relaxation of the thresholds was introduced because the purpose of the pre-alignment is not to provide an accurate estimate, but an estimate that is close enough to the true position to be significant for the next step of the algorithm. When the scan matching reached convergence, the mean and standard deviation of the error (estimated position - true position) were computed for the considered algorithms.

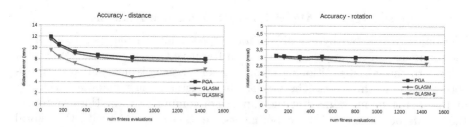

Fig. 5. Translation error (left) and rotation error (right) versus the number of fitness evaluations (right).

5.1   Genetic Pre-alignment: Performance Analysis

The performance of the genetic algorithm has been studied by varying the configuration of genetic parameters such as the size of the population and the number of generations. Figure 5 shows the translation (left) and rotation (right) errors versus the number of evaluations of the fitness function for the Polar Genetic Algorithm (PGA), GLASM and GLASM-g.

The calculation time of the algorithms based on the GLASM and GLASM-g lookup tables respectively represents the time taken only by the matching of the new scan with respect to the reference scan and not by the initialization phase of the lookup table. The figures show the times in milliseconds relating to various Success Ratios for *small* and *large* search spaces.

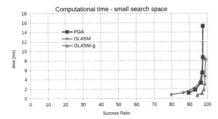 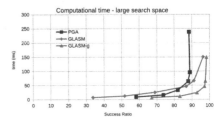

Fig. 6. Time needed to get various levels of success ratio for different sets of genetic algorithms with *small* (left) and *large* (right) search size.

As we can see in Fig. 6, the best Success Ratio results, calculation time and accuracy are always obtained with the fitness function used by the GLASM-g algorithm. In particular we see that the GLASM-g algorithm always offers better Success Ratio and calculation speed results than the PGA algorithm of Martinez *et al.*. The GLASM algorithm provides the same results as GLASM-g if the size of the population and the number of generations are both high.

This slower convergence toward high Success Ratios is explained by the nature of the GLASM fitness function. Actually, in addition to the smaller marking pattern, the GLASM fitness function also produces variations of the fitness score that are less refined than GLASM-g.

5.2   Proposed Hybrid Algorithms: Performance Analysis

The objectives of the hybrid algorithm proposed in this paper is to improve the robustness of the approaches based on Iterative Correspondence Point with respect to the initial positioning errors, to convergence issues and to the presence of noise in the scans. The increase in robustness must maintain accuracy and must not be introduced at the expense of computing time. Here we report the results of the third set of experiments, i.e. those obtained with the hybrid approach presented in this paper. The parameters of the genetic algorithm are described in the previous section and shown in Table 1 for the search spaces *small*, *medium* and *large*.

We describe now some results concerning the Success Ratios obtained with different values of initial position errors, both rotation and translation errors. On the other hand, the computation time slightly increases going from *small* to *large* sizes. But it is very important to observe that also the range of errors that can be corrected increases going from *small* to *large* sizes.

Figure 7 shows the Success Ratio curves versus different values of rotation and translation initial positioning errors, for *large* (left) and *medium* (right) search spaces. In addition to this, Fig. 8 (left) shows the same experimental patterns for *small* search space.

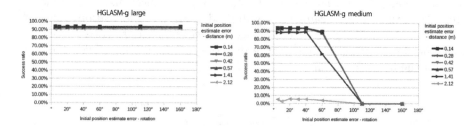

Fig. 7. Success ratio behavior versus rotation and translation errors for *large* (left) and *medium* (right) search space size.

As we can see in these figures, the Success Ratio for *large* search space size is about 90% for arbitrarily large rotation errors and all considered translation errors up to 2.12 meters. This large range of errors that can be corrected means that global positioning and tracking applications are possible. As the search space size decreases the range of errors that can be corrected is reduced, because the solution is searched in a reduced space. This allows to reduce the computation time in applications where a smaller search space size is sufficient.

For comparison we report in the following the Success Ratios obtained with the MbICP algorithm. Figure 8 (right) shows that the rotation and translation errors that can be corrected by the MbICP algorithm are much worse than that corrected by the hybrid algorithm described in this paper. Just to highlight a couple of results reported in Fig. 8 (right), consider the following initial positioning errors: 20 degree of rotation error and 0.57 meter of translation error. With these errors we have a Success Ratio of about 93% for large and medium search space sizes and about 85% for small search space size.

In the same condition the Success Ratio of the MbICP algorithm drops at about 20%. It is also important to consider the rotation error after which it is not possible to get significant Success Ratio. While HGLASM-g medium cannot get significant Success Ratio after 110 degrees and HGLASM-g small after 60 degree, the MbICP algorithm stops at 40 degrees.

If the IPE is inside the realignment margins of the local scan matcher, the addition of the pre-alignment step in theory would not be necessary since the ICP based algorithms is capable by itself of successful matching. On the first thought the addition of the pre-alignment step would then just add to execution time incurring speed penalty. However, in practice this often is not true because the addition of the pre-alignment step typically reduces the number of iterations for the ICP step. The reason for this speed-up lies in the iterative nature of ICP based algorithms. If the iteration starts from an initial point that is very far from the true position, it normally requires many iterations to improve the position. If we look at the last row of the table we see that if the search space is large, the proposed algorithm obtains a high Success Ratio in a very short time even if the initial position error is high.

An important aspect of the proposed algorithms is their robustness against uncertainty. Scan matching algorithms in general cannot adequately overcome singularity cases like navigation in long hallways. The uncertainty alongside the direction of the corridor cannot be addressed using the information contained in the scan alone. The proposed approach however manages to correct the error along the direction orthogonal to the direction of the hallways even for large errors of the initial position.

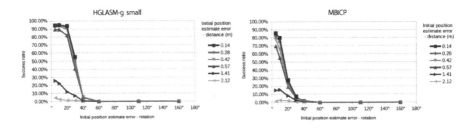

Fig. 8. Success ratio behavior versus rotation and translation errors for *small* search space size (left) and with respect to the MbICP algorithm (Right).

6   Conclusions and Future Work

In this paper we have presented a hybrid algorithm for the problem of scan matching. The algorithm better solves some scan-matching problems as compared to state-of-the-art algorithms, such as the problem of initial positioning errors and blocking the iterations in local minima. The proposed algorithm introduces a fitness function based on the look up tables whose content is used as a weight in the fitness calculation. Values in look up tables are gradually modified starting from the reference position. The algorithm was tested with a dataset consisting of laser scans obtained in various environments. The main contribution is that better results of the scan matching operation are obtained in terms of accuracy and robustness. The main reason is due to the new metric adopted that allows to compare not just single points but an entire scan. This permits to avoid the preliminary steps of point correspondence and the translation and rotation computation, that introduce errors in classical scan matching algorithms. Future works will be directed towards the extension of the described hybrid algorithms to the 3-D case, and higher dimensions. Also, we plan to make our comprehensive framework suitable to the emerging *big data trend* (e.g., [8–10,22–24]), as to make it able of dealing with specific features of such innovative settings, like also dictated by some recent studies (e.g., [17–19,25]).

References

1. Martínez, J.L., González, J., Morales, J., Mandow, A., García-Cerezo, A.J.: Mobile robot motion estimation by 2D scan matching with genetic and iterative closest point algorithms. J. Field Robot. **23**(1), 21–34 (2006)
2. Minguez, J., Montesano, L., Lamiraux, F.: Metric-based iterative closest point scan matching for sensor displacement estimation. IEEE Trans. Robot. **22**(5), 1047–1054 (2006)
3. Lenac, K., Mumolo, E., Nolich, M.: Robust and accurate genetic scan matching algorithm for robotic navigation. In: Jeschke, S., Liu, H., Schilberg, D. (eds.) ICIRA 2011. LNCS (LNAI), vol. 7101, pp. 584–593. Springer, Heidelberg (2011). https://doi.org/10.1007/978-3-642-25486-4_58
4. Pomerleau, F., Colas, F., Siegwart, R., Magnenat, S.: Comparing ICP variants on real-world data sets - open-source library and experimental protocol. Auton. Robots **34**(3), 133–148 (2013)
5. Rusinkiewicz, S., Levoy, M.: Efficient variants of the ICP algorithm. In: Proceedings of 3rd International Conference on 3D Digital Imaging and Modeling (3DIM 2001), pp. 145–152 (2001)
6. Bonarini, A., Burgard, W., Fontana, G., Matteucci, M., Sorrenti, D.G., Tardós, J.D.: RAWSEEDS: robotics advancement through web-publishing of sensorial and elaborated extensive data sets. In: Proceedings of IEEE/RSJ International Conference on Intelligent Robots and Systems Workshop Benchmarks in Robots and Systems, p. 5 (2006)
7. Ceriani, S., et al.: Rawseeds ground truth collection systems for indoor self-localization and mapping. Auton. Robots **27**(4), 353–371 (2009)
8. Cuzzocrea, A., De Maio, C., Fenza, G., Loia, V., Parente, M.: OLAP analysis of multidimensional tweet streams for supporting advanced analytics. In: Proceedings of ACM SAC 2016 International Conference, pp. 992–999 (2016)
9. Chatzimilioudis, G., Cuzzocrea, A., Gunopulos, D., Mamoulis, N.: A novel distributed framework for optimizing query routing trees in wireless sensor networks via optimal operator placement. J. Comput. Syst. Sci. **79**(3), 349–368 (2013)
10. Cuzzocrea, A.: Combining multidimensional user models and knowledge representation and management techniques for making web services knowledge-aware. Web Intell. Agent Syst. **4**(3), 289–312 (2006)
11. Chen, C.-C., et al.: A novel efficient big data processing scheme for feature extraction in electrical discharge machining. IEEE Robot. Autom. Lett. **4**(2), 910–917 (2019)
12. Zhu, D.: IOT and big data based cooperative logistical delivery scheduling method and cloud robot system. Future Gener. Comput. Syst. **86**, 709–715 (2018)
13. Huang, J., Zhu, D., Tang, Y.: Health diagnosis robot based on healthcare big data and fuzzy matching. J. Intell. Fuzzy Syst. **33**(5), 2961–2970 (2017)
14. Qian, C., Zhang, H., Tang, J., Li, B., Liu, H.: An orthogonal weighted occupancy likelihood map with IMU-aided laser scan matching for 2D indoor mapping. Sensors **19**(7), 1742 (2019)
15. Niu, X., Yu, T., Tang, J., Chang, L.: An online solution of LiDAR scan matching aided inertial navigation system for indoor mobile mapping. Mob. Inf. Syst. **2017**, 4802159:1–4802159:11 (2017)
16. Li, J., Zhong, R., Hu, Q., Ai, M.: Feature-based laser scan matching and its application for indoor mapping. Sensors **16**(8), 1265 (2016)

17. Jiang, Z., Zhu, J., Lin, Z., Li, Z., Guo, R.: 3D mapping of outdoor environments by scan matching and motion averaging. Neurocomputing **372**, 17–32 (2020)
18. Li, X., Du, S., Li, G., Li, H.: Integrate point-cloud segmentation with 3D LiDAR scan-matching for mobile robot localization and mapping. Sensors **20**(1), 237 (2020)
19. Deschaud, J.-E.: IMLS-SLAM: scan-to-model matching based on 3D data. In: Proceedings of ICRA 2018, pp. 2480–2485 (2018)
20. Wu, Y., Su, Q., Ma, W., Liu, S., Miao, Q.: Learning robust feature descriptor for image registration with genetic programming. IEEE Access **8**, 39389–39402 (2020)
21. Cannataro, M., Cuzzocrea, A., Pugliese, A.: XAHM: an adaptive hypermedia model based on XML. In: Proceedings of SEKE 2002, pp. 627–634 (2002)
22. Cuzzocrea, A., Mansmann, S.: OLAP visualization: models, issues, and techniques. In: Encyclopedia of Data Warehousing and Mining, 2nd edn., pp. 1439–1446 (2009)
23. Cuzzocrea, A., Song, I.Y.: Big graph analytics: the state of the art and future research agenda. In: Proceedings of DOLAP 2014, pp. 99–101 (2014)
24. Campan, A., Cuzzocrea, A., Truta, T.M.: Fighting fake news spread in online social networks: actual trends and future research directions. In: Proceedings of BigData 2017, pp. 4453–4457 (2017)
25. Budiharto, W., Irwansyah, E., Suroso, J.S., Chowanda, A., Ngarianto, H., Santoso Gunawan, A.A.: Mapping and 3D modelling using quadrotor drone and GIS software. J. Big Data **8**(1), 1–12 (2021)

A Stacking Approach for Cross-Domain Argument Identification

Alaa Alhamzeh[1,2]([✉]), Mohamed Bouhaouel[2], Előd Egyed-Zsigmond[1],
Jelena Mitrović[2], Lionel Brunie[1], and Harald Kosch[2]

[1] INSA de Lyon, 20 Avenue Albert Einstein, 69100 Villeurbanne, France
{Alaa.Alhamzeh,Elod.Egyed-zsigmond,Lionel.Brunie}@insa-lyon.fr
[2] Universität Passau, Innstraße 41, 94032 Passau, Germany
{Mohamed.Bouhaouel,Jelena.Mitrovic,Harald.Kosch}@uni-passau.de

Abstract. Argument identification is the cornerstone of a complete argument mining pipeline. Furthermore, it is the essential key for a wide spectrum of applications such as decision making, assisted writing, and legal counselling. Nevertheless, most existing argument mining approaches are limited to a single, specific domain. The problem of building a robust system whose models are able to generalize over heterogeneous datasets remains fairly unexplored. In this paper, we tackle the argument identification task on two different datasets (Student Essays and Web Discourse), following two approaches: a classical machine learning approach and a DistilBert-based approach. Moreover, this paper sheds light on a new direction for researchers in this domain since we validate the principle of ensemble learning. In other words, we show that combining multiple approaches via a well stacked model improves the system performance. The results are very promising with respect to the recent findings in the literature.

Keywords: Argument mining · Argument identification ·
Computational linguistics · Classical machine learning · Transfer
learning · Stacking

1 Introduction

Argumentation is a fundamental aspect of human communication, thinking, and decision making. It can be defined as the logical reasoning humans use to come to a conclusion or justify their opinions on a specific topic. It was first studied by ancient Greek philosophers in 6th century B.C., and they are known today as the first argumentation theorists. Later on, argumentation gained more attention from different domains like psychology, communication, linguistics and, more recently, computer science, in particular, as a Natural Language Processing (NLP) task.

An argument consists of two elementary components: one or more premises supporting one claim (conclusion) [1]. According to Missimer [2], *"The objective of argumentation is to convince an opponent of a certain **claim**. The claim is a*

© Springer Nature Switzerland AG 2021
C. Strauss et al. (Eds.): DEXA 2021, LNCS 12923, pp. 361–373, 2021.
https://doi.org/10.1007/978-3-030-86472-9_33

*perspective or belief that is justified through logical reasoning. The reasoning is an **inference relation** drawn from **supporting evidence** or reasons towards the claim. If the reasoning is valid, then the claim is a legitimate conclusion of the provided reasons. The manifestation of the application of this process is called an **argument**."*

Keeping this definition in mind, we can see the need for different argument diagrams and schemes. Each domain expert looks at the argument and the inference structure from a different angle considering the requirements of the task at hand. Thus, he tries to represent the relations between the premises and claims using a relative scheme. Actually, that leads to one of the main challenges in this domain - the problem of different annotation practices of available datasets, which we elaborate on in Sect. 2. Consequently, most studies have been concentrating only on one individual subtask of the following:

- Argument Identification: classification of text into Argument or non-argument.
- Argument Components classification: detection of premises and claims.
- Argument structure identification: consists of the argument components plus the relations between them [3].

Argument identification is, therefore, the first essential block in an argument mining pipeline. This phase is important because not every sentence in a text is a part of an argumentation process (some narrative parts serve as introduction or summary about the topic and are not relevant for the argument itself). Since a low performance in argument identification would eventually propagate further down to the next tasks, this step is currently the most refined subtask of argument mining and it will be our focus in this paper.

State-of-the-art literature reports mostly classical machine learning models and very few attempts at using deep learning to solve this problem [4–6].

In this paper, we strive to achieve the maximum profit of both approaches by combining a classical Support Vector Machine (SVM) model [7] with a Distil-BERT based model [8], using the concept of ensemble learning [9], which to the best of our knowledge has not been used yet in the domain of argument mining. This is a promising approach due to the fact that it gathers the power of both existing techniques and maximizes the accuracy of the final prediction. On the other hand, it makes a union between the high performance of deep learning models and the interpretability of classical machine learning models.

The contributions of this paper are:

(a) Implementation of a new transfer learning model for Argument Identification based on DistilBert [8], a distilled version of BERT [10]: smaller, faster, and lighter with 97% of its language understanding capabilities [8].
(b) Proposing of an ensemble learning model which combines the out of two different approaches and improve the overall system performance.
(c) We test our individual and ensembled models on two different corpora, and achieved a good improvement by the overall model.
(d) Our work is available publicly through the github repository.[1]

---

[1] Project source code is available at https://github.com/Alaa-Ah/Stacked-Model-for-Argument-Mining.

This paper is organized as follows: in Sect. 2, we take a close look at the conceptual background of our work as well as the state-of-the-art studies considering our particular task of argument identification. In Sect. 3, we come to our contribution details. We validate the results in Sect. 4. Finally, we discuss the overall research questions and future work in Sect. 5.

2   Related Work

In this section, we first concentrate on the state-of-the-art classical solutions regarding the argument identification task. Later on, we discuss the concept of transfer learning and come through the different argumentation tasks that it has been evaluated on. Finally, we introduce a conceptual background about ensemble learning and have a deeper look at its essence.

2.1   Argument Mining

The research in this domain witnesses a variance between the period before and after contextualized embeddings and transformers. Indeed, transformers have rapidly become the model of choice for NLP problems [11].

Traditional Machine Learning Methods: The problem of argument identification itself is a binary classification task. Many approaches have been conducted in the literature. Moens et al. [12] worked on detecting arguments from legal text. They investigated a set of textual features (word pairs, text statistics, verb features and keywords) on the Araucaria corpus, which contains arguments from various sources, and achieved the best results with a multinomial naïve Bayes classifier. They found that the lack and the ambiguity of the linguistic markers (e.g. should), seem to be a major source of errors. They revisited the topic in [13] and applied the same method on the ECHR (European Court of Human Rights) corpus as well as to the Araucaria corpus. The classifier performed significantly better on the ECHR corpus, which seems plausible considering that language cues in legal argumentation texts are more explicit and more restrictive. However, the AraucariaDB received additional annotations and modifications over the years. As of today, it does not include the original text anymore like it did in its first version. As a result, we could not use it for the argumentative text identification task.

Stab et al. [14] introduced the annotation of argument components in student essays corpus. They also identified the argumentative discourse structure of this corpus in [3,15]. Habernal et al. [16] worked on annotating and mining arguments from user-generated Web discourse. In our study, we investigate both of the latter corpora, therefore more details on this process will be stated in Sect. 3.

However, even an efficient domain-specific model was unable to reach a satisfying result in different domain corpora. Cross-domain identification of argument units is still a rarely tackled problem. One recent attempt is by J. Daxenbereger et al. [17] where the authors tried to overcome the problem of domain dependence only for claim identification. They found that in-domain and cross-domain experiments have few shared properties on the lexical level (like the word "should").

Transfer Learning Methods: In a traditional machine learning model, there is always an assumption that the training and testing data follow the same distribution and serve the same task. On the contrary, transfer learning seeks freedom from those constraints and searches for methods to adapt models trained on a given dataset to classify slightly different data. It goes towards the search of domain, task, and corpus agnostic models [18]. In principle, it aims to apply previous learned knowledge from one source task (or domain) to a different target one, considering that source and target tasks and domains may be the same or may be different but related.

So why is this useful in the argument mining domain? First of all, common knowledge about the language is obviously appreciable. Second, transfer learning can solve or at least help to solve one of the biggest challenges in the argument mining field, the lack of labeled datasets. Third, even available datasets are often of small size and very domain and task dependent. They may follow different annotations, argument schemes, and various feature spaces. This means that in each potential application for argument mining, we need argument experts to label a significant amount of data for the task at hand, which is definitely an expensive work in terms of time and human-effort. Hence, transfer learning will fine-tune a pre-trained knowledge on a big dataset to serve another problem.

Recently, in 2020, only two studies addressing transfer learning models for argumentation tasks were published. The first one is towards discriminating evidence related to Argumentation Schemes [19] where the authors train classifiers on the sentence embeddings extracted from different pre-trained transformers. The second one is by Wambsganss et al. [4] where the authors proposed an approach for argument identification using BERT (Bidirectional Encoder Representations from Transformers) [10]. Our transfer learning model is based on a distilled version of BERT, proposed by [8], which retains 97% of the language understanding capabilities of the base BERT model, with a 40% less in size, and being 60% faster.

2.2   Ensemble Learning

Ensemble learning is a machine learning research area where different models (i.e. learners) are trained to solve the same problem and combined to get better results [9]. The fundamental hypothesis behind it, is that when different models are correctly combined, the ensemble model tends to outperform each of the individual models in terms of accuracy and robustness [20].

This concept of ensemble learning usually comes to the scene with weak learners, so the overall model is highly improved (e.g., [21,22]). In our particular case, each individual model provides a considerable performance. Our goal, therefore, is to benefit from all the features a classical ML model uses, and the contextualized knowledge a deep learning model reveals, aiming to improve the performance and stability of the model.

To the best of our knowledge, this promising concept has never been used in argumentation tasks. Hence, we expect this paper to highlight its strength and potential.

3   Contribution

In this section, we present the setup of our experiments in addition to the methods that have been adopted to achieve the goal of extracting argumentative clauses from a natural text.

3.1   Problem Statement

The argument mining problem is broad and can be seen as a set of several subtasks. In this paper, we consider argument identification as a flat problem. Thus, a text contains only arguments and non-arguments. In our proposed approach, we decide to do **text classification** at **sentence level** in order to maintain the complete meaning of the sentence and ease the identification. For instance, given an argumentative passage, (1) we first apply sentence segmentation (i.e. split the text into sentences) and then we (2) classify each sentence as argument or non-argument by two individual models which we discuss in Sects. 3.3 and 3.4. (3) We finally combine their predictions by a stacked method that is capable to improve the performance on the two corpora as described in Sect. 3.5.

3.2   Corpora Description

In argument mining, finding a suitable annotated dataset for a specific task is very challenging due to, on one hand, the different scheme annotations of the available labeled datasets on the web. On the other hand, the annotation task itself is expensive. Moreover, labeling or annotating a new corpus needs typically domain experts to validate it. In our particular case, and in order to achieve the goal of cross-domain argument identification model, we searched for datasets that contain both argument and non-argument labels. Student Essays [14] and Web discourse [16] are public corpora which serve this purpose well.

The **Student Essays corpus** contains 402 Essays about 8 controversial topics. The annotation covers the following argument components: 'major claim', 'claim' and 'premise'. Moreover, it presents the support/attack relations between them. Thus, it could be used in several argument mining tasks.

The **User-generated Web Discourse corpus** is a smaller dataset that contains 340 documents about 6 controversial topics in education such as homeschooling. The document may refer to an article, blog post, comment, or forum posts. In other words, this is a noisy, unrestricted and less formalized dataset. The annotation has been done by [16] according to Toulmin's model [23]. Thus, it covers the argument components: 'claim', 'premise', 'backing' and 'rebuttal'.

In order to deduct a binary-labeled unified data for both corpora, we label any argument component as an 'argument', and the rest of the text sentences as 'non-argument'.

3.3   Classical Machine Learning Model - SVM

In terms of the first base model, we consider training a classical machine learning model. This model should be able to capture and learn textual features and patterns that identify argumentative sections of text. Inspired by the works of [12, 15], we defined a set of structural, lexical, and syntactic features in addition to discourse markers as shown in Table 1.

The structural features reflect the building of the sentence and its position in the document. For instance, *tokens count* or length of the sentence exploit the fact that premises tend to be longer than other sentences which can therefore contribute to the argument identification process. Likewise, *question mark ending* indicates that a sentence ending with a question mark is more likely to be a claim, and eventually an argument.

Table 1. Textual features, new added features are marked with '*'

	Features	Explanation
Structural features	Sentence position [3]	Indicates the index of the sentence in the document
	Tokens count [3, 12]	Indicates the count of tokens (words) in the sentence
	Question mark ending [3]	Boolean feature
	Punctuation marks count [3]	Indicates how many punctuation marks are there in the sentence
Lexical features	1–3 gram BoW [3, 12]	Unigrams, bigrams and trigram Bag of Words features
	1–2 gram PoS *	Unigram and bigram of Part of Speech features
	Named entity recognition *	Count of the present named entities in the sentence
Syntactic features	Parse tree depth [3, 12]	Indicates the depth of the sentence's parse tree
	Sub-clauses count [3, 12]	Indicates how many sub-clauses are in the sentence
	Verbal features *	Counts of [modal, present, past, base form] verbs in the sentence
Discourse markers	Keywords count [3, 12]	Number of existing keywords ('actually', 'because', etc.)
	Numbers count *	Indicates how many numbers are there in the sentence

In terms of lexical features, we found that *unigrams and bigrams of Part of Speech (PoS)* tags are very useful to capture the PoS patterns that are frequently observed in argument components. Moreover, *named entity recognition* is a sub-task of information retrieval that locates the named entities in unstructured text such as person names, organizations, quantities and time expressions. Such entities are usually used when stating a granted fact, reporting some incidents, or formulating a conclusion (i.e. in an argument component). Therefore, we take into account how many named entities appeared in the sentence as one feature to our model. We are using those two lexical features to improve the accuracy of the model and to make the identification of arguments more efficient and precise.

Furthermore, syntactic and grammatical features play an essential role for argument identification. In particular, the *depth of parse tree*, the *verbal features* and *count of sub-clauses* which clearly reflect the complexity of the sentence. This is important since evidence tend to appear in a complex sentence structure with more than one sub-clause. As far as we were able to find out in relevant literature [3] only the tense of the main verb of the sentence has been used to distinguish between claims and premises, however, the tense of the other verbs of the sentence is also helpful to make this identification more accurate. Indeed, sentences including several verbs in the past tense tend to be premises, whereas, the presence of many modal verbs and verbs in present tense makes the sentence more likely to be a claim.

Last but not least, we believe that the discourse markers present a direct indicator for argumentative text. For instance, the terms: 'consequently' and 'conclude that' are often followed by a claim, while the terms: 'for instance' and 'first of all', are mostly followed by a premise. Hence, we use a set of 286 discourse markers presented by A. Knott et al. [24] to generate the *keywords count* feature that reinforces argumentative text detection. Since statistics are generally used to support a claim, the existence of statistical numbers in a sentence (*numbers count* feature) makes it more likely to be identified as an argument. We chose to feed those features into an SVM model. This choice is justified by the fact that SVM performs effectively on small datasets and in high dimensional spaces.

3.4   Transfer Learning Model (DistilBERT-Based)

Among many existent transformers, BERT [10] has recently gained a lot of attention. It seems to achieve the state of the art results in several NLP tasks [4–6,25]. For our particular task, we performed different experiments using many BERT-based models (BERT base, RoBERTa-base [26], DistilRoBERTa, Distil-BERT)[2] and achieved very similar results. Hence, we finally decided to use the DistilBERT given that it is 40% less than BERT in size with a relevant in-line performance and a faster training/testing time [8].

Figure 1 describes the adopted pipeline to perform the text classification using DistilBERT. The first block is the Tokenizer that takes care of all the BERT input requirements: (1) It transforms the sentence's words into an array of DistilBERT tokens. (2) It adds the special starting token ([CLS] token). (3) It adds the necessary padding to have a unique size for all sentences (we set 128 as a maximum length). The second block is the DistilBERT fine-tuned model, that outputs mainly a vector of length of 768 (default length). Our mission now is to

---

[2] We used Transformers from huggingface.co for our experiments.

Fig. 1. Transfer learning model architecture

adapt the output of this pre-trained model to our specific task. We achieve this by adding a third block, which is a linear layer applied on top of the DistilBERT transformer, and outputs a vector of size 2. The index of the maximum value in this vector represents the predicted class id. We trained the model for 3 epochs, using AdamW [27] as an optimizer and Cross Entropy for the loss calculation.

3.5 Overall Model (SVM + DistilBERT)

At this step, we have two models based on two completely different approaches. One is based on textual features while the other is based on the NLP transformer's ability of language understanding. Since they are two heterogeneous learning models, we chose to use the **stacking** ensemble method to combine their predictions.

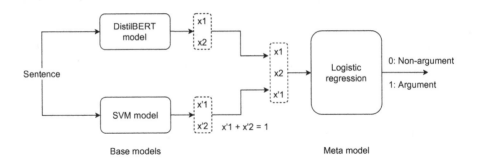

Fig. 2. Stacked model architecture

Figure 2 presents the stacked model architecture, consisting of two main components: 1. the base models, that include the trained SVM model and the trained transformer based model (DistilBERT) in parallel, 2. the meta model, that will learn from the outputs of the two models to produce the final prediction of a sentence. In order to have an array of independent features for the meta-model, and since SVM outputs two probabilities $x'1$ and $x'2$ (i.e. $x'1 + x'2 = 1$), we consider only $x'1$. Whereas, $x1$ and $x2$ are two independent raw logits so both of them are considered. Given that we are dealing with a binary classification problem where the input features are independent, logistic regression serves well as a meta-model to accomplish the task. For the training/testing steps, we split first the combined dataset into 75% training and 25% for the overall testing.

This testing data remains unseen for all the models and it is used only for the final validation of the overall model. The base models are trained on the 75% training data. The training data of the meta model is prepared by 5-folds cross validation of the two base models. In each fold, the out-of-fold predictions are used as a part of the training data for the meta-model.

4   Evaluation

In this section, we discuss the performance of each of the individual learners apart and the final stacked model. In addition, we state some comparison with the most recent previous work [4] tackling the same problem of argument identification on the same datasets. We will further discuss the results shown in Table 2, 3 and Table 4.

In terms of SVM model, we can see that it works very well on the Essays corpus with an accuracy of 90.95% and F1-score of 83.75%. On the other hand, it seems less efficient on the Web Discourse corpus, where the transfer learning model provides better measurements. This can be interpreted by the formal structure of Student Essays compared to Web Discourse as we mentioned in Sect. 3.2. We conducted this evaluation in the context of a cross-domain argument identification, in the sense that we apply the model on different merged corpora. In these settings, SVM achieved an accuracy of 85.42%, using the textual features that learn a set of patterns in argument identification. In some cases[3], SVM fails to classify an argumentative sentence as an argument due to the absence of such necessary features. These limitations might be handled by understanding the meaning of the sentence using the characteristic of transfer learning through the pre-trained DistilBERT model. Of course, there are other cases where the contrary happens - SVM classifies correctly and DistilBert model fails. Hence, we have decided to combine these models.

Table 2. Achieved results on **Student Essays** corpus

Model	Accuracy	Precision	Recall	F1-score
SVM	0.9095	0.8730	0.8116	0.8375
DistilBERT	0.8727	0.8016	0.7477	0.7697
Stacking model	0.9162	0.8890	0.8195	0.8483

As we can see in the normalized confusion matrices (Fig. 3), SVM model reaches a higher percentage than DistilBERT-based model in terms of True Positive (TP) whereas the latter performs better than SVM for True Negative

---

[3] Here is an example (from Essays dataset) of an argument sentence that SVM fails to identify while DistilBERT succeeds: *"Personally, I think both government and common people should have the responsibility for the environment, but we need to analyze some specific situations."*

(TN). Therefore, the stacked model is getting the most out of both of them in terms of TN and TP, and thus it records a better classification accuracy, precision and recall as shown in Table 4.

Table 3. Achieved results on **Web Discourse** corpus

Model	Accuracy	Precision	Recall	F1-score
SVM	0.7437	0.7051	0.5882	0.5874
DistilBERT	0.7799	0.7718	0.6484	0.6655
Stacking model	0.7855	0.7449	0.6958	0.7113

In recent work [4], the authors implemented a BERT-based transfer model on different corpora including the two datasets we use in this paper. Our stacked model overcomes theirs on the Student Essays achieving an accuracy of 91.62% and F1-score of 84.83% compared to their accuracy of 80.00% and F1-score of 85.19%. On the Web Discourse corpus, we have similar accuracy values (78.5% to 80.00%) while on the level of the combined model, our approach achieved better performance even though they have investigated on more training corpora.

Table 4. Achieved results on the merged corpora (**Student Essays and Web Discourse**)

Model	Accuracy	Precision	Recall	F1-score
SVM	0.8542	0.8037	0.7012	0.7331
DistilBERT	0.8587	0.7887	0.7529	0.7683
Stacking model	0.8780	0.8326	0.7659	0.7921

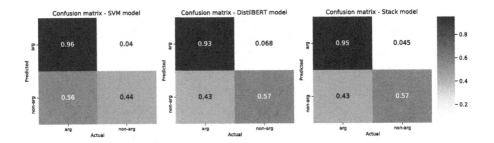

Fig. 3. Normalized confusion matrices

Furthermore, we emphasize that the idea of combining two different approaches is not only about the improvement of results, but also a step forward for the model's interpretability. On one hand, deep learning models (transformers in our case) reduce the task of feature engineering. Yet, it is difficult for humans to fully understand their behaviour. On the other hand, the direct feature engineering involved in classical ML makes those models more interpretable and understandable.

5   Conclusion

In this paper, we present a novel approach in the field of argument mining. Our model leverages the ensemble learning stacking method to combine a classical machine learning model and a deep transfer learning model to benefit from the advantages of both. The evaluation of the presented model shows good performance in the task of argument identification with an accuracy of 0.8780 and F1-score of 0.7921 on the two merged corpora. The proposed approach is a first insight into combining different argument mining techniques and methods to build a more general and robust model. In future work, we consider to extend our approach to other corpora from different domains. We plan also to use the model stacking method to cover additional argument mining tasks such as argument components classification and argument structure identification.

Acknowledgement

SPONSORED BY THE

Federal Ministry
of Education
and Research

This work was supported by the French Ministry of Higher Education and Research. It has been also co-funded by the German Federal Ministry of Education and Research (BMBF) under the funding code 01—S20049.

References

1. Govier, T.: A Practical Study of Argument. Wadsworth, Belmont (2001)
2. Missimer, C.A.: Good Arguments: An Introduction to Critical Thinking. Prentice Hall, Englewood Cliffs (1995)
3. Stab, C., Gurevych, I.: Identifying argumentative discourse structures in persuasive essays. In: Proceedings of the 2014 Conference on Empirical Methods in Natural Language Processing (EMNLP), pp. 46–56 (2014)
4. Wambsganss, T., Molyndris, N., Söllner, M.: Unlocking transfer learning in argumentation mining: a domain-independent modelling approach. In: 15th International Conference on Wirtschaftsinformatik (2020)
5. Reimers, N., Schiller, B., Beck, T., Daxenberger, J., Stab, C., Gurevych, I.: Classification and clustering of arguments with contextualized word embeddings. arXiv preprint arXiv:1906.09821 (2019)
6. Niven, T., Kao, H.-Y.: Probing neural network comprehension of natural language arguments. arXiv preprint arXiv:1907.07355 (2019)

7. Cortes, C., Vapnik, V.: Support-vector networks. Mach. Learn. **20**(3), 273–297 (1995)
8. Sanh, V., Debut, L., Chaumond, J., Wolf, T.: Distilbert, a distilled version of bert: smaller, faster, cheaper and lighter. arXiv preprint arXiv:1910.01108 (2019)
9. Sagi, O., Rokach, L.: Ensemble learning: a survey. Wiley Interdisc. Rev. Data Mining Knowl. Disc. **8**(4), e1249 (2018)
10. Devlin, J., Chang, M.-W., Lee, K., Toutanova, K.: BERT: pre-training of deep bidirectional transformers for language understanding. arXiv preprint arXiv:1810.04805 (2018)
11. Wolf, T., et al.: Transformers: state-of-the-art natural language processing. In: Proceedings of the 2020 Conference on Empirical Methods in Natural Language Processing: System Demonstrations, pp. 38–45 (2020)
12. Moens, M.-F., Boiy, E., Palau, R.M., Reed, C.: Automatic detection of arguments in legal texts. In: Proceedings of the 11th International Conference on Artificial Intelligence and Law, pp. 225–230 (2007)
13. Palau, R.M., Moens, M.-F.: Argumentation mining: the detection, classification and structure of arguments in text. In: Proceedings of the 12th International Conference on Artificial Intelligence and Law, pp. 98–107 (2009)
14. Stab, C., Gurevych, I.: Annotating argument components and relations in persuasive essays. In: Proceedings of COLING 2014, the 25th International Conference on Computational Linguistics: Technical Papers, pp. 1501–1510 (2014)
15. Stab, C., Gurevych, I.: Parsing argumentation structures in persuasive essays. Comput. Linguist. **43**(3), 619–659 (2017)
16. Habernal, I., Gurevych, I.: Argumentation mining in user-generated web discourse. Comput. Linguist. **43**(1), 125–179 (2017)
17. Daxenberger, J., Eger, S., Habernal, I., Stab, C., Gurevych, I.: What is the essence of a claim? Cross-domain claim identification. In: Proceedings of the 2017 Conference on Empirical Methods in Natural Language Processing, pp. 2055–2066, Copenhagen, Denmark, September 2017. Association for Computational Linguistics
18. Pan, S.J., Yang, Q.: A survey on transfer learning. IEEE Trans. Knowl. Data Eng. **22**(10), 1345–1359 (2009)
19. Liga, D., Palmirani, M.: Transfer learning with sentence embeddings for argumentative evidence classification (2020)
20. Van der Laan, M.J., Polley, E.C., Hubbard, A.E.: Super learner (2007)
21. Goubin, R., Lefeuvre, D., Alhamzeh, A., Mitrovic, J., Egyed-Zsigmond, E., Fossi, L.G.: Bots and gender profiling using a multi-layer architecture. In: CLEF (Working Notes) (2019)
22. Ciccone, G., Sultan, A., Laporte, L., Egyed-Zsigmond, E., Alhamzeh, A., Granitzer, M.: Stacked gender prediction from tweet texts and images notebook for pan at CLEF 2018. In: CLEF 2018-Conference and Labs of the Evaluation, p. 11p (2018)
23. Toulmin, S.E.: The Uses of Argument. Cambridge University Press, Cambridge (2003)
24. Knott, A., Dale, R.: Using linguistic phenomena to motivate a set of rhetorical relations, August 1997
25. Caselli, T., Basile, V., Mitrović, J., Kartoziya, I., Granitzer, M.: I feel offended, don't be abusive! Implicit/explicit messages in offensive and abusive language. In: Proceedings of LREC (2020)

26. Liu, Y., et al.: Roberta: a robustly optimized BERT pretraining approach. arXiv preprint arXiv:1907.11692 (2019)
27. Loshchilov, I., Hutter, F.: Decoupled weight decay regularization. arXiv preprint arXiv:1711.05101 (2017)

Property Analysis of Stay Points for POI Recommendation

Junjie Sun[✉], Yuta Matsushima, and Qiang Ma

Kyoto University, Kyoto, Japan
{jj-sun,matsu_shiba}@db.soc.i.kyoto-u.ac.jp, qiang@i.kyoto-u.ac.jp

Abstract. Stay points extracted from trajectories are often treated as points of interest (POI) in the data preprocessing of POI recommendations. Popularity (i.e., the number of visits) is one of the important features to distinguish the value of different POIs, especially for tourists traveling in an unfamiliar city. However, popularity could not reveal a user visits the point due to its attractiveness (e.g., views) or convenience for transport (e.g., railway station). In this paper, we introduce two properties: authority and hub, to further express the popularity of stay points. We apply a weighted HITS-based algorithm to calculate the scores of the two properties and conduct various experiments to demonstrate its effectiveness. The experimental results show the potential of the introduced properties that can be applied to various POI recommendation tasks.

Keywords: Link analysis · Trajectory mining · Sightseeing mining

1 Introduction

With the rapid development of location-based social networks (LBSNs) in recent years, various POI recommendation models and applications have been proposed. For instance, a sequence of POIs (i.e., a tour) can be recommended according to the user's query in [6]. To apply these POI recommendation models, user check-in data (i.e., user-location pair) are required. One of the major ways to obtain such data is to mining stay points from users' GPS trajectories, and these stay points usually are treated as a POI in POI recommendation services [11]. To distinguish the value of different stay points, a popularity-based score (i.e., the number of visiting users) is often used as an important indicator, especially for tourists traveling in an unfamiliar city.

However, the popularity score is not enough to assess the value of a stay point. Some of the highly popular stay points are due to the value of attraction or experience (e.g., landmarks and shopping mall), but there are still some because of the transitional function and are served as hub points to access to other stay points (e.g., transport station). For example, there are five typical trajectories of stay points shown in Fig. 1a. $Traj_1$, $Traj_2$, and $Traj_4$ may be the trajectories of daily life, and visits occurred between residence area, restaurant, and sports gym. Trajectories $Traj_3$ and $Traj_5$ seem to be typical travel trajectories where

© Springer Nature Switzerland AG 2021
C. Strauss et al. (Eds.): DEXA 2021, LNCS 12923, pp. 374–379, 2021.
https://doi.org/10.1007/978-3-030-86472-9_34

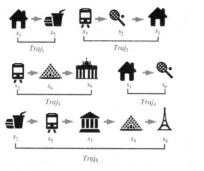

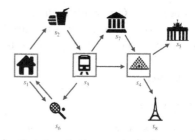

(a) Trajectories of stay points

(b) Graph of the trajectories, points marked with red rectangle show a good property of hub

Fig. 1. An example of trajectories of stay points and its graph.

users moved from public transport to landmarks. We can construct a graph of the trajectories according to incoming and outgoing links and shown in Fig. 1b. As a result, we can observe that some points are popular due to they are attractive such as s_4 and s_6. But for some points such as s_3, users visit it for transporting to other points.

In this paper, we study the links in the trajectories of stay points, especially the relation of incoming and outgoing links. We introduce two properties of stay points to better express popularity:

- **Authority:** The value of attraction or experience of a location.
- **Hub:** The ability to access to other stay points.

High authority points attract users to set them as destinations such as s_4 and s_6 shown in Fig. 1b. High hub points make users more easily move to other points, such as s_1, s_3, and s_4 that near several high authority points. Note that the two properties are not opposed to each other, a good authority point can also be a good hub point.

2   Related Work

Link analysis algorithms such as HITS are widely used to rank web pages [3]. These algorithms extract the features of nodes by analyzing the structure of the web page graph. Similar to the graph of web pages, LBSN data such as user-location can also be constructed as graphs. Social relationships were incorporated into a HITS-based POI recommendation model in [7]. A tensor decomposition model that incorporated with weighted HITS algorithm was proposed in [9].

The problem of stay point detection was first introduced in [4]. It is usually considered as a data pre-processing step in POI and tour recommendation [6]. Suzuki et al. studied the problem assign POI to users' GPS trajectory, which built a mapping between the POI database and raw GPS data [8]. Transportation mode was inferred from raw GPS data in [10]. Data integration methods were applied in [1] to construct a POI database from multiple data sources by linked data technologies. Our work focuses on analyzing the trajectories of stay points and no data from other sources are used, which can be treated as a supplement of data pre-processing for POI and tour recommendation.

3   Method

3.1   Preliminaries

From a user u's GPS device, we obtain spatial points from the log data, which record the locations that user u has been visited. For simplicity, we directly apply the mean shift clustering algorithm on all spatial points' coordinates to obtain stay points $S = (s_1, s_2, ..., s_m)$. Each stay point s_i is also associated with its latitude/longitude coordinates and the visit duration Dur_{s_i}.

In this way, we extract the trajectories of stay points of all users as $Traj$. Then we can obtain a transition matrix $Tran$ of size $m \times m$, where each entry $Tran_{s_i,s_j}$ represents the transition times from one location to another that can be calculated from trajectories of all users $Traj$. Similarly, an adjacency matrix A of size $m \times m$ is computed, where each entry a_{s_i,s_j} represents whether there is a link from one location to another. Finally, we compute the number of visit times for all stay points as a vector Pop.

3.2   Weighted HITS-Based Algorithm

Inspired by Hyperlink-Induced Topic Search (HITS) which originally is a link analysis algorithm for ranking web pages [3], we take the transitions between two stay points as links and construct a graph of stay points. A good authority point represents a stay point that is attractive and linked by many different hub points, while a good hub point represents a stay point that pointed to many other attractive points.

Unlike the graph of web pages or documents, a trajectory of stay points often generates both incoming and outgoing links on the location. HITS is difficult to distinguish the property of the stay point since it gives the same weight to all the stay points during the computation. We apply a weighted HITS-based algorithm [5] which gives different weights for links so that different links have different contributions to the score calculation. The detail is showed in Algorithm 1.

In order to figure out how different weights affect the results of the algorithm, we consider the following three kinds of weights in this paper:

- **Popularity Weight:** Intuitively, the higher the popularity of the stay point that the link points to, the greater its weight. The popularity weight is computed as: $w_{s_i,s_j}^{Pop} = \frac{Pop_{s_j}}{max(Pop)}, \forall s_i, s_j \in S$.
- **Transition Weight:** To distinguish the contribution of different links, we use the number of transitions and compute the transition weight: $w_{s_i,s_j}^{Tran} = \frac{Tran_{s_i,s_j}}{max(Tran_{s_i,:})}, \forall s_i, s_j \in S$.
- **Duration Weight:** Based on an idea that attractive places will make users stay longer, we introduce the duration weight: $w_{s_i,s_j}^{Dur} = \frac{Dur_{s_j}}{max(Dur)}, \forall s_i, s_j \in S$.

Algorithm 1: Weighted HITS-based Algorithm

Input : Adjacency matrix A, weight matrix W, and number of iterations
Output: Authority score vector $\mathbf{x}$ and hub score vector $\mathbf{y}$
Initialize $\mathbf{x}$ and $\mathbf{y}$;
while *Iterations still left* **do**
    for $s_i, \in S$ **do**
        for $s_j, \in S$ **do**
          | $x_{s_i} \mathrel{+}= a_{s_j,s_i} w_{s_j,s_i} y_{s_j}$;
        end
    end
    for $s_i, \in S$ **do**
        for $s_j, \in S$ **do**
          | $y_{s_i} \mathrel{+}= a_{s_i,s_j} w_{s_i,s_j} x_{s_j}$;
        end
    end
    Normalize $\mathbf{x}$ and $\mathbf{y}$;
end

---

4   Experiments and Results

Our experiments are conducted on two datasets. We share our source code for reference[1].

- **Kyoto dataset.** The dataset contains 1004 GPS trajectories of foreign tourists and school trip students who traveled around the Kansai area, Japan.
- **Melbourne dataset.** It comprises 1000 trajectories of stay points that were extracted from geo-tagged photos through Flickr API in earlier work [2].

Property Analysis. The Algorithm 1 outputs two scores namely authority score and hub score, respectively, which stand for the property of the stay point. We divide the score into three levels (i.e., low, middle, high) by λ and 2λ, where λ is the median of the score.

Figure 2 shows different groups of stay points in the Kyoto dataset by dividing different levels of authority and hub scores which were calculated by transition weight. The points marked in blue represent the stay points that belong to its group combination. The lines connecting different points represent the incoming and outgoing links of the users, and the more links the thicker the line in the figure.

For instance, Fig. 2a depicts the group of points with high authority and high hub scores. Most of these points are located in the center of Kyoto city, which contains many views and convenient public transportation. Different from points in Fig. 2a that can easily access other places, points with high authority and middle hub scores are attractions away from the center area. In contrast, most of the points in Fig. 2c and 2d are railway stations and crossroads, mainly served as transport points in the trajectories.

---

[1] https://github.com/singreen/poi-links.

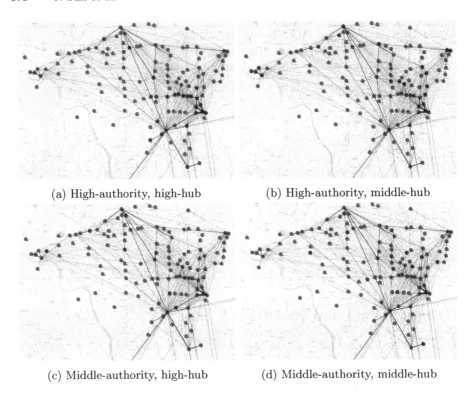

(a) High-authority, high-hub   (b) High-authority, middle-hub

(c) Middle-authority, high-hub   (d) Middle-authority, middle-hub

Fig. 2. Different groups of stay points in Kyoto dataset.

Tour Recommendation. We use the authority score as the reward for the optimization objective function to recommend tours. We strictly follow the settings in [6] and please refer to it for the detail. The performance in terms of tour precision, recall, and F_1-Score is showed in Table 1. **WPHITS, WTHITS** and **WDHITS** are methods that incorporated with popularity weight, transition weight, and duration weight, respectively.

WDHITS with duration weight outperforms all other methods on both datasets on all metrics. Although HITS performs not as well as POP, WPHITS

Table 1. Performance of tour recommendation (bold means the best).

Methods	Kyoto			Melbourne		
	Precision	Recall	F_1-Score	Precision	Recall	F_1-Score
POP	.253 ± .350	.151 ± .233	.181 ± .262	.109 ± .148	.225 ± .318	.141 ± .188
HITS	.203 ± .279	.126 ± .185	.150 ± .208	.094 ± .136	.210 ± .308	.125 ± .176
WPHITS	.273 ± .351	.152 ± .232	.183 ± .251	.121 ± .162	.235 ± .324	.154 ± .202
WTHITS	.254 ± .328	.139 ± .209	.170 ± .228	.102 ± .152	.204 ± .308	.131 ± .191
WDHITS	**.326 ± .346**	**.184 ± .225**	**.224 ± .249**	**.153 ± .199**	**.245 ± .339**	**.179 ± .231**

that incorporated with popularity weight outperforms POP and HITS on all metrics. This verified our idea that using authority and hub score can further represent the popularity score to get a better result.

5 Conclusion

In this paper, given trajectories of stay points, we introduced two properties, namely authority, and hub, to further explain the popularity score for POI recommendation services. A weighted HITS-based algorithm was applied to calculate the scores of the two properties. The experiment showed the potential that can be applied in various applications.

Acknowledgment. This work is partly supported by MIC SCOPE (201607008).

References

1. Athanasiou, S., et al.: Big POI data integration with Linked Data technologies. In: EDBT, pp. 477–488 (2019)
2. Chen, D., Ong, C.S., Xie, L.: Learning points and routes to recommend trajectories. In: Proceedings of the 25th ACM International on Conference on Information and Knowledge Management, pp. 2227–2232 (2016)
3. Kleinberg, J. M. Authoritative sources in a hyperlinked environment. In: Proceedings of the ninth Annual ACM-SIAM Symposium on Discrete Algorithms, pp. 668–677 (1998)
4. Li, Q., Zheng, Y., Xie, X., Chen, Y., Liu, W., Ma, W.Y.: Mining user similarity based on location history. In: Proceedings of the 16th ACM SIGSPATIAL International Conference on Advances in Geographic Information Systems, pp. 1–10 (2008)
5. Li, L., Shang, Y., Zhang, W.: Improvement of HITS-based algorithms on web documents. In: Proceedings of the 11th International Conference on World Wide Web, pp. 527–535 (2002)
6. Lim, K.H., Chan, J., Leckie, C., Karunasekera, S.: Personalized tour recommendation based on user interests and points of interest visit durations. In: Twenty-Fourth International Joint Conference on Artificial Intelligence (2015)
7. Long, X., Joshi, J.: A HITS-based POI recommendation algorithm for location-based social networks. In: Proceedings of the 2013 IEEE/ACM International Conference on Advances in Social Networks Analysis and Mining, pp. 642–647 (2013)
8. Suzuki, J., Suhara, Y., Toda, H., Nishida, K.: Personalized visited-poi assignment to individual raw GPS trajectories. ACM Trans. Spatial Algorithms Syst. (TSAS) 5(3), 1–28 (2019)
9. Ying, Y., Chen, L., Chen, G.: A temporal-aware POI recommendation system using context-aware tensor decomposition and weighted HITS. Neurocomputing 242, 195–205 (2017)
10. Zheng, Y., Liu, L., Wang, L., Xie, X.: Learning transportation mode from raw GPS data for geographic applications on the web. In: Proceedings of the 17th International Conference on World Wide Web, pp. 247–256 (2008)
11. Zheng, Y.: Trajectory data mining: an overview. ACM Trans. Intell. Syst. Technol. (TIST) 6(3), 1–41 (2015)

Beacon Technology for Retailers - Tracking Consumer Behavior Inside Brick-and-Mortar-Stores

Alexander Voelz[✉], Andreas Mladenow, and Christine Strauss

University of Vienna, Oskar-Morgenstern-Platz 1, 1090 Vienna, Austria
{alexander.voelz,andreas.mladenow,christine.strauss}@univie.ac.at

Abstract. For many years, brick-and-mortar retail has been confronted with an ever-growing e-commerce as a strong competing alternative, and challenging customer expectations. As a result, retailers need to increasingly focus on integrating emerging technologies to keep pace with online retailers and improve the customer's shopping experience. In this context, beacon technology has been credited with the potential to fundamentally transform the retail industry. However, to date, widespread adoption by retailers has failed to materialize. One reason for the failed adoption could be false expectations regarding the applications of the beacon technology in the retailing context. To investigate this assumption in depth, we analyzed relevant literature to identify possible applications that beacon technology enables for brick-and-mortar retailers. We categorized the identified applications into three groups: (i) *indoor navigation and localization*, (ii) *tracking and analyzing customer behavior* and (iii) *personalized and location-based advertising*. Our work aims at revealing these challenges to make the true potential of beacon technology for brick-and-mortar stores transparent.

Keywords: Beacon technology · Brick-and-mortar · Retail · Customer data

1 Introduction

The share of e-commerce in total retail sales has been growing steadily for many years, and the growth is predicted to continue [38]. As a result, brick-and-mortar retailers are confronted not only with shrinking market shares, but also with rising and sophisticated customer expectations based on offers and services on the Internet [11,28]. The ongoing COVID-19 pandemic and related temporary lockdowns have further reinforced this trend, as consumers are taking advantage of e-commerce offerings to an even greater extent [10,16]. Therefore, it is of big importance for retailers to meet the rising expectations of their customers. Considering these challenges, technical and management-oriented scientific communities have increasingly discussed ways in which retailers may improve their customers' shopping experience by integrating recent technologies.

© Springer Nature Switzerland AG 2021
C. Strauss et al. (Eds.): DEXA 2021, LNCS 12923, pp. 380–390, 2021.
https://doi.org/10.1007/978-3-030-86472-9_35

Applications, such as digital signage [5, 7] or contactless payment using near field communication (NFC) [17], already have been established successfully in practice.

In 2013, beacon technology was presented for the first time by Apple at the annual World Wide Developers Conference under the name iBeacon [4]. Only shortly after its introduction the technology started to generate attention, as Google and Facebook launched their own beacons as well. Various media reported that beacon technology has the potential to fundamentally transform the retailing industry. The value proposition was that by using beacons, retailers would gain insights into customer behaviors and preferences, which in turn could be used for personalized and location-based advertising. The direct proximity of customers was said to help target them exclusively with relevant advertising content, which would subsequently lead to increased sales for retailers.

This view is supported by the scientific literature, where beacon technology is often said to have the potential "to reach the right person at the right time with the right message" ([2] p. 226). The interest for this kind of advertising is motivated by the fact that advertising messages are more effective if they are perceived as relevant by the customer [6].

In recent years, however, the initially euphoric predictions have not persisted and further press releases highlighting the benefits of beacons became rare. This development raises the question of what has become of this promising technology, and why it has not transformed retail industry in the way it was expected to.

The structure of the paper is as follows: Sect. 2 covers the theoretical background relevant for this study, which includes the current challenges of brick-and-mortar stores, the elements of beacon technology, and its various applications for brick-and-mortar retailers. In Sect. 3, we extract and aggregate relevant information and insights based on hands-on experience in the field from an in-depth interview conducted with an expert from industry. His know-how is used to verify and complement the insights from literature. The study and its results are summarized in Sect. 4.

2   Theoretical Background

The following section provides an overview of current challenges brick-and-mortar stores are facing. Furthermore, the functionality of beacon technology is presented. Finally, we present the results of a systematic literature review to demonstrate the state of the art and provide an overview of possible applications that beacons can enable for retailers.

2.1   Challenges of Brick-and-Mortar Stores

Understanding the behavior of customers within stores is of great importance for retailers. On the one hand, such knowledge about the motivations behind customer behavior is necessary to support an ideal shopping experience [15]. On the other hand, a large proportion of purchase decisions are unplanned and impulsive, which is why deeper insights into the causes underlying these unplanned and spontaneous decisions are of great value for marketers and retailers [24].

In recent years, the understanding of customer behavior has become even more important due to the modern consumer who is almost constantly online and uses his smartphone in stores for price comparisons or product reviews [26]. Furthermore, customers are increasingly using digital channels for shopping [11]. As a result, retailers are facing shrinking market shares and at the same time rising customer expectations due to ubiquitous offers and services available on the Internet [9]. Among many others, the lack of data on customer behavior in brick-and-mortar stores stems from limited alternatives to observe and track customers physically and poses a severe problem and a serious disadvantage of physical retail compared to e-commerce.

Collecting data on the Internet has become ubiquitous [20]. Online retailers thus gain detailed insights regarding their customers' navigation, dwell times, and other behaviors [34]. In contrast to physical retail, personalized offers or services based on collected data have become the prevailing strategy for targeting customers on the Internet [3]. Due to the lack of customer data, retailers introduced loyalty cards early on [13]. In exchange for voluntarily disclosing data about their shopping behavior, customers receive extended access to special offers or can redeem collected points for discounts or rewards [23,43]. Nowadays, loyalty cards are often integrated into retailers' app [27].

Besides loyalty cards, however, there are limited opportunities for retailers to collect high-quality sensor data at given physical locations, as accurately recording individual customer behavior is cost-intensive [34]. One of these cost-intense methods to monitor customer behavior is using cameras, software for facial recognition and algorithms for image processing to identify as well as analyze the characteristics in the individual behavior of different customers [32].

As retail industry is digitizing too, more cost-efficient ways of collecting data have emerged, which also promise an increased shopping experience for customers [21]. One technology that has gained more and more attention in the literature over the past years are beacons.

2.2   Beacon Technology

Not only in press releases, but also in scientific publications, the terms beacons and beacon technology are often seen as synonyms. However, it seems important to distinguish between these two terms.

The elements of beacon technology can be divided into the beacon itself, a corresponding application on a mobile device, and a processing server [44]. Beacons are small and rather inexpensive transmitters that send small amounts of data at a continuous interval via Bluetooth Low Energy (BLE) to all receptive mobile devices that are within a certain range of the respective beacon. The most important component of the transmitted data is the so-called Universally Unique Identifier (UUID), which is needed for the identification of each beacon [18]. The UUID can be retrieved by the addressed devices - provided BLE is activated, and a corresponding application is installed [2]. Nowadays, a vast majority of retailers already offers individual apps [25]. The apps form the link between the beacons and the server, by receiving the UUID emitted by the beacons and forwarding it to the server via Wi-Fi or mobile data connection. The server in turn is needed to store and analyze the data transmitted by the app [44].

BLE is the key property that makes beacon technology superior over GPS. With beacons, it is possible to locate smartphones indoors precisely, which GPS is unable to do due to weak signal strength and resulting inaccuracy [18,35,40]. Relative to WLAN or RFID, implementing beacons is also far less expensive. In addition, BLE benefits from the already existing penetration of Bluetooth [14].

2.3   Applications of Beacon Technology in Brick-and-Mortar Stores

To identify possible and useful applications of beacon technology for retailers discussed in literature, we conducted a literature search in the databases of ACM, IEEE and Scopus, using combinations of different key words in a search string (see Table 1). The conducted search all in all generated 464 items, which were reduced to 187 after eliminating duplicates and publications before 2013. Among the remaining 187 total items, we identified in the end nine relevant publications, in which different applications of the beacon technology in the retailing context are suggested and discussed.

Table 1. Combined keywords and filtering to identify relevant literature discussing applications of beacon technology in retail.

Search string	Total results	Abstract analysis	Full text analysis
Beacon OR ibeacon OR ble AND retail OR shopping	187	54	9

We divided the identified applications discussed in the literature into three areas: (i) indoor navigation and localization; (ii) tracking and analyzing customer behavior; (iii) personalized and location-based advertising. Table 2 shows which area of application is discussed among the nine publications, that we identified during our literature review.

Navigation: Indoor Navigation and Localization. The category "Navigation" referring to indoor localization and navigation is frequently used in literature in the form of product localization [2,25,42]. In combination with a digital shopping list this function can be further used to guide customers and assist them navigating through a whole shopping trip through a store [39]. Such navigation features, as well as recommendations based on shopping lists, are mentioned as desirable functions of the beacon technology by consumers [42].

While navigation features are not yet very common in retail applications, they have already been successfully implemented and used in the context of museums for several years [12,22,37].

Table 2. Application areas of the beacon technology in the literature.

Publications	Navigation	Tracking	Advertising
Thamm et al. (2016)	✓	–	–
Inman and Nikolova (2017)	✓	✓	✓
Pugaliya et al. (2017)	–	–	✓
Shende et al. (2017)	–	✓	✓
Adkar et al. (2018)	✓	–	✓
Betzing (2018)	–	✓	–
van de Sanden et al. (2019)	✓	–	–
Zaim et al. (2019)	–	✓	✓
Zualkernan et al. (2020)	–	✓	–

Tracking: Tracking and Analyzing Customer Behavior. Tracking and analyzing customer behavior is of particular importance in the context of retailing, as it provides the basis for personalized recommendations and offers [36,45]. Such services are based on large data collections, which are indispensable prerequisites to offer such services at an acceptable quality level. To date, this has mainly been performed by the use of loyalty cards that record data on the past purchasing behavior of the respective customer. Nowadays, beacons can be used to track the customer behavior in more detail [39].

Complementing past purchases, beacons can be used to collect sensor data on customers' movements and dwell times inside of stores [44,45]. Other studies have even designed a collaborative platform for high street retailers that uses the beacon technology to enable comprehensive collection of spatio-temporal data on customer behavior, inside as well as outside of stores [9]. In addition, data mining methods can be engaged to identify correlations and patterns within the collected data [44]. The application of data mining methods in combination with customer profiling is said to improve predicting customers' future behaviors and needs [19]. In summary, beacon technology can be used in the retailing context to track and analyze sensor data about consumer behavior, which would not be available in terms of quantity, quality, and diversity if only loyalty cards are used for data collection.

Advertising: Personalized and Location-Based Advertising. Collected data on customer behavior plays a crucial role in personalized advertising. Most commonly, this involves sending out personalized offers or providing personalized recommendations [2,25,36,42,44]. In [25], both past purchase behavior data and location-based data are used to personalize offers. Data on past purchase behavior is available through loyalty cards integrated into apps, while location-based data is collected in-store through beacon technology.

The combination of individual purchase behavior patterns and location-based data within the store, commonly form the basis for personalizing offers and recommendations [2, 36, 44]. However, several studies have suggested that personalized offers are often not preferred by consumers [42] and may lead to negative reactions [25, 30].

2.4   Challenges of Beacon Technology in Brick-and-Mortar Stores

Our literature review has revealed that most publications on the topic of beacon technology in retailing focus on personalized advertising as area of applications, while tracking of consumer behavior is mostly just mentioned as necessary prerequisite [2, 25, 31, 36, 44]. Only few publications focus on the potential of tracking and analyzing customer behavior that goes beyond the purpose of personalizing advertisements [9, 45].

The misuse of beacon technology by retailers in the past has created the serious problem of disappointed consumer expectations. Considering this, customer privacy concerns and trust issues are mentioned in literature as neglected factors when it comes to the adoption of new retail technologies [15, 25, 30]. In these cases, privacy concerns can take different forms, as for example concerns about potential harassment from too many notifications, [39] or the fear of manipulative behavior by retailers [42]. Accordingly, it is not surprising that some studies rank customers privacy and the protection of personal data as a decisive prerequisite for customers to accept a new retailing technology [30]. In this context, perceived fairness and trust are mentioned as important factors, which may absorb data protection concerns of customers.

Perceived fairness can be defined as the extent to which an individual (in this case: a customer) perceives the relationship with another party (in this case: the retailers), as balanced and fair [29]. Conversely, such a relationship is perceived as unfair if customers value the resulting benefits to be disproportionate to their investment, i.e. the disclosure of their data [15, 25]. Therefore, excessive use of push notifications by retailers without providing equivalent benefits has led to the perception of an unfair relationship from the perspective of customers.

3   Recommendations for Implementation

The following recommendations are divided into two groups. The first group deals with prerequisites that should be met before an adaption of beacon technology. The second group presents recommendations that need to be considered after the implementation of beacon technology to ensure customers' acceptance.

3.1   Prerequisites for Adapting Beacon Technology

The basic requirements for the adaption of beacon technology by retailers are an established app [2, 25] and the existence of a customer database [42]. An adaptation of the beacon technology is only recommended when these prerequisites have been considered in a diligent and timely planning.

The acceptance of an existing app poses particular problems for retailers, since customers generally have concerns about the need to install too many apps and wish for a single app that can be used across several retailers [15, 25].

Furthermore, it is important for retailers to build up and maintain trust in their relationship with customers. Several studies have confirmed, that the trustworthiness of retailers plays a crucial role regarding the adoption and acceptance of beacon technology by customers [15, 25, 30].

3.2 Recommendations for Establishing Beacon Technology

Even after the implementation of beacon technology, there are a number of recommendations to be followed in order to increase the chances of success for such projects. Of utmost importance is the provision of applications according to the preferences of the customers so that they agree to the collection of data about their behavior. The collected data in return is necessary to provide personalized and individualized customer service, thus laying the foundation for sustainable, long-term customer loyalty.

In general, it is recommended to use a pull strategy for the provided applications of beacon technology, so that customers actively consent to the collection of their data in the form of opt-ins, which have a positive effect on the assessment of perceived fairness and trustworthiness [1, 41]. This is especially important, since generating customer loyalty requires long-term trust building to convince customers of retailers' trustworthiness and honest intentions [30, 42]. Keeping this in mind, retailers need to make sure that the resulting benefits of any new technology outweigh the "investment" of the customers, which is the consent to the tracking and analysis of their disclosed data.

Benefits for the customers can be provided through indoor navigation and product localisation, which create value for customer through saved time and effort [25]. In addition, the tracking and analyzing of customer behavior can be used to provide personalized and individualized customer service [33].

4   Conclusion

The initial euphoria following the release of beacons has faded, and the technology failed to bring the transformation of the retail sector that it was originally credited with. As a result, this paper raised the fundamental question of the underlying reasons for the failure of this transformation.

It can be concluded that the overuse of location-based push notifications is one of the major reasons for the failed transformation. Customers often perceive push notifications as annoying and intrusive. While many publications discuss this application as one of the big advantages for retailers, it can also be seen a disruptive form of advertising, which could be responsible for the failed widespread adoption. Primarily, because a majority of consumers perceive personalized and locations-based advertising as intrusive, which in return negatively affects the perceived fairness and the trust in the relationship with retailers.

In summary, the true potential among the different applications of beacon technology lies within the tracking and analyzing of customer behavior inside brick-and-mortar stores. Beacons are able to provide retailers with temporal and spatial sensor data, which in return can be combined with data on past purchase behaviors to reveal previously unknown information about the consumers. Nevertheless, the expectations that have already been disappointed in the past create a difficult basis for the future widespread adoption of the beacon technology in retail. As a result, beacons might quietly establish themselves as one part in a bigger system of interconnected retailing technologies, while their performed functions will no longer be apparent to customers.

Future research should aim at interviewing a larger number of experts from industry to acquire more reliable insights on various challenges underlying the adoption of beacon technology. First steps in this direction haven been taken by van de Sanden et al. [42]. However, a more systematic approach is needed to develop and justify recommendation for the successful implementation of beacon technology by brick-and-mortar retailers in the future. Furthermore, it would be interesting for future work to investigate the upcoming importance of appropriate application of data mining methods to support retailers in managing, analyzing and utilizing the collected data on the behavior of customers inside brick-and-mortar stores.

References

1. Achadinha, N.M.J., Jama, L., Nel, P.: The drivers of consumers' intention to redeem a push mobile coupon. Behav. Inf. Technol. **33**(12), 1306–1316 (2014)
2. Adkar, N., Talele, A., Mundhe, C., Gunjal, A.: Bluetooth beacon applications in retail market. In: 2018 International Conference on Advances in Communication and Computing Technology (ICACCT), Sangamner, India, pp. 225–229. IEEE (2018)
3. Aguirre, E., Mahr, D., Grewal, D., De Ruyter, K., Wetzels, M.: Unraveling the personalization paradox: the effect of information collection and trust-building strategies on online advertisement effectiveness. J. Retail. **91**(1), 34–49 (2015)
4. Apple Developer. https://developer.apple.com/videos/play/wwdc2013/307/. Accessed 1 May 2021
5. Bauer, C., Dohmen, P., Strauss, C.: Interactive digital signage - an innovative service and its future strategies. In: International Conference on Emerging Intelligent Data and Web Technologies (EIDWT), Tirana, Albania, pp. 7–9. IEEE (2011)
6. Bauer, C., Strauss, C.: Location-based advertising on mobile devices. Manag. Rev. Q. **66**(3), 159–194 (2016). https://doi.org/10.1007/s11301-015-0118-z
7. Bauer, C., Garaus, M., Strauss, C., Wagner, U.: Research directions for digital signage systems in retail. Procedia Comput. Sci. **141**, 503–506 (2018)
8. Betzing, J.H.: Beacon-based customer tracking across the high street: perspectives for location-based smart services in retail. In: Proceedings of the 24th Americas Conference on Information Systems (AMCIS 2018), pp. 1–10. AIS, New Orleans (2018)

9. Betzing, J.H., Hoang, A.Q.M., Becker, J.: In-store technologies in the retail servicescape. In: Multikonferenz Wirtschaftsinformatik (MKWI 2018) - Band IV, Lüneburg, Germany, pp. 1671–1682 (2018)

10. Bhatti, A., Akram, H., Basit, H.M., Khan, A.U., Raza, S.M., Naqvi, M.B.: E-commerce trends during COVID-19 Pandemic. Int. J. Future Gener. Commun. Netw. **13**(2), 1449–1452 (2020)

11. Bollweg, L., Lackes, R., Siepermann, M., Sutaj, A., Weber, P.: Digitalization of local owner operated retail outlets: the role of the perception of competition and customer expectations. In: Proceedings of the 20th Pacific Asia Conference on Information Systems (PACIS 2016), p. 348. AISeL, Chiayi (2016)

12. Choi, H.S., Kim, S.H.: A content service deployment plan for metaverse museum exhibitions-centering on the combination of beacons and HMDs. Int. J. Inf. Manag. **37**(1), 1519–1527 (2017)

13. Coll, S.: Consumption as biopower: governing bodies with loyalty cards. J. Consum. Cult. **13**(3), 201–220 (2013)

14. Conte, G., De Marchi, M., Nacci, A.A., Rana, V., Sciuto, D.: BlueSentinel: a first approach using ibeacon for an energy efficient occupancy detection system. In: Proceedings of the 1st ACM Conference on Embedded Systems for Energy-Efficient Buildings (BuildSys 2014), pp. 11–19. ACM, Memphis (2015)

15. Cordiglia, M., Van Belle, J.P.: Consumer attitudes towards proximity sensors in the South African retail market. In: 2017 Conference on Information Communication Technology and Society (ICTAS), Umhlanga, South Africa, pp. 1–6. IEEE (2017)

16. Dannenberg, P., Fuchs, M., Riedler, T., Wiedemann, C.: Digital transition by COVID-19 pandemic? The German food online retail. J. Econ. Hum. Geogr. **111**(3), 543–560 (2020)

17. Demeter, P.: Near field communication im Handel: Expertenbefragung mittels Delphi-Methode. HMD Praxis der Wirtschaftsinformatik **52**(2), 240–248 (2015). https://doi.org/10.1365/s40702-015-0122-8

18. Fard, H.K., Chen, Y., Son, K.K.: Indoor positioning of mobile devices with agile iBeacon deployment. In: 28th IEEE Canadian Conference on Electrical and Computer Engineering (CCECE), Halifax, NS, Canada, pp. 275–279. IEEE (2015)

19. Fayyad, U., Piatetsky-Shapiro, G., Smyth, P.: From data mining to knowledge discovery in databases. AI Mag. **17**(3), 37–54 (1996)

20. Goldfarb, A., Tucker, C.E.: Privacy regulation and online advertising. Manag. Sci. **57**(1), 57–71 (2011)

21. Hagberg, J., Sundstrom, M., Egels-Zandén, N.: The digitalization of retailing: an exploratory framework. Int. J. Retail Distrib. Manag. **44**(7), 694–712 (2016)

22. He, Z., Cui, B., Zhou, W., Yokoi, S.: A proposal of interaction system between visitor and collection in museum hall by iBeacon. In: 2015 10th International Conference on Computer Science & Education (ICCSE), Cambridge, UK, pp. 427–430. IEEE (2015)

23. Im, H., Ha, Y.: Is this mobile coupon worth my private information? Consumer evaluation of acquisition and transaction utility in a mobile coupon shopping context. J. Res. Interact. Mark. **9**(2), 92–109 (2015)

24. Inman, J.J., Winer, R.S., Ferraro, R.: The interplay among category characteristics, customer characteristics, and customer activities on in-store decision making. J. Mark. **73**(5), 19–29 (2009)

25. Inman, J.J., Nikolova, H.: Shopper-facing retail technology: a retailer adoption decision framework incorporating shopper attitudes and privacy concerns. J. Retail. **93**(1), 7–28 (2017)

26. Jung, K., Cho, Y.C., Lee, S.: Online shoppers' response to price comparison sites. J. Bus. Res. **67**(10), 2079–2087 (2014)
27. Ketelaar, P.E., et al.: "Opening" location-based mobile ads: how openness and location congruency of location-based ads weaken negative effects of intrusiveness on brand choice. J. Bus. Res. **91**, 277–285 (2018)
28. Kryvinska, N., Van Thanh, D., Strauss, C.: Integrated management platform for seamless services provisioning in converged network. Int. J. Inf. Technol. Commun. Converg. **1**(1), 77–91 (2010)
29. Maxham, J.G., III., Netemeyer, R.G.: Firms reap what they sow: the effects of shared values and perceived organizational justice on customers' evaluations of complaint handling. J. Mark. **67**(1), 46–62 (2003)
30. Pizzi, G., Scarpi, D.: Privacy threats with retail technologies: a consumer perspective. J. Retail. Consum. Serv. **56**, 102160 (2020)
31. Pugaliya, R., Chabhadiya, J., Mistry, N., Prajapati, A.: Smart shoppe using beacon. In: 2017 IEEE International Conference on Smart Technologies and Management for Computing, Communication, Controls, Energy and Materials (ICSTM), Chennai, India, pp. 32–35. IEEE (2017)
32. Quintana, M., Menéndez, J.M., Alvarez, F., Lopez, J.P.: Improving retail efficiency through sensing technologies: a survey. Pattern Recogn. Lett. **81**, 3–10 (2016)
33. Roy, S.K., Balaji, M.S., Sadeque, S., Nguyen, B., Melewar, T.C.: Constituents and consequences of smart customer experience in retailing. Technol. Forecast. Soc. Change **124**, 257–270 (2017)
34. Rykowski, J., Chojnacki, T., Strykowski, S.: In-store proximity marketing by means of IoT devices. In: Camarinha-Matos, L.M., Afsarmanesh, H., Rezgui, Y. (eds.) PRO-VE 2018. IAICT, vol. 534, pp. 164–174. Springer, Cham (2018). https://doi.org/10.1007/978-3-319-99127-6_15
35. Saranya, K., Fathima, S.S.A., Ismail, M.N.A.: Enhancing customer engagement using beacons. Int. J. Recent Technol. Eng. **7**(4S), 390–393 (2018)
36. Shende, P., Mehendarge, S., Chougule, S., Kulkarni, P., Hatwar, U.: Innovative ideas to improve shopping mall experience over E-commerce websites using beacon technology and data mining algorithms. In: 2017 International Conference on Circuit, Power and Computing Technologies (ICCPCT), Kollam, India, pp. 1–5. IEEE (2017)
37. Spachos, P., Plataniotis, K.N.: BLE beacons for indoor positioning at an interactive IoT-based smart museum. IEEE Syst. J. **14**(3), 3483–3493 (2020)
38. Statista. https://www.statista.com/statistics/534123/e-commerce-share-of-retail-sales-worldwide/. Accessed 1 May 2021
39. Thamm, A., Anke, J., Haugk, S., Radic, D.: Towards the omni-channel: beacon-based services in retail. In: Abramowicz, W., Alt, R., Franczyk, B. (eds.) BIS 2016. LNBIP, vol. 255, pp. 181–192. Springer, Cham (2016). https://doi.org/10.1007/978-3-319-39426-8_15
40. Uitz, I., Koitz, R.: Consumer acceptance of location based services in the retail environment. Int. J. Adv. Comput. Sci. Appl. **4**(12), 124–131 (2013)
41. Unni, R., Harmon, R.: Perceived effectiveness of push vs. pull mobile location based advertising. J. Interact. Advert. **7**(2), 28–40 (2007)
42. van de Sanden, S., Willems, K., Brengman, M.: In-store location-based marketing with beacons: from inflated expectations to smart use in retailing. J. Mark. Manag. **35**(15–16), 1514–1541 (2019)
43. Wright, C., Sparks, L.: Loyalty saturation in retailing: exploring the end of retail loyalty cards? Int. J. Retail Distrib. Manag. **27**(10), 429–440 (1999)

44. Zaim, D., Benomar, A., Bellafkih, M.: Developing a geomarketing solution. Procedia Comput. Sci. **148**, 353–360 (2019)
45. Zualkernan, I.A., Pasquier, M., Shahriar, S., Towheed, M., Sujith, S.: Using BLE beacons and machine learning for personalized customer experience in smart Cafés. In: 2020 International Conference on Electronics, Information, and Communication (ICEIC), Barcelona, Spain, pp. 1–6. IEEE (2020)

Author Index

Printed in the United States
by Baker & Taylor Publisher Services